Currents of Genius

E.B Cohen

First edition, 2026

Contents

Currents of Genius

Edison, Tesla, Westinghouse, and the Epic Battle to Electrify the World

E.B Cohen

Currents of Genius: Edison, Tesla, Westinghouse, and the Epic Battle to Electrify the World

First edition, 2026.

This is a work of narrative non-fiction. It is drawn from the public historical record, from the scientific and biographical literature, and from primary sources including the Edison Papers, the patent record, and contemporary reporting. Where the record is uncertain, contested, or shaped by legend, the text says so. Any errors that remain are the author's own.

Published by E.B Cohen.

For everyone who has ever flicked a switch without thinking — and for the handful of restless minds who made that thoughtlessness possible.

"Let there be light."

— the oldest wish of humankind

"I do not think there is any thrill that can go through the human heart like that felt by the inventor as he sees some creation of the brain unfolding to success."

— Nikola Tesla

"Genius is one percent inspiration and ninety-nine percent perspiration."

— Thomas Edison

Author's Note

For most of human history, night was a wall. When the sun went down, the world contracted to the small trembling circle of a fire or a candle, and beyond it lay the dark. Work stopped. Streets emptied. The rich burned whale oil and the poor went to bed. This was true for the pharaohs and true for Napoleon; it was true, in fact, for very nearly everyone who ever lived, right up until the lifetimes of people whose grandchildren are alive today.

Then, in the span of a single human generation — roughly the forty years between 1880 and 1920 — that wall came down. Cities that had gone dark for ten thousand years began to glow. A farmer who had been born reading by tallow smoke died listening to a radio powered by a dam a hundred miles away. It is, without much competition, the fastest and most complete transformation of daily life in the human record, and we have grown so used to it that we notice it only when it fails.

This book is the story of how it happened. It is a story with a long, slow beginning — a philosopher rubbing a piece of amber, a monk building a jar that could store a shock, an American flying a kite in a thunderstorm — and a sudden, violent climax, when three of the most remarkable men of the nineteenth century collided over a single question: not *whether* the world would be electrified, but *how*. Their fight had a body count, a propaganda war, electrocuted animals, a financier's fortune, and the first death in an electric chair. It ended at a waterfall. And out of it came the humming, invisible nervous system that now carries the modern world.

I have tried to tell it as it deserves to be told: as a human drama and as a genuine education. You do not need to know any physics to read this book. By the end, I hope, you will understand not only what Edison, Tesla, and Westinghouse did, but *why* it worked — why a changing magnetic field makes electricity, why alternating current could cross a continent when direct current could barely cross a city, and why the light comes on when you touch the switch. Where the popular legend has drifted from the truth — and with Tesla especially, it has drifted a long way — I have gone back to the record and said so plainly.

Come to the edge of the dark, then, and watch the lights come on.

— E.B Cohen

Prologue: The Night the Lights Came On

At three o'clock in the afternoon on Monday, the fourth of September, 1882, a man in a frock coat stood in the offices of the banker J. P. Morgan at 23 Wall Street, in lower Manhattan, and waited for a signal. His name was Thomas Alva Edison. He was thirty-five years old, already the most famous inventor in the world, and he was about to do something no one had ever done before: switch on a city.

A few streets away, in a squat brick building at 255–257 Pearl Street, an engineer stood beside a machine the size of a locomotive. The machine was a dynamo — a generator — and Edison, with the grandiose affection he brought to his machines, had nicknamed it "Jumbo," after the famous circus elephant. When the moment came, a switch was thrown, steam drove the great flywheel, and along a web of copper wires buried beneath the streets of the First District, an invisible river began to flow.

In hundreds of rooms across a square mile of Manhattan — in the counting-houses and law offices and the editorial rooms of the newspapers — pear-shaped glass bulbs, each holding a slender loop of carbonized bamboo, began to glow. They did not flicker like gas. They did not smoke, or smell, or need a match, or threaten to burn the building down. They simply came on, warm and steady and clean, and the men working beneath them looked up.

The next morning's *New York Times*, whose own offices were among the eighty-five customers on Edison's new circuit, reported the event with the slightly stunned tone of people describing a magic trick they could not explain. The light, the paper said, was "soft, mellow, and grateful to the eye," and "without a particle of flicker." It was, in every way that mattered, the beginning of the modern world.

And yet the light that came on that afternoon carried within it the seed of a war.

Edison's system was built on *direct current* — electricity that flows steadily in one direction, like water down a pipe. It was simple, it was safe at the low voltages he used, and it worked beautifully. It had only one fatal flaw, and Edison, for all his genius, either could not see it or would not admit it: direct current could not travel. The current that lit Pearl Street grew weaker with every yard of wire it crossed, so that a customer more than half a mile from the station got a feeble, orange glow, and a customer a mile away got nothing at all. To light a city, Edison would have to build a power station every few blocks — a dynamo on every corner, a coal fire in every neighborhood, forever.

There was another way. Across the ocean, and soon in a laboratory a few miles from Pearl Street, other minds were bending toward *alternating current* — electricity that reverses direction many times a second, and that, through a device called a transformer, can be pushed up to enormous voltages, sent across a hundred miles of thin wire with almost no loss, and then gently stepped back down for safe use at the far end. Alternating current could do the one thing Edison's system could not. It could carry power over the horizon.

The problem was that in 1882, no one had built a practical motor that could run on it — and a lighting system that could not also turn wheels, pump water, and drive machines was only half a system. That missing piece existed only in the mind of a strange, brilliant, penniless young Serb who was, at that very moment, working in one of Edison's own European offices and dreaming of a motor with no moving contacts at all: a machine driven by a spinning ghost of magnetism. His name was Nikola Tesla, and within two years he would sail into New York harbor with four cents in his pocket and a letter of introduction to the very man he was destined to fight.

The third figure in the drama was not yet paying attention. George Westinghouse, of Pittsburgh, was in 1882 a wealthy industrialist known for a single brilliant invention — an air brake that had made the railroads safe — and for a restless appetite for the next big thing. When he turned that appetite toward electricity a few years later, he would see instantly what Edison refused to see, buy the patents that Edison had let walk out his door, and become the champion of the current that could conquer distance.

So the stage was set. On one side stood Edison: American, self-made, deaf, relentless, the greatest inventor of the age, defending the system he had built and the fortune he had staked on it. On the other stood Westinghouse, the courtly Pittsburgh magnate, armed with Tesla's patents and a better idea. And between them, half genius and half prophet, moved Tesla himself, the man whose invention would decide the whole contest and who would, in the end, be crushed by the very forces he had set in motion.

The fight they were about to have has come down to us with a misleadingly tidy name: the War of the Currents. It sounds like a schoolroom topic, a matter of engineering committees. It was nothing of the kind. It was a brawl fought with propaganda and patents, with bribed newspapermen and staged public executions, with electrocuted dogs and horses and, at its lowest point, with the first human being ever strapped into an electric chair. Fortunes were made and broken. Reputations were poisoned. And the stakes could not have been higher, because whoever won would not simply win a market. He would decide the shape of the electrical age — the voltage in your walls, the hum of the grid, the reach of the light.

This book is the story of that war and of the ten thousand years of human curiosity that led up to it. It begins a very long way from Wall Street, in the sunlit market of an ancient Greek city, with a man holding a lump of fossilized tree sap and wondering why it clings to a feather. It ends in our own century, with the grid straining under a load its founders could never have imagined and with Edison's banished current, of all things, quietly staging a comeback. In between lie the kite and the frog's leg, the battery and the field, the invention factory and the rotating magnetic field, the White City and the thunder of Niagara.

Along the way I will ask you to understand a few beautiful and genuinely simple ideas — what electricity *is*, why a moving magnet can make it, why one kind of current could cross a continent and the other could not — because the drama makes no sense without them, and because they are among the most elegant things the human mind has ever worked out. You will not need mathematics. You will only need the willingness to look, for a moment, at the invisible thing running through your walls, and to feel a little of the wonder that gripped a Greek philosopher, a Philadelphia printer, and a boy on a railroad platform in Michigan who would grow up to switch on a city.

The lights are about to come on. Let us go back to the dark and begin.

Part I

The Amber and the Spark

From a Greek philosopher's lump of fossil resin to the first battery — the twenty-four centuries in which electricity was a curiosity, a parlor trick, and a terror, and slowly, haltingly, became a science.

Chapter 1: The Stone That Loved Straw

In the Greek city of Miletus, on the sun-baked coast of what is now western Turkey, sometime around the year 600 before the common era, a man sat turning a small golden-brown stone over in his hands. The stone was amber — not a mineral at all, but the fossilized resin of pine trees that had died millions of years before, hardened into a warm, translucent jewel the color of dark honey. The Greeks prized it, traded for it, carved it into beads. But the man holding it, whose name was Thales, was not admiring its beauty. He had rubbed it against a piece of wool or fur, and now, when he held it near a few wisps of straw or a downy feather, the straw leapt up and clung to the stone as if it were alive.

We do not have Thales's own words about this; almost nothing he wrote survives, and what we know of him comes secondhand from later Greeks who treated him as the first of the philosophers. But the observation was remembered and repeated for centuries, and it deserves to be remembered still, because it is the oldest recorded human encounter with the force that would one day light the world. The Greek word for amber was *ēlektron*. Twenty-four centuries later, an Englishman would reach back to that word to name the mysterious property Thales had stumbled upon, and from *ēlektron* we would get *electric*, *electricity*, *electron*, and every switch and socket and glowing screen that follows from them. The whole electrical age is, in a sense, named after a warm yellow stone that a curious Greek rubbed on his sleeve.

It is worth pausing on how strange this moment really was, because familiarity has drained the wonder out of it. Imagine you had never heard of electricity — that no one you knew had ever heard of it, that the very idea did not exist. Now imagine you rub a stone on your coat, and dead straw jumps toward it across empty air. There is no wind. Nothing touches the straw. It simply *moves*, pulled by an invisible hand. To a Greek, for whom the world was full of hidden souls and animating spirits, the most natural explanation was that the amber was, in some sense, alive — that it possessed a *psyche*, a soul, that reached out and grasped. Thales, according to Aristotle, may have used the amber and the lodestone (a naturally magnetic rock) as evidence that even apparently dead matter could contain soul. He was wrong about the soul. But he was asking exactly the right question: *what is this force, and where does it come from?*

To understand what was really happening in Thales's hand, we can borrow an everyday experience that almost everyone has had. On a dry winter day, you pull a wool sweater over your head and your hair stands up and crackles; you shuffle across a carpet and touch a doorknob and a tiny blue spark stings your finger. That is the amber effect, alive in your own home. When two different materials rub together — wool and amber, hair and wool, your shoes and the carpet — one of them steals a vanishingly small number of tiny particles from the other. We now call those particles *electrons*, and we call the imbalance they leave behind an *electric charge*. The amber, robbed or enriched, becomes charged, and a charged object pulls on light things nearby. Rub, and you build up the charge. Touch a conductor, and it leaps away as a spark. Thales had, without any possible way of knowing it, performed the first electrical experiment. He had made static electricity.

Here it is worth defining our first real term, plainly, because everything else in this book is built on it. **Electric charge** is a basic property of matter, like mass. It comes in two kinds, which we label positive and negative, and the fundamental rule — the rule that runs through the entire story from here to the modern grid — is that *opposite charges attract and like charges repel.* The straw was pulled to the amber because the two carried opposite charges. Your hair stands up because every strand carries the *same* charge and they all shove one another apart. That single rule, attract and repel, is the seed from which the lightning rod, the electric motor, and the power station all eventually grow.

The ancient world noticed the amber effect and, for the most part, left it there. This is one of the quietly astonishing facts of the history of electricity: for roughly two thousand years after Thales, essentially nothing happened. The Greeks catalogued it as a curiosity. The Romans wore amber and admired its glow; Pliny the Elder wrote about it in his great encyclopedia of natural history without adding much to Thales. The lodestone got more attention, because a sliver of it, floated on water or hung from a thread, would swing to point north — the basis of the compass, which by the medieval period was guiding ships across oceans. But magnetism and the amber effect were treated as two separate marvels, unrelated to each other and to anything useful. That they were secretly two faces of a single force — the deepest and most consequential discovery in this whole book — lay more than two millennia in the future.

Why the long silence? Partly it was that the amber effect is feeble and fickle. It works best on cold, dry days; humidity kills it. It requires the right materials, rubbed the right way. It produces nothing you can obviously *use* — no light, no heat you can feel, no motion beyond a twitching feather. And the intellectual tools to investigate it did not yet exist. The Greeks were brilliant geometers and subtle philosophers, but the idea of the controlled experiment — of teasing nature apart by deliberately changing one thing at a time and measuring the result — was not yet part of the human toolkit. It would take the scientific revolution of the sixteenth and seventeenth centuries to supply it.

Here is a fact worth carrying forward, because it corrects a misconception that quietly shapes how we imagine progress. We tend to assume that once a phenomenon is noticed, understanding follows quickly, as if curiosity were a fuse leading inevitably to discovery. The amber effect is the great refutation of that assumption. It was noticed, reliably and repeatedly, in the sixth century BCE. It was not *understood*, even crudely, until the seventeenth century CE. The practical exploitation of electricity — the battery, the motor, the light — came a further two hundred to three hundred years after that. The distance from Thales's straw to Edison's bulb is roughly twenty-five centuries, and for the first twenty-two of them the human race made almost no progress at all. Genius, it turns out, is not enough. You also need the right questions, the right tools, and the accumulated groundwork of generations. This book is, among other things, the story of how all three finally came together.

A second surprising fact: the very word we use for the force betrays how backward the discovery ran. We did not name amber after electricity; we named electricity after amber. The Greeks had the substance and the word centuries before anyone had the concept. It is as though we had named fire after the particular kind of stick first used to make it — a reminder that human beings often live with a phenomenon for a very long time, and even give it a name, before they have the faintest idea what it is.

Consider, finally, how much of the modern world descends from that one small observation. Every time you peel plastic wrap and it clings to your hand, every time lightning splits the sky, every time your phone charges or a spark plug fires or a screen lights up, you are watching a more elaborate version of straw leaping to amber. The force is the same. Only our command of it has changed. Thales could make a feather twitch. Twenty-five centuries later, his intellectual descendants would send that same force howling through copper cables to power entire nations. Between those two points lies everything in this book.

But before the philosophers of Miletus could even begin to make sense of the clinging stone, humankind had to reckon with electricity in its wildest and most terrifying form — the form that fell from the sky, split trees, and killed men where they stood. Long before anyone rubbed amber on wool, people had been staring up at the lightning and calling it the anger of the gods.

Chapter 2: Lightning and the Gods

On a summer evening in the highlands of ancient Anatolia, a shepherd watched a storm build over the mountains and understood, with perfect certainty, that he was watching a god lose his temper. The clouds had turned the color of a bruise. The air had gone strange and heavy, the way it does before a storm, so that even the sheep grew restless. Then the sky cracked open with a white slash of fire, a single jagged line drawn from heaven to earth in less time than it takes to blink, and an instant later the thunder arrived — a sound so vast and physical that the shepherd felt it in his chest, in the ground beneath his feet, in his teeth. A tree on the far ridge stood suddenly wrapped in flame. To this man, and to nearly every human being who lived before the eighteenth century, there was no mystery about what had happened. A god had thrown a spear of fire. The only question worth asking was *why*, and whether the god could be appeased.

We began this book with amber — with electricity in its mildest, gentlest, most easily overlooked form, a stone that made straw twitch. But there was another face of the same force, and it was the first one humanity ever met. Long before any Greek philosopher rubbed a warm yellow stone on his sleeve, our ancestors had been staring up in terror at the lightning. Here was electricity at its wildest: a discharge of unimaginable power, hotter than the surface of the sun, capable of splitting an oak or killing a man where he stood. That the delicate clinging of amber and the annihilating violence of a lightning bolt were the *same phenomenon*, differing only in scale, is one of the great counterintuitive truths of this whole story — and it would not be proven until Benjamin Franklin flew a kite in a thundercloud, some twenty-four centuries after Thales. For all the ages before that, lightning belonged not to the philosophers but to the gods.

And nearly every people who ever looked up gave the lightning to their mightiest god. This is one of the quietly astonishing patterns in the history of human belief. Cultures that never met, separated by oceans and millennia, independently arrived at the same idea: that the being who ruled the sky and hurled the thunderbolt was the king of all the gods. The Greeks had Zeus, lord of Olympus, whose very emblem was the *keraunos*, the thunderbolt, forged for him — the myth said — by the one-eyed Cyclopes. When Zeus was angry, he threw lightning; when he wished to seal an oath or punish a liar, the bolt was his instrument. The Romans took him over as Jupiter, and a place struck by lightning became sacred ground, walled off and consecrated, because the god had reached down and touched it. The Norse peoples of the cold north had Thor, red-bearded and enormous, whose hammer Mjölnir made the thunder and the lightning as he rode his chariot across the sky; the very word *Thursday* is Thor's day, so that every week, without knowing it, we still name a day after the thunder god. In the Hindu tradition, Indra wielded the *vajra*, a thunderbolt weapon, and rode to battle against the serpent of drought. The Baltic peoples had Perkūnas, the Slavs had Perun, the Finns had Ukko, whose name simply meant "old man" and whose thunderbolts were struck, in some tellings, when he and his wife embraced. The Yoruba of West Africa had Shango, a thunder god whose double-headed axe split the sky. Across the whole spread of human cultures, the same instinct repeats: the thunder is the voice of the greatest power there is.

Why should so many peoples, with no contact between them, land on the same story? The answer tells us something important about how the human mind works, and it is worth pausing on, because it corrects a misconception we tend to carry — the flattering assumption that ancient people were simply foolish or credulous. They were nothing of the kind. Faced with lightning, they were reasoning as well as anyone could reason without the tools of modern science. Consider what they had to explain. Here was a force that came from the sky, which was clearly the domain of something vast and beyond reach. It was overwhelmingly powerful, capable of killing in an instant. It was unpredictable, striking here and not there for no reason a person could discern. And it seemed, above all, *intentional* — it picked out a single tree, a single ship, a single man, and destroyed him while sparing his neighbor. Human beings are pattern-finding, agency-detecting animals. When we see something powerful, selective, and unpredictable, our deepest instinct is to suspect a *mind* behind it, a will, a chooser. To an ancient watching a bolt strike one house and leave the next untouched, the hypothesis of an angry god was not stupidity. It was the most reasonable explanation available, and it would remain so for a very long time.

The fear was entirely rational, because lightning genuinely was — and remains — one of nature's great killers. Here is a fact that ought to give any modern reader pause: a single bolt of lightning heats the narrow channel of air it travels through to roughly 30,000 degrees Celsius, which is about five times hotter than the visible surface of the sun. That superheated air explodes outward faster than the speed of sound, and the shockwave of that explosion is the thunder we hear. The lightning and the thunder happen at the very same instant; the reason we see the flash before we hear the boom is simply that light travels to our eyes almost instantaneously while sound crawls along at roughly a fifth of a mile per second. This is the origin of the old trick, still taught to children, of counting the seconds between the flash and the thunderclap: every five seconds of counting means the strike was about a mile away. That homely piece of folk wisdom is, in fact, perfectly good physics, and it was being practiced for centuries before anyone knew what lightning actually was.

Certain places drew the fire again and again, and this deepened the sense that the bolt was aimed by an intelligent hand. Tall structures — temples, towers, the masts of ships — were struck far more often than the low ground around them, for reasons no one understood but everyone noticed. The temples were often the tallest buildings a community possessed, which meant that the house of the god was, with grim regularity, the very thing the lightning destroyed. This produced some strange and revealing episodes. There is a famous case involving the great bell towers and churches of medieval and early modern Europe, whose lofty spires were struck over and over. It was widely believed that ringing the church bells during a storm would drive away the lightning and the evil spirits riding it — some bells were even inscribed with the Latin motto *fulgura frango*, "I break the lightning." The tragic result was that when a storm came, bell-ringers were sent up into the very highest and most exposed part of the tallest building in town, hauling on ropes attached to a large piece of metal, precisely where the danger was greatest. Over one thirty-three-year stretch in the eighteenth century, by one much-cited count, lightning struck hundreds of church towers across Germany and killed scores of bell-ringers at their ropes. The custom meant to ward off the bolt was quietly delivering men to it.

Ships suffered the same way. A wooden sailing vessel on the open sea presented a tall mast to the sky with nothing higher for miles, and sailors dreaded a storm not only for its wind and waves but for the fire that might come down the mast, splinter it, ignite the tarred rigging, and — worst of all — reach the powder magazine. Naval records from the age of sail are grim with accounts of ships crippled or destroyed by lightning, of masts shattered and men flung to the deck. The sailors, like the bell-ringers, had noticed the pattern without understanding the cause. They knew the height of the mast invited the bolt; they simply had no idea *why*, and no means to defend against it, until the lightning rod arrived in the eighteenth century and gave the fire a safe path to follow.

What, then, did the ancients think lightning actually *was* — beyond the weapon of a god? A few daring minds tried to reach past the mythology. The Greek philosopher Anaximander, and later the great atomist thinkers, proposed that thunder and lightning came from wind bursting out of clouds, or from clouds colliding and tearing. Aristotle, whose ideas would dominate European thought for nearly two thousand years, taught that the sun drew up a dry, smoky "exhalation" from the earth into the clouds, and that when this hot dry stuff was violently squeezed out of the cold wet cloud, it burst into flame — the flame was the lightning, and the noise of the bursting was the thunder. It was wrong, but it was a genuine attempt at a natural, mechanical explanation, framed without recourse to Zeus. The Roman philosopher Seneca devoted a whole book of his *Natural Questions* to lightning, cataloguing its varieties with real care. And here is a detail that captures the strange half-light in which these early thinkers worked: Seneca recorded, apparently as a plain observation, that lightning could melt the blade of a sword while leaving its wooden scabbard untouched, or melt the coins inside a purse without scorching the leather. He had no framework to explain it, but he had noticed, with startling accuracy, that this force did something peculiar and selective to metal — a hint, unrecognized, that lightning had a special relationship with the very materials we would one day call *conductors*.

That word points to the deepest puzzle of all, the one that no ancient could solve and that turns out to be the hidden thread running through this entire book. Lightning, the amber effect, the shock you get from a doorknob, the spark that jumps a gap — these look like utterly different things. One is a stone that tugs at straw; another is a bolt that levels a tree. What could they possibly have in common? The ancients had no way even to ask the question, because they had no reason to suspect a connection. It would take the slow construction of an entire experimental science — the friction machines, the storage jars, the careful measurements of the seventeenth and eighteenth centuries — before anyone could look at a laboratory spark an inch long and a lightning bolt a mile long and dare to say: *these are the same thing*. The whole intervening story is, in a sense, the story of shrinking that gulf, of taming the god's weapon down to a spark you could make on a workbench, and then recognizing, with a shock of its own, that the tame spark and the terrible bolt were one.

You still meet this ancient fear, and its ancient logic, every single day, though you have wrapped it in the language of engineering. When a storm rolls in and someone tells you to stay off the golf course, out of the open field, away from the lone tree and the tall metal flagpole, they are repeating — in the vocabulary of statistics and physics rather than gods — exactly what the bell-ringers and sailors learned at such cost: that height and metal attract the fire. When your airliner flies serenely through a thunderhead and is struck by lightning, as commercial aircraft are on average once or twice a year, and nothing happens at all, it is because the metal skin of the plane carries the current harmlessly around the passengers and lets it flow off the wingtips — the lightning rod's principle, perfected. And every building you have ever entered that stands taller than its neighbors carries, somewhere on its roof, a metal spike wired to the ground, silently offering the lightning an easier path than through the walls. The god's thunderbolt has not grown any weaker. We have simply, at last, learned where it wants to go, and gotten out of its way.

For all of human history until the dawn of the modern age, then, electricity wore two faces that no one knew were connected: the harmless amber that pulled at straw, and the murderous lightning that fell from the sky. Bringing those two faces together into a single science required a new kind of thinker entirely — a man who would refuse to take the old marvels on faith, and who set out, at the court of Queen Elizabeth, to test them one by one.

Chapter 3: Gilbert and the Birth of a Science

In a paneled room in London, sometime in the last years of the sixteenth century, an aging physician bent over a workbench cluttered with the strangest instruments a doctor ever owned. There were lumps of amber and jet, chips of gemstone and glass, a spherical lodestone he called a *terrella* or "little earth," and a curious lightweight needle, balanced on a pin so that it could spin freely like the pointer of a compass. William Gilbert was in his fifties, a man of high reputation and higher standing — he would soon be named personal physician to Queen Elizabeth I herself — and by every expectation of his age he should have spent his evenings with his patients, his classical texts, and his fees. Instead he spent them here, rubbing bits of amber, watching a needle twitch, and quietly assembling the demolition of two thousand years of received wisdom. When he brought a freshly rubbed piece of amber near his balanced needle, the needle swung to follow it, and Gilbert did something that almost no one before him had thought to do with such a marvel. He did not marvel. He measured. He tried another material, and another, and wrote down exactly what each one did.

Gilbert's world took its science almost entirely on the authority of the ancients, and above all of Aristotle. For roughly two thousand years, to answer a question about the natural world was to ask what the great authorities had written, not what nature itself might show you if you asked it directly. The amber effect had been sitting in the books, unexamined, since Thales — noted, repeated, embroidered with legend, but never systematically tested. Amber attracts straw; everyone knew it. So does the lodestone attract iron; everyone knew that too. And because the two effects looked superficially alike — both were invisible attractions exerted by special stones across empty space — the learned tradition had quietly lumped them together, or explained both away with the same vague talk of sympathies and occult virtues, of amber giving off a sticky "effluvium," a kind of invisible glue, that seized the straw. Gilbert's revolutionary act was not any single discovery. It was a method. He resolved to trust nothing he had not seen with his own eyes, to test every claim by experiment, and to throw out whatever failed — even if Aristotle himself had written it.

The book in which he laid out the results appeared in 1600 under the title *De Magnete — On the Magnet* — and it is one of the founding documents of experimental science, published a full generation before Galileo's famous clashes with the Church and decades before the word *scientist* would even exist. Most of *De Magnete* is about magnetism, and its central argument was breathtaking in its ambition: Gilbert proposed, correctly, that the whole Earth is itself a giant magnet, and that this is why a compass needle points north. His little spherical lodestone, the *terrella,* was his model of the planet; he showed that a tiny compass moved across its surface behaved exactly as a real compass does across the face of the Earth, dipping and swinging just so. It was a stunning leap — to explain the mysterious northward pull that had guided ships for centuries not by appeal to the Pole Star or some magnetic mountain at the top of the world, as legend had it, but by the sober claim that we all live on the surface of an enormous magnet. He was right, and it remains true today: the compass in your phone works because the Earth's iron core generates a magnetic field, precisely as Gilbert argued four centuries ago.

But it is one section of *De Magnete*, easy to overlook amid the magnetism, that earns Gilbert his place at the head of *our* story. For Gilbert did something no one had done before: he set out to determine, by experiment, whether the amber effect and the magnetic effect were truly the same thing, as the tradition assumed — or whether they were two entirely different forces that merely looked alike. And to do it, he built the first scientific instrument ever devised for the study of electricity. He took a light metal needle, delicately balanced on a sharp point so that the slightest force could turn it, free to pivot in any direction — much like a compass needle, but responsive to the amber effect rather than to magnetism. He called it a *versorium*, from the Latin for "to turn." Bring a rubbed piece of amber near it, and the needle swings toward the amber. It was, in effect, the world's first electroscope: a detector sensitive enough to register the feeble pull that Thales had felt in his hand, and to compare one material against another. With this simple, elegant device, the study of electricity graduated from anecdote to measurement.

What Gilbert found with his versorium overturned the old confusion completely. First, he discovered that amber was not remotely unique. The tradition had treated the clinging power as a special virtue belonging to amber almost alone; Gilbert, patiently testing material after material, found that a whole family of substances behaved the same way once rubbed. Jet, sulphur, sealing wax, hard resins, various gemstones, and above all glass — all of them, when rubbed, would swing his little needle. He had discovered that the amber effect was a *general* property of a broad class of materials, not a quirk of one honey-colored stone. And he needed a word for this shared property, a single name for "the power to behave the way rubbed amber behaves." Reaching back, as scholars did, to the Greek, he took *ēlektron*, the Greek word for amber, and coined from it the Latin adjective *electricus* — "amber-like," "of the nature of amber." From that coinage, over the following decades, English drew the words *electric* and, a little later, *electricity*. It is a fact worth savoring: the entire vocabulary of the electrical age — every *electric* motor, *electric* light, and *electronic* device — descends from a single word that an Elizabethan physician minted in 1600 to mean, quite literally, "acting like rubbed amber." When you flip a light switch, you are speaking Gilbert's Latin without knowing it.

Second, and more important still, Gilbert proved that electricity and magnetism were *not* the same force, contrary to two thousand years of loose assumption. He drew up a careful list of the ways they differed, and the differences were undeniable once you looked. The lodestone attracted only iron and a few related metals; rubbed amber attracted nearly anything light — straw, feathers, bits of paper, even droplets of water and thin wisps of smoke. The lodestone worked all the time, needing no preparation; amber worked only after being rubbed, and its power faded. The lodestone's pull had direction, a north and a south that oriented a needle along a definite line; the amber's pull simply drew light things toward it from any side. Magnetism passed straight through paper, wood, and even the human hand; the amber effect was blocked and killed by the slightest breath of moisture or the interposition of certain materials. Point by point, Gilbert showed that here were two distinct phenomena, each with its own laws, which had merely been mistaken for one because both involved invisible attraction. Drawing that line — *electricity is one thing, magnetism is another* — was an essential act of intellectual housekeeping. The strange and beautiful irony, which will occupy much of this book, is that Gilbert was in the deepest sense wrong: electricity and magnetism turn out to be two faces of a single unified force, as a Danish professor would begin to reveal in 1820 and as Maxwell would prove with mathematical finality in the 1860s. But before science could earn its way to that grand reunion, someone first had to separate the two cleanly and study each on its own terms. Gilbert's "error" was the necessary first step; you cannot rejoin what has never been properly told apart.

It is worth correcting a misconception that clusters around this era. We often imagine that the scientific method sprang, more or less fully formed, from a single famous mind — Galileo, perhaps, or Francis Bacon, the philosopher who wrote so eloquently about learning from nature by experiment. But Bacon's great manifesto for experimental science appeared decades after *De Magnete*, and Bacon, though he admired Gilbert's diligence, actually criticized him for building too grand a theory on his magnet. The truth is that the experimental method had no single inventor; it was assembled, gropingly, by many hands. Gilbert was one of the earliest and clearest of those hands. He spent, by his own account, years and a considerable fortune on his investigations, refusing to publish until he had tested his claims himself. Galileo, who read *De Magnete* closely, praised Gilbert warmly and adopted his magnetic ideas. It is entirely fair to say that when Gilbert set his amber against his versorium and wrote down what he saw, he was practicing, in the year 1600, exactly the discipline that would eventually give us everything from the light bulb to the microchip: ask nature a precise question, and let its answer, not the authority of the ancients, decide.

There is a detail about Gilbert's standing that deserves telling, because it corrects the flat picture we tend to carry of "early scientists" as marginal cranks tinkering in obscurity. Gilbert was nothing of the sort. He was among the most eminent physicians in England, president of the Royal College of Physicians, and personal doctor to the aging Queen Elizabeth I in the last years of her reign; when James I came to the throne he retained Gilbert in the same office. This was a man at the very summit of the Elizabethan establishment, who could have rested comfortably on his practice and his patronage. That he chose instead to spend his fortune and his evenings rubbing amber and pivoting needles, and that Elizabeth is said to have taken an interest in his experiments and even watched some of his demonstrations at court, tells us something about the temper of that extraordinary age. The queen who sent Drake around the world and defeated the Armada also, it seems, peered curiously at a little turning needle as her physician showed her that rubbed glass could reach across empty air. The new experimental spirit was not confined to obscure workshops; it had reached the throne itself.

Gilbert also gave us, almost in passing, the first genuine theory of *why* the amber effect happens — and while it was wrong, it was wrong in an interesting and honest way. He proposed that rubbing warmed the amber and caused it to release a subtle, invisible material effluvium, a fine emanation that streamed out, reached the light object, and then, cooling and contracting, drew the object back toward the amber as it returned home, like an outstretched arm gathering something in. This mechanical picture — attraction by an invisible fluid or emanation reaching across the gap — would dominate thinking about electricity for the next two centuries, right through Franklin's day. It was replaced only when scientists learned to speak of charge and, eventually, of electrons. But notice what Gilbert had done even in being wrong: he had insisted that the invisible attraction must have a *physical cause*, something real passing between the objects, rather than a magical sympathy or the will of a god. He had taken electricity out of the temple and put it on the workbench.

You are surrounded, right now, by the descendants of Gilbert's versorium. Every device that senses an electric charge — the touchscreen that knows where your finger is, the humidity sensor in a thermostat, the little grounding straps that technicians wear to bleed static safely away before touching a delicate circuit board — is a distant, refined child of that balanced needle swinging toward a lump of rubbed amber. More fundamentally, the very idea that you can build an *instrument* to detect and measure an invisible force, rather than merely marveling at it, is Gilbert's bequest. The dry crackle of static when you pull off a wool hat is the same effect he catalogued; the difference is that we now have four centuries of instruments, born from his needle, that can measure it to a precision he could not have dreamed of.

Gilbert died in 1603, only three years after his great book appeared, most likely of the plague that swept London that year, and the same year his royal patron Elizabeth died. He had drawn the crucial line between electricity and magnetism, given the phenomenon its enduring name, and, above all, shown that the amber effect could be studied by patient experiment rather than swallowed on authority. But he had left a glaring practical problem for those who came after. To study electricity at all, you first had to *make* it, and Gilbert's only method was the ancient one: rub a stone by hand and hope for a feeble, fading charge on a dry day. What the young science needed next was a machine — some engine that could conjure electricity reliably, powerfully, and on demand — and in the workshops of a German burgomaster, the first such machine was about to be born.

Chapter 4: Machines to Make Sparks

In the German city of Magdeburg, around the middle of the seventeenth century, the mayor was doing something no mayor had any business doing. Otto von Guericke had rebuilt his town from the ashes of one of the most savage sacks of the Thirty Years' War, in which tens of thousands of its people had died; he was a diplomat, an engineer, a man of weighty public affairs. And in his spare hours he retreated to a workshop to melt sulphur and pour it, molten and reeking, into a hollow glass globe the size of a large melon. When the sulphur cooled and hardened, he broke away the glass, leaving a solid yellow sphere of sulphur pierced through its center by an iron rod, so that it could be mounted on a frame and spun like a grindstone. Then he set it turning, laid his dry palm against the whirling globe, and watched what happened. The sphere crackled. It clung to bits of fluff and drew them dancing after it. In the dark, it glimmered with a faint, eerie light. And when Guericke brought his knuckle close, a spark leapt the gap with a tiny, distinct snap. He had built the first machine in history for manufacturing electricity on demand.

To feel why this mattered, recall the bind that Gilbert had left the young science in. Electricity, in 1600, could be produced only by hand-rubbing amber or glass, a laborious business that yielded a weak and fleeting charge, and only on a dry day. It was rather like trying to study fire when your only method of making it was to strike two flints together and blow desperately on a spark — you could prove fire existed, but you could hardly do anything ambitious with it. What the field needed was the equivalent of a reliable furnace: a device that would produce electricity steadily, strongly, and whenever the experimenter wished. Guericke's spinning sulphur globe was that device, or its rough ancestor. By turning a crank and holding a hand to the sphere, he could keep the amber effect going continuously and build it up to a strength no hand-rubbed lump had ever reached — strong enough to draw visible sparks and to make feathers hover and drift in the air above the globe as it repelled and attracted them by turns.

Guericke, it must be said, did not fully understand what he had made, and he is far better remembered for a different and gaudier experiment. This was the same man who, to demonstrate the power of a vacuum and the pressure of the atmosphere, had two teams of horses harnessed to pull apart a pair of empty copper hemispheres held together by nothing but the air pressing on them from outside — the famous "Magdeburg hemispheres," which the horses, straining, could not separate. He was a showman as much as a scientist, and he interpreted his electrical globe partly through the lens of his cosmic speculations, imagining its powers of attraction and repulsion as a little model of how the Earth held the Moon and drew objects to itself. He was reaching, without the tools to grasp it, toward something like a unified idea of the forces of the world. He missed the mark. But the machine remained, and machines have a way of outliving the theories of the men who build them.

Across the Channel, at the dawn of the eighteenth century, an Englishman named Francis Hauksbee took up the friction machine and made it glow. Hauksbee was an instrument-maker and demonstrator who worked under the eye of Isaac Newton himself at the Royal Society in London, performing experiments for the assembled gentlemen-scientists at their weekly meetings. He improved the machine by using a spinning glass globe rather than sulphur, and then he made a discovery that must have stopped the room. He partly emptied the air out of one of his glass globes, creating a near-vacuum inside, and set it spinning against his hand in a darkened room. The globe filled with a soft, flickering purple-blue light, bright enough, it was said, that one could read large print by it. Hauksbee had produced a cold electric glow inside a nearly airless glass — and though neither he nor anyone in that room could have known it, he was looking at the direct ancestor of every neon sign, every fluorescent tube, and every glowing screen that would one day fill the modern night. The gas discharge that lights a "neon" sign is the same phenomenon Hauksbee coaxed out of his spinning globe in 1705: electricity passing through thin gas, making it shine. He had lit the first electric light, two full centuries before Edison, though he had no way to sustain it or to grasp what it was.

The next great step belonged to a man who had spent most of his life in obscurity and poverty, and it is one of the most appealing stories in the whole history of the science. Stephen Gray was an aging, self-taught dyer's son from Canterbury, a man of no fortune and little formal standing, who by his sixties had been admitted, poor and dependent on charity, into the Charterhouse, a London almshouse for gentlemen who had fallen on hard times. There, with borrowed and improvised equipment, this humble pensioner made a discovery that ranks among the most important in the early history of electricity. Gray was experimenting, around 1729, with a glass tube that he rubbed to electrify — and he had corked the ends of the tube to keep out dust. He noticed, almost by accident, that the cork itself, though he had never rubbed it, seemed to have become electrified: it would attract a feather. The "electric virtue," as he called it, had somehow traveled from the rubbed glass into the cork that touched it.

This nagging observation set Gray on the trail of a truly fundamental discovery: that electricity could be made to *travel*, to move from one object to another along a connecting path — and that some materials would carry it while others stopped it dead. He stuck a slender stick of wood into the cork, and the electric virtue traveled to the far end of the stick. He tied a length of packthread — ordinary twine — to the stick, hung an ivory ball on the end of the thread, and found that the ball, yards away from the rubbed glass, would now attract bits of leaf. The electricity was running down the string. Gripped by the possibilities, he set out to see how far the virtue could be sent. In a series of experiments in a great barn and a gallery, he strung packthread across long distances, suspending it from the ceiling on loops of silk thread so that it hung in the air, and he found he could transmit the electric virtue hundreds of feet — by some accounts, when he and a collaborator ran lines across a garden and between buildings, on the order of several hundred feet, and in later trials farther still. He had sent electricity down a wire, so to speak, generations before there were any electric wires. Every telegraph line, every power cable, every cord in your home descends from Gray's packthread.

But the truly decisive part of the discovery lay in the failures. When Gray first tried to hold his transmitting line up with ordinary hemp or wire supports, it did not work — the electric virtue drained away and vanished. It was only when he happened to hang the line from loops of *silk* that the electricity ran all the way to the end. The reason, Gray gradually realized, was that some materials let the electric virtue pass through them and away, while others refused to carry it at all and so bottled it up on the line. Silk stopped it; that was why hanging the packthread from silk loops kept the charge from leaking into the ceiling. We now divide the world's materials, following exactly this discovery, into two great families. **Conductors** are substances that let electricity flow through them easily — metals above all, and also damp string, the human body, and moist air. **Insulators**, sometimes called nonconductors, are substances that block the flow and hold electricity in place — silk, glass, dry wood, resin, rubber, and most plastics. Gray had discovered the single most practically important distinction in all of electrical engineering. Every wire in the world is a conductor wrapped in an insulator: a copper core to carry the current where you want it, a rubber or plastic sheath to keep it from leaking where you don't. When you hold the insulated grip of a plugged-in appliance in perfect safety, you are trusting your life, quite literally, to Stephen Gray's distinction between the string that carries and the silk that blocks.

There is one more experiment of Gray's that deserves telling, because it is at once charming, faintly horrifying, and scientifically profound. To prove that the human body was itself a conductor, Gray suspended a boy — an orphan, a "charity boy," by the accounts — horizontally in the air on silk cords, like a hammock, so that the insulating silk cut him off from the ground. Then Gray brought a rubbed glass tube up to the boy's feet and electrified him. The boy's whole body became charged, and when the experimenters held bits of gold leaf and feathers near his face and hands, the light objects flew up and clung to him; sparks could even be drawn from his nose. The famous "flying boy" demonstration made a sensation, and it makes a genuine point: it proved that a person is a conductor, that the electric virtue flows through living flesh as readily as through metal, and that only the silk insulation kept the charge from running out of him into the earth. It is a discovery whose darker implication — that the human body is an excellent conductor, and therefore terribly vulnerable to electric current — would haunt the last part of this book, when the safety of alternating current became the deadliest weapon in a corporate war.

Across the water in France, a customs official and gifted experimenter named Charles François de Cisternay du Fay took Gray's results and pushed them one crucial step further, uncovering a hidden duality that had lurked in the phenomenon since Thales. Du Fay noticed something puzzling. When he electrified two light objects — say, two small pieces of gold leaf or two bits of cork — by touching them to the *same* electrified rod, they would push each other apart. Yet an object electrified by rubbed *glass* would attract an object electrified by rubbed *amber* or resin. Following the clues with care, du Fay concluded around 1733 that there were, in fact, **two different kinds of electricity**. One he called *vitreous*, from the Latin for glass, because it was the kind produced by rubbing glass; the other he called *resinous*, produced by rubbing amber, resin, or sealing wax. And he found the governing rule: two objects carrying the *same* kind of electricity repel each other, while two objects carrying *opposite* kinds attract. This is the very rule we met in the first chapter, "like repels, opposite attracts" — but now it had a physical basis in two distinct sorts of electric charge. Du Fay's *vitreous* and *resinous* would, a generation later, be rechristened by Benjamin Franklin as *positive* and *negative*, the names we still use, and the two-fluid picture du Fay proposed would compete with Franklin's one-fluid picture for decades. Between them, Gray and du Fay had established the essential grammar of static electricity: it flows through conductors and not insulators, and it comes in two opposing kinds.

The machines and discoveries of this restless century — Guericke's spinning globe, Hauksbee's glowing vacuum, Gray's traveling virtue, du Fay's two electricities — had transformed the amber effect from an isolated curiosity into a genuine science with instruments, rules, and vocabulary. But every one of these experiments shared a maddening limitation. The electricity they made was fleeting: you could generate it, watch it leap and cling and glow, but you could not *keep* it. The charge drained away the instant you stopped cranking. What the experimenters longed for, without quite knowing how to ask for it, was some vessel that could hold electricity the way a jug holds water — a bottle you could fill with the invisible fire and stopper up for later. In a Dutch university town, and very nearly at the same moment in a German cathedral city, that bottle was about to be invented, and it would bite the first men who made it hard enough to knock them senseless.

Chapter 5: The Jar That Bit Back

In January of 1746, in the Dutch university town of Leiden, a respected professor of physics named Pieter van Musschenbroek sat down to write a letter to a colleague in Paris, and what he wrote was very nearly a warning. He had received, he reported, a shock so violent that he thought his life was over. He would not, he declared, take such a blow again for the whole kingdom of France. This from a sober, middle-aged Dutch professor, not given to melodrama — a man who had spent his career carefully measuring the behavior of heat and light and the strength of materials. What had reduced him to this state was a simple glass jar, partly filled with water, that he had been trying to charge with electricity. He had grasped the jar in one hand while its water was connected to a friction machine, and when he reached to touch the apparatus with his other hand, something in the jar had discharged its whole store of electricity through his arms and chest at once. His body convulsed. His arm and shoulder ached for days. He had just become one of the first human beings ever knocked flat by a stored electric charge, and in the process he had helped invent the single most important instrument in the early history of electricity.

To understand what had happened to poor Musschenbroek, we need to appreciate the problem that he and every experimenter of his day had been wrestling with. The friction machines could *make* electricity, in ever greater quantities, but nobody could *keep* it. The charge sat on the surface of the electrified globe or the suspended wire, and it bled away into the air and the ground almost as fast as it was produced. It was, in the frustrated experimenters' minds, rather like trying to collect a precious liquid that evaporated the instant it appeared. Musschenbroek and his assistant, a lawyer named Andreas Cunaeus who was helping in the laboratory, had been trying to solve this by, in effect, pouring the electricity into water. The reasoning, dim as it was, went like this: perhaps if you ran the electricity from the machine down a wire into water sealed in a glass bottle, the water would trap the electric fluid and hold it, the way a jug holds a liquid. So they had a glass jar with water in it, a wire or brass rod dipping into the water and leading up out of the mouth of the jar to the friction machine, and Cunaeus — or Musschenbroek, accounts differ on who suffered first — held the jar in his bare hand while it charged. That grip was the fatal detail.

Almost simultaneously, and quite independently, a German cleric named Ewald Georg von Kleist, a dean in Pomerania, had done much the same thing a few months earlier, in late 1745. He had put electricity into a medicine bottle partly filled with liquid, holding it in his hand, and had likewise been staggered by the discharge when he touched the nail sticking out of the cork. Von Kleist's reports were garbled and hard for others to reproduce, so history has tended to attach the device's name to Leiden — it became known across Europe as the **Leyden jar**, after the town, though in fairness von Kleist has as good a claim to have discovered it first. What matters is that within a single year, two men in two countries had stumbled onto the same astonishing fact: a glass jar, wet inside and held in the hand outside, could soak up and store a quantity of electricity vastly greater than anything a friction machine could put on a bare wire — and could then release the whole of it in one savage instant.

What they had built, without any idea of how it worked, was the first **capacitor** — a device for storing electric charge. The principle, which took decades to understand and can be stated simply now, is this. Electricity likes to accumulate on the two sides of a thin insulating barrier when those two sides are held at opposite charges. In the Leyden jar, the glass wall of the jar was the barrier — an insulator, glass being one of the substances Stephen Gray had shown would not carry electricity. On the *inside* of the glass sat the water, connected through the rod to the machine, piling up one kind of charge. On the *outside* of the glass sat the experimenter's warm, damp, conducting hand, connected through his body to the ground, and there the opposite charge gathered. The two clouds of opposite charge faced each other across the thin glass, each holding the other in place by their mutual attraction, so that far more electricity could be crammed onto the surfaces than either could hold alone. The jar was, in effect, a reservoir under pressure. And the moment the experimenter closed a path from the inside to the outside — by reaching his free hand over to touch the rod, completing a circuit through his own chest — the whole pent-up store rushed across in a single convulsive discharge, and the reservoir emptied itself through his body.

A homely comparison helps. Think of the Leyden jar as a drawn bow, or a compressed spring. The friction machine does the work of slowly drawing the bow back, storing up energy little by little; the jar holds that tension, the string quivering at full draw; and the discharge is the instant of release, when everything stored so patiently is dumped out at once. A friction machine on its own is like an archer who can only ever nudge the arrow forward by hand — a continuous but feeble push. The Leyden jar let the experimenters draw the bow fully back and then loose it, delivering in a single flashing instant a blow out of all proportion to the gentle cranking that had charged it. That is why a machine that had only ever produced tingles and small sparks could suddenly, once its output was stored in a jar, deliver a shock that flung a grown man backward. The *rate* of delivery had changed everything.

Word of the Leyden jar swept across Europe with a speed that tells you how starved the age was for wonders it could touch. Here, at last, was electricity you could *feel* — dramatically, unforgettably, and on demand. What followed was one of the strangest chapters in the history of science: for roughly a decade, electricity became a form of popular entertainment, a parlor spectacle and a fairground marvel. Traveling showmen and natural philosophers alike toured the courts and salons of Europe charging up Leyden jars and administering shocks to volunteers, who yelped and jumped to the delight of onlookers. And the most famous demonstration of all was staged in Paris by the Abbé Jean-Antoine Nollet, a French clergyman and the leading electrical showman of his day, before an audience that, by the popular account, included the King of France himself.

Nollet's experiment was designed to answer a real scientific question — how fast does the electric shock travel? — while also being pure theater. He assembled a long line of people, in the celebrated version some *two hundred monks* of a Carthusian monastery, and had each man grip a length of iron wire connecting him to the monk on either side, so that the whole community stood joined in a single human chain said to stretch the better part of a mile. Then Nollet connected a charged Leyden jar to the two ends of the chain. Every monk in the line leapt into the air, cried out, and jerked at the same instant — the entire mile-long ring of holy men convulsing together as the discharge ran through them all. It was, at once, a magnificent piece of showmanship and a genuine measurement: the near-simultaneity of the monks' reactions demonstrated that electricity traveled through the human chain with astonishing, seemingly instantaneous speed. It is one of the enduring images of the whole history of the science — a religious brotherhood transformed, for one shocking moment, into a living circuit. The story is likely embellished in the retelling, as such crowd-pleasing tales always are, and the exact numbers vary from account to account, but the demonstrations of chains of shocked people were real and repeated, and the point they proved — that electricity moves through conductors with a speed near the limit of what the eye can register — was sound.

The craze had a serious side that is easy to miss beneath the theatrics, for the Leyden jar quickly became the first electrical device that ordinary physicians tried to turn to healing. Within a few years of its invention, doctors and hopeful quacks across Europe were administering measured shocks from Leyden jars to patients suffering from paralysis, palsy, rheumatism, and a hundred other complaints, in the belief that the electric fire might rouse dead limbs and stir sluggish nerves back to life. Much of this was wishful and useless, and some of it was dangerous. But it was not entirely foolish, and here is a genuinely surprising thread that runs from those crude jars straight to the modern hospital: the instinct that a jolt of electricity could act upon the nerves and muscles of the body was fundamentally *correct*. The same principle — a stored charge released in a single controlled pulse to shock living tissue — is exactly what a cardiac defibrillator does today when it stops and restarts a fibrillating heart. The eighteenth-century "medical electricians," fumbling with their jars in the salons, were the distant and clumsy ancestors of the emergency-room paddles. They had the right idea two centuries too early, with none of the understanding needed to use it safely.

There are two persistent misconceptions worth clearing away here. The first is the assumption that the men who invented the Leyden jar understood it; they emphatically did not. Musschenbroek could not explain his own device and, badly frightened, wanted little more to do with it. For years afterward, brilliant experimenters argued over where in the jar the electricity was actually *stored* — in the water, in the glass, in the wire? — and performed ingenious experiments pouring the charged water from one jar to another only to find, to their bewilderment, that the electricity stayed behind in the original glass. It was Benjamin Franklin, a few years later, who would work out that the charge resided on the surfaces of the glass barrier itself, and who would build flat versions — panes of glass coated with metal foil on both faces — that made the point plain. The jar worked long before anyone knew *why* it worked, which is a recurring pattern in this story and a useful corrective to the tidy way science is often taught.

The second misconception is that the Leyden jar was a mere toy, all spectacle and no substance. In fact it was the indispensable tool that made the next century of discovery possible. Before the jar, electricity was too weak and fleeting to do much with; after it, experimenters could store a serious charge and release it in a controlled, powerful pulse, which is exactly what they needed to probe what electricity could do. It was a charged Leyden jar, in essence, that let Franklin capture and study the electricity he drew from the sky. It was Leyden jars that let later researchers pass strong discharges through gases, liquids, and living tissue, opening the studies of chemistry, physiology, and eventually the electron. The entertainment and the science were the same thing, viewed from different chairs in the room.

You carry the great-great-grandchildren of the Leyden jar with you at every moment, by the billions. A **capacitor** — a device that stores charge on two conducting surfaces separated by a thin insulator, exactly the jar's principle in miniature — is one of the most common components in all of electronics. There are dozens, often hundreds, inside your phone; there are capacitors smoothing the power in your laptop charger, storing the sudden burst of energy that fires a camera flash, and holding the memory in some computer chips. The camera flash is perhaps the most faithful modern echo of Musschenbroek's jar: a capacitor charges up quietly over a second or two, storing energy little by little, then dumps its entire store in a single brilliant instant to fire the flashtube — a drawn bow loosed, precisely as the jar loosed its charge through the Dutch professor's arm. The touchscreen you tap works by sensing the tiny capacitance of your fingertip. Every one of these is Leyden's jar, tamed, shrunk, and multiplied beyond counting.

The Leyden jar gave the eighteenth century a way to store and unleash electricity, and it turned the invisible force into something people could feel in their bones and see convulse a chain of monks. But the jar also sharpened an old and tantalizing suspicion. That sudden crack of the discharge, the blue-white spark leaping the gap, the sharp sting and the smell of it — did it not resemble, in miniature, the crack and flash of lightning splitting the sky? In Philadelphia, a printer, postmaster, and self-made philosopher had begun to think exactly that, and he was preparing an experiment that would drag the god's thunderbolt down out of the heavens and prove it was nothing more, and nothing less, than the same fire he could bottle in a jar.

Chapter 6: Franklin's Kite

On a gray and gusty afternoon in June of 1752, on the outskirts of Philadelphia, a stout, balding man of forty-six stood under the shelter of a shed doorway, holding a length of twine that ran up into a storm. Beside him was his twenty-one-year-old son, William. At the far end of the string, high in the roiling clouds, rode a homemade kite — two crossed sticks and a large silk handkerchief, with a sharp pointed wire fixed to the top of the frame, thrusting up into the wet sky. Where the wet twine met the doorway, the man had tied a short length of dry silk ribbon, which he kept carefully under cover so that it stayed dry, and to the junction of the two he had fastened an ordinary iron key. Benjamin Franklin was waiting, with the patience of a man who had thought the thing through, to see whether the storm above would send its fire down the string to the key in his hand. When at last he brought his knuckle near the key, a spark jumped the gap to his finger with the same small, familiar snap he had drawn a hundred times from a Leyden jar on his workbench. In that instant, one of the oldest terrors of the human race was quietly demoted from a divine mystery to a laboratory phenomenon. The lightning was electricity. Franklin had proven it.

Before we go further, honesty requires a caution, because

the kite experiment is wrapped in as much legend as almost any event in the history of science. Franklin himself wrote a brief, matter-of-fact account of it only months afterward, but he left maddening gaps: he did not say exactly when he did it, or precisely who witnessed it, and the fullest version we have was written down years later by the chemist Joseph Priestley, presumably from Franklin's own telling. Some modern historians have wondered whether the experiment happened exactly as the famous story has it, or at all. What is *not* in doubt is the science and Franklin's central role in it. He had designed and published the crucial experiment before he ever flew the kite, and by the time the kite story circulated, other men had already carried out his plan and gotten the result he predicted. So while we should tell the kite tale with a raised eyebrow — the popular image of Franklin joyfully flying a kite in a raging thunderstorm is certainly overdramatized, and had the kite been struck by a full lightning bolt he would very likely have been killed — the underlying achievement is rock solid. Franklin proposed a way to test whether lightning was electrical, the test was performed, and it passed.

The experiment Franklin actually designed first, and published in 1750, was not the kite at all but the "sentry-box" experiment, and it is worth understanding because it shows how carefully he reasoned. Franklin had noticed, in his own words, that electricity seemed to be drawn off most readily by sharp points — a pointed conductor would silently pull the charge out of a nearby electrified object at a distance, where a blunt one would not. From this he made a daring leap. If the storm clouds overhead were themselves electrified, as he suspected, then a tall pointed iron rod, thrust up into the sky, ought to draw electricity down out of the clouds — and one could then test that captured electricity to see whether it behaved exactly like the electricity of the laboratory. He proposed erecting such a rod on a high tower or steeple, with a man standing on an insulating platform inside a little sentry-box shelter at its base, who could draw sparks from the rod and prove the sky's fire was ordinary electricity. Franklin published this proposal, and in May of 1752 a French experimenter named Thomas-François Dalibard, following Franklin's printed instructions, erected a forty-foot iron rod in a garden at Marly, near Paris. When a thunderstorm passed over, an old soldier left to watch the apparatus drew crackling sparks from the rod's base. Franklin's prediction had come true in France a month before he is said to have flown his kite in Philadelphia. It was one of the first occasions in history when an American's scientific idea was tested and confirmed by European experimenters — a small but telling sign that the New World had begun to contribute to the great conversation of science.

The kite, then, was Franklin's own more portable version of the same test, devised — the story goes — because he was tired of waiting for a tall enough steeple to be finished in Philadelphia. A kite could reach higher than any tower, and it needed no scaffolding. The physics of what he did is worth spelling out plainly, because it is often misunderstood. Franklin did *not* intend to be struck by lightning, and the experiment does not depend on a lightning bolt hitting the kite. Rather, a thundercloud carries an enormous electric charge, and a sharp point thrust up into the electrified air beneath it will quietly draw off some of that charge — not the whole violent bolt, but a steady trickle. That trickle ran down the wet, conducting twine (wet string conducts; this was Stephen Gray's discovery at work) to the key. The dry silk ribbon that Franklin held, kept under the shed out of the rain, was an *insulator* — it would not carry the charge — and so the electricity, unable to escape through Franklin to the ground, accumulated on the key, from which he could draw it off as sparks. That dry silk ribbon was quite literally the thread on which his life hung: it was the insulating barrier between him and the charge, and had it become soaked, the story might have ended very differently. Indeed it very nearly did end that way for others: within a couple of years a Swedish-born professor in St. Petersburg named Georg Wilhelm Richmann, attempting a version of the experiment during a storm, was struck and killed by a discharge that leapt from his apparatus to his head — a sobering reminder that Franklin had been playing with the real thunderbolt, and had been at least as lucky as he was clever.

Having proven that lightning was electricity, Franklin did something that reveals the deepest cast of his mind: he immediately turned the discovery to *use*. If lightning was merely electricity, and if electricity was drawn off harmlessly by sharp metal points and conducted safely to the ground, then a building could be protected from the god's thunderbolt by a simple device — a pointed metal rod fixed to the highest part of the roof and connected by a metal wire running all the way down into the moist earth. When lightning threatened, the rod would either quietly bleed the charge away or, if a bolt did strike, offer it an easy metal path straight down to the ground, carrying the destroying fire around the building instead of through it. This was the **lightning rod**, and it was arguably the first time in human history that a genuine, life-saving technology emerged directly from a piece of pure scientific discovery. Franklin published the design freely, refusing to patent it, and within a few years pointed rods were sprouting from the rooftops, steeples, and ships of America and Europe. For the first time in the whole span of human existence, people had a real defense against the lightning. The bell-ringers and their useless, deadly bells could stand down; the sailors could fit a rod to the mast. It is difficult to overstate what a psychological revolution this was. For millennia lightning had been the untouchable weapon of the gods, to be met only with prayer. Now a Philadelphia printer had shown that a length of iron wire could turn it aside. Some clergymen actually objected, complaining that the rod impiously thwarted the will of heaven — to which Franklin's supporters could reasonably reply that God, presumably, had also given us the wit to come in out of the rain.

Franklin's contributions to electrical science went far beyond the drama of the kite, and here we meet a fact that surprises many people: much of the everyday vocabulary of electricity was coined by Benjamin Franklin. Wrestling with the confusion left by du Fay's "vitreous" and "resinous" electricities, Franklin proposed a simpler and more powerful picture. He suggested that electricity was a single invisible fluid present in all matter, and that electrical effects came from having too much or too little of it. An object with an excess of the fluid he called **positive**, or plus; an object with a deficiency he called **negative**, or minus. Rubbing did not *create* electricity, in Franklin's view; it merely transferred the one fluid from one body to the other, so that whatever one gained, the other lost in exactly equal measure. This was the germ of one of the most important principles in all of physics — the **conservation of charge**, the rule that electric charge is never made or destroyed but only moved around. From Franklin's pen, in his letters and papers on electricity, came *positive* and *negative*, *charge*, *conductor*, and *battery* (a word he borrowed from artillery, where a "battery" was a group of cannon, to describe a bank of Leyden jars linked together), along with terms like *condenser* and *electrician*. When a modern engineer speaks of a positive terminal or a negative charge, she is speaking Franklin's language.

His single-fluid theory was elegant, largely correct in its bookkeeping, and wrong in one detail that we have been stuck with ever since — which makes for one of the most delicious "did-you-know" facts in the whole subject. Franklin had to guess which way his invisible fluid flowed, and he guessed that it flowed *from* the positive body *to* the negative one — that current runs from plus to minus. He had a fifty-fifty chance, and he got it backward. We now know that in a metal wire the things that actually move are electrons, and electrons are negatively charged, so the real flow of charge-carriers is from minus to plus — the opposite of what Franklin assumed. By the time this was understood, in the late nineteenth century, Franklin's convention was so deeply embedded in every textbook, every equation, and every engineer's habits that the world simply decided to keep it. To this day, "conventional current" is defined as flowing from positive to negative, in the direction Franklin guessed, even though the electrons themselves are trooping the other way. Every electrical engineer alive works with a pair of arrows pointing in opposite directions, and cheerfully lives with a mistake made by a Philadelphia printer in the 1740s. It is a small monument to how history, once it sets, resists correction.

There is a final dimension to Franklin that matters for the shape of this book, and it is captured in a Latin motto later coined in his honor: *Eripuit caelo fulmen sceptrumque tyrannis* — "He snatched the lightning from the sky and the scepter from tyrants." For this same man who tamed the thunderbolt went on, in the very next decade, to help lead a revolution and found a nation; the scientist became a statesman, drafting and signing the Declaration of Independence and negotiating the treaties that secured American independence. Franklin embodied a new and thoroughly modern idea: that the same rational, experimental, practical intelligence that could unriddle nature could also be turned to the improvement of human affairs. He is the first great figure in this book who was, at once, a serious scientist and a maker of the modern world in the fullest sense — a foreshadowing of the inventor-entrepreneurs, Edison above all, who would later fuse discovery and enterprise into a single restless drive.

You meet Franklin's legacy every time you see a metal spike on a rooftop, or fly safely through a storm, or read the little plus and minus signs stamped on the ends of a battery. That battery in your remote control is labeled with Franklin's words, wired according to Franklin's convention, storing the same fire he pulled down from a Philadelphia thundercloud. He had shown that the tame spark of the laboratory and the terrible bolt of the sky were one and the same, and he had given humanity, for the first time, a measure of command over it. But everything Franklin and his predecessors had studied was still *static* electricity — charge that built up, sat still, and then discharged in a single flash. No one could yet make electricity flow steadily and continuously, like water in a stream. That breakthrough would begin, of all places, with the twitching leg of a dead frog on a dissecting table in Italy.

Chapter 7: Twitching Frogs

In a laboratory in the university city of Bologna, in the 1780s, an anatomy professor named Luigi Galvani stood over a dissecting table on which lay the lower halves of several dissected frogs — skinned legs, pale and glistening, their nerves exposed, prepared for study. Galvani was a careful, pious, meticulous man, a physician and professor of anatomy with a special interest in the nervous system, and a scientist not much given to wild speculation. Yet what he was about to observe would set off one of the most famous scientific disputes of the age, seed a whole pseudoscience, and, by a strange path, help inspire one of the most enduring monsters in all of literature. As the story is usually told, an assistant happened to touch a scalpel to the exposed crural nerve of a frog's leg at the very moment that a nearby electrical friction machine threw off a spark — and the dead leg kicked, sharply, as though the frog were briefly, horribly alive again. Galvani, watching, felt the hair rise on the back of his neck. Something about electricity and the spark had reached across the room and animated dead flesh.

Galvani was gripped. Here was a possible bridge between two of the deepest mysteries in nature: electricity, that invisible fire the philosophers had been chasing since Thales, and *life* itself, the animating principle that made muscles contract and hearts beat. For years he pursued the connection with the patience of a born anatomist, and his experiments grew more precise and more strange. He found that he could make the frog legs twitch not only when a spark flew from a machine, but during a thunderstorm — he hung prepared frog legs from brass hooks on an iron railing outside his window and watched them kick each time lightning flashed, confirming Franklin's insight that the storm's fire and the laboratory's spark were one and the same, now reaching into living tissue. But the most consequential observation came on a day with no storm and no machine at all. Galvani hung a frog's leg from a hook of one metal — brass — and let it touch a plate or railing of a different metal — iron. And the leg twitched, all on its own, with no external source of electricity anywhere in sight. Merely completing a circuit of two dissimilar metals through the moist flesh was enough to make the dead leg jerk.

From this Galvani drew a bold and, as it turned out, largely mistaken conclusion. He decided that the electricity making the leg move came from *within the animal itself* — that living creatures generated and stored their own special electric fluid, which he called **animal electricity**. In his picture, the muscle was like a living Leyden jar, charged with a vital electricity secreted by the animal, and the nerve was the conducting wire that carried the discharge; the metals merely provided a path to release the animal's own stored electricity and trigger the contraction. It was an intoxicating idea. It seemed to say that the life force and electricity were, at bottom, the same thing — that the spark of life might be a literal spark. Galvani published his findings in 1791, and they caused a sensation across Europe. The phenomenon and the whole line of research took his name: to this day we speak of **galvanism** for the effect of electric current on muscle and nerve, of a *galvanized* reaction for a sudden jolt into action, and of a *galvanometer* for the instrument that detects electric current. Galvani's name survives, quietly, in a dozen everyday words, even though his central theory was wrong.

For it did not go unchallenged. Some two hundred miles away, at the University of Pavia, a brilliant, worldly, and combative physicist named Alessandro Volta read Galvani's paper with mounting skepticism. Volta was in almost every respect Galvani's opposite. Where Galvani was a modest, devout anatomist absorbed in the mysteries of living tissue, Volta was a confident, cosmopolitan physical scientist, more interested in the behavior of electricity itself than in frogs, and temperamentally inclined to distrust anything that smacked of a mysterious "vital force." At first Volta accepted Galvani's animal-electricity idea and even praised it warmly. But as he repeated the experiments, a nagging suspicion grew into a conviction. Volta fixed his attention on a detail Galvani had treated as incidental: the *two different metals*. In the experiments that most impressed Volta, the leg twitched precisely when it was touched by an arc made of two dissimilar metals — brass and iron, say, or silver and zinc — joined together. What, Volta asked, if the metals were not merely a passive path for the frog's own electricity, as Galvani thought, but were themselves the *source* of it? What if the electricity was generated by the contact of the two different metals, and the frog's leg was merely a sensitive detector — a delicate meter that registered the current by twitching?

This was the crux of one of the great scientific quarrels of the eighteenth century, and it is worth dwelling on because it is a model of how science actually advances — not by one genius being simply right and another simply wrong, but by a productive collision of interpretations, each seeing part of the truth. Galvani insisted the electricity came from the animal. Volta insisted it came from the junction of the metals. The two men, both honest and both able, marshaled experiment after experiment. Galvani, pressed hard, devised an ingenious counterstroke: he made a frog leg twitch by touching the exposed nerve of one leg directly to the muscle of another, using *no metal at all*, only living tissue completing the circuit. This was a genuinely important result — it showed that living tissue really does generate its own electrical signals, and in this narrow but crucial sense Galvani was right, more right than he could prove. Volta, for his part, could reply that in most of the dramatic demonstrations the two metals were doing the real work.

Here is the resolution that history has settled on, and it does honor to both men. *Both were partly right, and the dispute gave birth to two whole sciences.* Galvani had genuinely discovered **bioelectricity** — the fact that nerves and muscles do produce and transmit real electrical signals, that the body runs on tiny currents. Every heartbeat, every thought, every twitch of a finger is triggered by minute pulses of electricity racing along nerve fibers, and this is the direct descendant of Galvani's animal electricity. When a modern doctor reads the electrical trace of your heartbeat on an electrocardiogram, or measures the electrical storms of your brain on an electroencephalogram, or when a defibrillator shocks a stopped heart back into rhythm, she is working in the science Galvani founded. His instinct that electricity and life were intimately linked was, at the deepest level, correct. But Volta had also seen something true and, for the immediate future of technology, even more consequential: that two different metals in contact with a moist conductor will, by themselves, generate a steady electric current. Galvani's "animal electricity" was real, but it was not what powered most of his experiments; Volta's metallic electricity was doing that. In pursuing his side of the argument to its conclusion, Volta was driven to build a device that would prove metals alone could make electricity without any animal at all — and that device, as we will see in the next chapter, changed the world.

It is worth correcting the common notion that Galvani was simply a deluded man clinging to a mystical "life force" while the rational Volta swept him aside. The reality is subtler and more interesting. Galvani was a rigorous experimenter who made a foundational discovery about the electrical nature of the nervous system, one that anticipated the entire field of electrophysiology by more than a century. His error was one of interpretation, not of observation, and it was an error that led directly, by way of Volta's rebuttal, to the invention of the battery. Science needed both the man who saw the electricity of life and the man who saw the electricity of metals; their argument was more fruitful than either man's being right alone.

It is worth pausing, too, on the human dimension of the quarrel, because it is easy to imagine such disputes as bloodless clashes of ideas when they are nearly always clashes of people. Galvani and Volta never met to thrash the matter out; they fought at a distance, through published papers, and the fight took a heavy toll on the older man. Galvani was, by temperament, retiring and devout, deeply attached to his wife Lucia — herself the daughter of one of his teachers, and by some accounts a partner in his laboratory work — and when she died in 1790 he was left grief-stricken just as the controversy over his life's work was reaching its height. Worse was to come. When Napoleon's armies swept into northern Italy and established a new republic, Galvani, a devout Catholic, refused on principle to swear the required oath of loyalty to the new regime, and for this he was stripped of his university chair and his income. He died in 1798, impoverished and largely forgotten, in the house where he had been born, not living to see the vindication that history would grant him. Volta, by contrast, prospered: honored across Europe, courted by Napoleon himself, who made him a count and a senator. That the retiring, principled Galvani should have died in obscurity while his rival was showered with titles is one of the quiet injustices of scientific history — the more so because time has shown that Galvani, too, had seen a genuine and profound truth. The frog's leg really did contain its own electricity.

The idea that electricity might be the very spark of life did not stay quietly in the laboratories. It leaked out into the culture and took on a lurid life of its own, and this is where the story of the twitching frogs acquires its most famous and eerie afterlife. Galvani's nephew, a flamboyant experimenter named Giovanni Aldini, took galvanism on tour as a kind of grim theater. In the most notorious of his demonstrations, in London in 1803, Aldini applied electrodes from a powerful battery to the corpse of a recently executed murderer named George Forster. Before a horrified and fascinated audience, the dead man's jaw quivered, his face contorted into ghastly grimaces, one eye opened, and — most dreadfully — a leg kicked and a fist clenched and rose, so that some onlookers feared the criminal was about to sit up and return to life. Aldini's public reanimations of animal and human corpses with electricity made galvanism a byword for the tantalizing, forbidden possibility that death itself might be reversed by the electric fire.

It was into this charged atmosphere, a decade later, that a young Englishwoman named Mary Shelley absorbed the talk of galvanism, of currents that made dead limbs move, of the boundary between life and death dissolving under the influence of electricity. When, in 1816, she conceived the story of a scientist who assembles a creature from dead flesh and shocks it into hideous life, the science of Galvani and the spectacles of Aldini stood plainly behind it. *Frankenstein; or, The Modern Prometheus*, published in 1818, never specifies the exact method by which Victor Frankenstein animates his creature, but Mary Shelley herself, in a later preface, invoked galvanism as part of the inspiration, recalling talk of experiments in which a corpse or a fragment might be "re-animated" — and generations of readers and filmmakers have ever since imagined the monster jolted to life by lightning and crackling electrodes, an image drawn straight from Galvani's frogs and Aldini's convulsing corpses. It is a remarkable thing to consider: that the most famous fictional monster in the English language, and our whole cultural anxiety about scientists overreaching to steal the fire of life, traces back to a professor in Bologna watching a dead frog's leg kick on his workbench. The twitching frog became the reanimated man, and the reanimated man became a permanent fixture of the modern imagination.

You meet Galvani's legacy far more often than Frankenstein's, though you rarely notice it. It is in the electrode patches a nurse presses to your chest, in the pacemaker regulating a faltering heart with tiny timed pulses, in the nerve-conduction test that measures how fast a signal races down your arm, in the very fact — strange when you stop to think of it — that your every sensation and decision is carried by threads of electricity through your flesh. Galvani was closer to a profound truth than his rival ever admitted: life really does run on electricity.

But it was his rival's answer to the quarrel that broke open the future of technology. Convinced that the electricity came from the metals and not the frog, Alessandro Volta set out to prove it by building a machine of metal and moisture that would pour out electricity endlessly, with no living thing anywhere near it — and in doing so, he was about to give humanity the first steady, continuous electric current, and end forever the age of the fleeting static spark.

Chapter 8: Volta's Pile

In the spring of 1800, in his study at the University of Pavia, Alessandro Volta sat down to build a tower. It was not much to look at — a neat column of small metal discs, perhaps as tall as a stack of coins built up hand over hand, rising on the table before him. He laid down a disc of silver, then on top of it a disc of zinc, then a round of cardboard or cloth that he had soaked in salt water or dilute acid; then silver again, then zinc, then wet cloth; and so on, up and up, dozens of layers, a patient sandwich of two metals with brine between. When he had built his column high enough, Volta touched a wire to the bottom disc and another to the top, and brought the two wire-ends together near his hand. There was no spark of the old dramatic kind, no convulsion, no thunderclap. There was something far stranger and, as it would prove, far more important: a steady, continuous, unrelenting flow of electricity, pouring out of the metals without pause, for as long as he cared to leave the wires connected. Volta had built the first battery. And with it he ended, in a single stroke, the entire two-thousand-year age of static electricity and opened the age of the current — the age that would end with the whole world wired and humming.

To grasp why this humble stack of discs deserves to stand among the great inventions of history, we have to appreciate the profound limitation of everything that had come before it. From Thales's amber to Franklin's kite, every kind of electricity human beings had ever produced was *static* — charge that built up, sat still, and then discharged all at once in a brief, violent flash. A friction machine, a Leyden jar, even a lightning bolt: all of them delivered their electricity in a sudden burst, here and gone in a fraction of a second. It was, to use a comparison that will serve us well, like the crack of a whip or the splash of a bucket of water thrown all at once against a wall — a single dramatic release, impossible to sustain and hard to control. You could study electricity with such tools, and marvel at it, and shock a chain of monks with it, but you could not make it *work*, because work of any steady kind requires a steady supply. What no one had ever produced was electricity that flowed *continuously*, on and on, like water running from a tap or a stream turning a mill wheel. That continuous flow is what we call an **electric current**, and until Volta, it simply did not exist as anything a person could make and hold in their hands.

Volta arrived at his tower by way of the great quarrel we met in the last chapter. Determined to prove, against Galvani, that the twitching of the frog's leg came from the contact of two different metals rather than from any "animal electricity," Volta had set out to show that metals alone, with no living tissue whatsoever, could generate electricity. He experimented with pairs of dissimilar metals separated by moist materials, and found that each such pair produced a tiny, feeble electric push — far too weak to notice easily, but real. His masterstroke was to reason that if one pair produced a little, then many pairs stacked in series, each adding its small push to the total, should produce a great deal. So he piled them up, silver and zinc and brine, over and over, and the little pushes added together into a current strong enough to feel, to spark, and to do chemistry with. Because it was a *pile* of discs, the device was called, in Italian and then in every European language, the **voltaic pile** — the first battery. He described it in a letter to the Royal Society in London in 1800, and the news detonated across the scientific world.

How does it actually work? The plain-language explanation is one of the most satisfying in all of science, and it rewards a moment's patience. The secret is that electricity here is being made not by rubbing, not by friction, but by a *chemical reaction* — the two different metals and the salty liquid between them are quietly reacting with one another, and that chemical reaction pushes electric charge around a circuit. Think of it this way. The two metals have different appetites for their own electrons; one of them, the zinc, holds its electrons more loosely and is more eager to react with the surrounding liquid, giving up electrons in the process. Those liberated electrons, if you give them a path — a wire connecting the top of the pile to the bottom — will flow through that wire from the metal that has a surplus to the metal that has a hunger for them, and that steady flow of electrons through the wire *is* the electric current. As long as the chemicals have not been used up, the reaction continues, and so does the flow. The battery is, in essence, a device that converts stored chemical energy into a steady stream of electricity — a little chemical engine, quietly burning its metals to pump electric charge around and around.

Two everyday analogies make the leap from static to current clearer. First, imagine the difference between a bolt of lightning and a garden hose. The lightning delivers a colossal amount of energy in a single instant and is then gone; you cannot wash your car with a lightning bolt, however powerful, because it will not stay. The garden hose delivers far less power at any one moment, but it delivers it *continuously*, and precisely because the flow is steady and sustained, you can actually do useful work with it — fill a bucket, turn a water wheel, run a fountain. Volta's pile was the first electrical garden hose: modest in its instantaneous power, but able to run and run, and therefore able, for the first time, to *do things*. Second, think of the old static charge as money you find on the ground now and then — a windfall, unpredictable, quickly spent — whereas the battery is a steady wage, a small amount arriving reliably and continuously, which is what actually lets you build a life, or in this case a technology. The whole electrical revolution to come depended on having a wage rather than a windfall: a dependable, controllable, continuous current.

The consequences came almost immediately, and they were spectacular, because scientists suddenly had a tool that could do what no fleeting spark ever could — deliver energy steadily into a substance and hold it there. Within weeks of Volta's announcement, two English experimenters used a voltaic pile to pass a current through water and split it into its component gases, hydrogen and oxygen, bubbling up from the wires — the birth of the science of **electrochemistry**. A few years later the brilliant Humphry Davy, at the Royal Institution in London, built a gigantic battery of hundreds of pairs of plates and used its powerful current to tear apart chemical compounds that had resisted every previous method, isolating for the first time the pure elements sodium, potassium, calcium, magnesium, and others — new metals, some so reactive they burst into flame on water, pried out of nature by the steady force of Volta's current. Davy also connected his great battery to two rods of charcoal and drew between them a blindingly bright, sustained electric arc — the first electric light bright enough to be useful, the ancestor of the arc lamps that would later light the streets before Edison's bulb lit the parlors. In a handful of years, the battery had launched whole new sciences and technologies. It had given the nineteenth century its indispensable tool: a reliable, portable source of continuous electric current, on tap in any laboratory.

It is worth correcting a couple of misconceptions here, gently. The first is the loose modern habit of imagining that Volta "discovered electricity." He did nothing of the kind; electricity had been known and studied for over two millennia by the time he built his pile. What Volta discovered was how to generate a *continuous current* of it — a very specific and world-changing thing, but not electricity itself. The second misconception concerns the very word "battery." In everyday speech we call a single cell — the little cylinder in a flashlight — a "battery," but strictly a *battery* means a group of cells joined together, just as Franklin's "battery" of Leyden jars meant a bank of them, borrowing the artillery sense of a row of cannon. Volta's pile, being many pairs of metals stacked in series, was a true battery in the original sense: a whole company of tiny electrical cells firing in concert. A third small correction worth making is that Volta's chemistry was not quite right — he believed, wrongly, that the mere *contact* of the two metals was the whole source of the electricity, and underrated the role of the chemical reaction in the liquid. As with Galvani and Gilbert before him, the man who built the world-changing device did not fully understand the thing he had built. The pile worked, and later chemists explained why.

So great was the achievement that when, decades later, the community of scientists came to name the units by which electricity is measured, they reached for Volta's name to honor the quantity at the very heart of what his pile produced. The **volt** — the unit of electrical "push," of the pressure or force that drives a current along a wire — bears his name to this day. Every time you read that a household outlet supplies 120 volts, or that a car battery is a 12-volt battery, or that a phone charger delivers 5 volts, you are invoking Alessandro Volta and the tower of discs he stacked in Pavia in 1800. To carry forward the garden-hose picture: if the current is the *amount* of water flowing, then the voltage is the *pressure* pushing it — and Volta lent his name to the pressure, the driving force, which was exactly what his pile supplied that no static machine ever could sustain. It is a fitting memorial. Few people in this book have their name in more pockets: there is a Volta in every battery on Earth.

You are, of course, surrounded by Volta's descendants at every moment of your life, more thoroughly than by almost any other invention in this book. The battery in your phone, your laptop, your watch, your car, your smoke detector, your child's noisy toy; the vast banks of batteries now being built to store the output of solar farms and steady the electrical grid; the cell that will one day, perhaps, power most of the world's automobiles — every one of them is a direct heir of the voltaic pile, the same basic idea of two dissimilar materials and an intervening chemistry, generating a continuous current from stored chemical energy. The chemistry has grown vastly more sophisticated, and modern cells are lighter, stronger, and rechargeable in ways Volta could not have dreamed, but the principle is his. When you unplug your phone in the morning and carry its charge with you through the day, you are carrying a little tower of Pavia in your pocket.

With Volta's pile, the long prehistory of electricity reaches its hinge. For twenty-four centuries, from Thales's amber to Franklin's kite, electricity had been a thing of sparks and shocks and fleeting static marvels — fascinating, frightening, but essentially useless for any sustained purpose. Now, at last, it could be made to *flow*, steadily and controllably, and everything changes from this point on. Almost every device in the rest of this book — the motor, the generator, the telegraph, the light, the whole electrified world — depends on the steady current that Volta was first to produce. The static age was over; the age of the current had begun. And the very first thing that steady current would reveal, twenty years later, in the hands of a Danish professor lecturing to his students, was the secret that Gilbert had gotten exactly backward: that electricity and magnetism were not two separate forces at all, but two faces of a single power — a discovery that would arrive, like so many in this story, entirely by accident.

Part II

The Age of the Field

From a twitching compass needle to Maxwell's four equations — the half-century in which electricity and magnetism were revealed as one force, and humanity learned that a moving magnet could make current flow.

Chapter 9: The Compass and the Wire

On an evening in April of 1820, in a lecture room at the University of Copenhagen, a bearded Danish professor stood before a small audience of advanced students and did something that, by his own later admission, he had not entirely planned. His name was Hans Christian Ørsted, he was forty-two years old, and he was in the middle of demonstrating the newest marvels of physics — the flow of electric current from a voltaic pile, the battery that Alessandro Volta had unveiled just twenty years earlier. On the table beside his apparatus, almost incidentally, sat a compass: an ordinary mariner's compass, its slender magnetized needle floating quietly on its pivot, pointing north as compass needles had pointed north for a thousand years. Ørsted connected a wire across the terminals of his battery so that a current flowed through it, and he held the live wire near the compass. The needle twitched. It swung away from north, hesitated, and settled at a new angle, as though some invisible hand had reached out and nudged it aside.

To almost everyone in that room, including perhaps most of the students, it may have looked like a small thing. A needle moved a few degrees. But Ørsted understood at once that he had seen something that was not supposed to be possible, and the ground shifted quietly beneath the whole of physics. For two thousand years, since Thales and his amber and the ancient sailors with their lodestones, electricity and magnetism had been regarded as two entirely separate powers of nature — two neighbors who lived on the same street but never spoke. The clinging of amber and the pointing of a compass were assumed to have nothing to do with one another. Now, in a single flick of a needle, Ørsted had caught them in the act of speaking. An electric current — pure electricity, flowing through a copper wire — had reached out through empty air and moved a magnet. The two forces were not strangers after all. They were somehow the same thing, or two faces of one deeper thing, and the word we now use for that unified force is **electromagnetism** — the single power that binds electricity and magnetism together, the force that would eventually spin every motor and fill every generator in the world.

Before we go further, we should be honest about a legend that has grown up around this moment, because it is one of the most repeated stories in the history of science and it is only half true. The tidy version says that Ørsted made his discovery purely by accident — that he was doing something else entirely, that the wire happened to lie near the compass, and that the twitch surprised him completely, a bolt of dumb luck. The truth is more interesting and does more credit to the man. Ørsted had spent years convinced, on almost philosophical grounds, that the forces of nature must ultimately be unified — that electricity and magnetism *ought* to be connected, if only someone could catch the connection in the act. He had been hunting for exactly this effect. What was accidental was the timing and the setting: the discovery came during a live demonstration, not in the quiet of a planned experiment, and Ørsted himself admitted he was as startled as anyone when it finally appeared. So it was neither pure accident nor pure design. It was a prepared mind meeting a lucky moment — which, as it happens, is how a great many discoveries are actually made.

To feel why the twitch was so shocking, you have to understand how strange the *direction* of the force was. When you push something, it moves in the direction you push. When gravity pulls an apple, the apple falls straight down, toward the earth. Every force anyone had ever studied acted along the line between the two objects — a straight line, a shove or a pull. But the compass needle did not swing *toward* the wire, and it did not swing *away* from it. It swung so as to set itself *at right angles* to the wire, crosswise, sideways, as if the force curled around the wire in circles rather than reaching out in a straight line. Imagine holding a garden hose with water rushing through it, and finding that nearby leaves did not blow away from the hose or toward it, but instead arranged themselves in rings around it, all circling the stream. That is roughly how bizarre this was to physicists trained on straight-line forces. The magnetic influence of a current, it turned out, wraps around the wire like the grain of wood around a knot. This circling, sideways character of magnetism is not a minor detail; it is the very feature that later makes electric motors turn, because a force that pushes crosswise is a force that can spin a wheel.

Here it helps to define our terms plainly, since they run through everything that follows. An **electric current** is simply a flow of electric charge — think of it as electricity in motion, the way a river is water in motion rather than water sitting still in a pond. The static electricity of Thales's amber was charge at rest, piled up and waiting; a current is that same charge streaming steadily along a wire, and it took Volta's battery to produce it reliably for the first time. A **magnetic field**, meanwhile, is the region of influence around a magnet — the invisible zone within which a compass needle feels a push. You can make that field visible, as generations of schoolchildren have, by scattering iron filings on a sheet of paper over a bar magnet: the filings line up into graceful looping curves, mapping out the shape of the field like grass bending in a wind you cannot see. What Ørsted proved is that a current-carrying wire generates one of these magnetic fields all by itself, with no magnet present at all. Electricity in motion *makes* magnetism. That sentence, only four words long, is the foundation of the modern electrical world.

The news traveled with a speed that astonished even its own age. Ørsted wrote up his result in a short Latin pamphlet in July 1820 and mailed copies to scientists across Europe, and within weeks the twitching needle was being reproduced in laboratories from Paris to London to Geneva. Part of the reason it spread so fast is a happy accident of the experiment's own simplicity: anyone with a battery, a length of wire, and a compass could see it for themselves in under a minute. There was no need to take Ørsted's word for it. A physicist in Paris could read the pamphlet in the morning and hold a trembling needle in his own hand by the afternoon — and, as we will see in the next chapter, at least one of them did exactly that, and within a single week had begun to build the whole thing into mathematics. Rarely in the history of science has a discovery been so instantly, so widely, and so easily confirmed.

A few surprising facts deserve their place here. The first is that Ørsted was not, by trade, only a physicist. He was a broad Enlightenment intellectual, a friend of the fairy-tale author Hans Christian Andersen, and a chemist of real distinction — he was the first person to isolate the metal aluminum, in 1825, an element so common in the earth's crust and yet so hard to purify that it was once considered more precious than gold. The man who linked electricity and magnetism also gave us the metal of soda cans and airplanes. The second surprising fact is one of pure historical bad luck: there is evidence that an Italian jurist and amateur scientist named Gian Domenico Romagnosi had noticed a current disturbing a compass needle back in 1802, nearly two decades before Ørsted, and had even published a notice of it. But he published in an obscure newspaper, framed it poorly, and used a setup that muddied the effect, and the observation sank without a trace. Romagnosi had the fact but not the fame; the world simply did not notice. It is a reminder that in science, a discovery half-made and badly announced is, for all practical purposes, no discovery at all. The third fact is that Ørsted's name lives on as a scientific unit — the *oersted,* a measure of magnetic field strength — one of that select company of researchers whose surnames became the vocabulary of the very phenomena they uncovered.

It is worth lingering on *why* the twitching needle stunned the scientific world as thoroughly as it did, because to a modern reader, raised on the casual knowledge that electricity and magnetism are related, the shock can be hard to feel. For two millennia the two phenomena had lived in entirely separate mental compartments. Magnetism belonged to the lodestone and the compass, to navigation and the mysterious pull of iron toward north; electricity belonged to rubbed amber, to sparks and shocks, to Volta's battery. No one had any reason to suspect a link, and several great minds had actively looked for one and failed. William Gilbert, the founder of the science, had insisted two centuries earlier that the amber effect and the magnetic effect were fundamentally different and unrelated, and his verdict had hardened into orthodoxy. To overturn a two-hundred-year-old consensus of that weight, backed by a figure of Gilbert's stature, took more than a clever argument; it took a demonstration so simple, so direct, and so instantly repeatable that no amount of authority could argue it away. A needle either moved or it did not, and in Ørsted's hands, and then in a hundred other hands across Europe, it moved. That is the peculiar power of a good experiment: it can topple a settled belief in an afternoon.

There is a further reason the discovery carried such force, and it lies in the intellectual climate Ørsted came from. He had been steeped in a German-flavored philosophy of nature — sometimes called *Naturphilosophie* — that held, almost as an article of faith, that all the powers of the universe were secretly manifestations of a single underlying unity. Light, heat, electricity, magnetism, chemical affinity: to a believer in this romantic worldview, these were not separate forces but different expressions of one great cosmic power waiting to be revealed. Most hard-nosed experimental physicists of the day dismissed such talk as mystical fog. Ørsted's twitching needle handed the mystics an unexpected vindication, because here was the unity they had preached, caught red-handed in a laboratory. It is one of the stranger episodes in the history of science: a discovery of the first rank, guided in part by a half-mystical philosophy that most working scientists rejected, yet proven by the most concrete and hard-edged of experiments. Ørsted, uniquely, had a foot in both camps — the dreamer who expected unity and the careful experimenter who could catch it — and that double vision may be exactly why he, and not someone else, saw what the needle was telling him.

It is worth correcting one more misconception before we move on, because it distorts how people imagine this whole era. We tend to assume that once electricity and magnetism were shown to be linked, the practical machines followed almost immediately — that the motor and the generator were just around the corner. They were not. Ørsted's twitch in 1820 revealed *that* a current makes magnetism, but it would take another decade of hard work by other hands before anyone discovered the reverse and infinitely more useful truth — that a moving magnet can make a current — and several decades more before any of it powered a single lamp. The distance from the twitching needle to the working power station was more than sixty years. Ørsted opened a door; he did not walk through to the other side. What he provided was the first fact of the electrical age proper, the seed fact, the one from which every dynamo and transformer would eventually grow.

You meet Ørsted's discovery, quite literally, every waking moment of a modern day, though you never see the twitching needle. Every time an electric motor turns — in the fan cooling your laptop, in the pump circulating water through your refrigerator, in the little vibrating motor that buzzes your phone, in the windshield wipers of a car, in the whir of a washing machine — you are watching Ørsted's principle at work. All of those motors run on the same brute fact he stumbled into on that April evening: send a current through a wire and it becomes, in effect, a magnet, able to push and pull on other magnets. Reverse the direction of the push a few thousand times a minute, arrange the geometry cleverly, and you have rotation, and from rotation you have the mechanical labor of the industrial world. The needle that jumped on Ørsted's table was the first faint stirring of every motor that has spun since.

What Ørsted had not done was turn his discovery into law. He had a phenomenon — vivid, repeatable, world-changing — but he had no equations to describe how strong the force was, how it depended on the current, how it fell off with distance. He had caught the fish; he had not yet weighed or measured it. That task fell to a grieving, brilliant, and famously unlucky Frenchman who read Ørsted's pamphlet in Paris and, within a single astonishing week, began turning the twitch of a needle into the mathematics of force.

Chapter 10: Ampère's Mathematics of Force

In the first week of September 1820, a slightly disheveled Frenchman of forty-five sat in a meeting of the Academy of Sciences in Paris and listened, electrified, as a colleague described the news arriving from Copenhagen: a Danish professor named Ørsted had made a compass needle swing by running a current through a nearby wire. Most of the academicians nodded, filed the report away as an interesting curiosity, and went to dinner. André-Marie Ampère went home and could not sleep. Within days he was back in front of the Academy with experiments of his own, and within a week — a single week — he had leaped past Ørsted's discovery to a deeper and stranger one. If a current could move a magnet, Ampère reasoned, then perhaps a current was, in some hidden sense, *itself* a kind of magnet. And if that were true, then two wires each carrying a current ought to push and pull on each other directly, with no magnet involved at all. He set up two parallel wires, sent current through both, and watched. When the currents flowed the same way, the wires drew together. When they flowed in opposite directions, the wires shoved apart. Two ordinary copper wires, carrying nothing but invisible electricity, were attracting and repelling each other across empty space like a pair of magnets.

This was a discovery every bit as profound as Ørsted's, and in some ways more so. Ørsted had shown that electricity could talk to magnetism. Ampère had shown that electricity could talk to *itself* — that the magnetism was not some separate ingredient but was generated by the moving charge alone, so that two flows of current would grip one another with a force that could be seen, measured, and, crucially, calculated. He gave this new science a name that we still use: **electrodynamics**, the study of electric currents in motion and the forces between them. If electrostatics was the physics of charge sitting still — Thales's amber, the crackle of a wool sweater — then electrodynamics was the physics of charge on the move, and it was in that moving world that all the useful machines of the future would live.

To picture what Ampère found, imagine two ropes hung side by side, and suppose that whenever you send a pulse of energy down both ropes in the same direction, they mysteriously swing toward each other and cling; send the pulses in opposite directions, and they fling apart. That is essentially the behavior of two current-carrying wires. Or think of it another way, closer to the truth: each wire, once a current runs through it, wraps itself in a circling magnetic field, exactly the kind Ørsted's needle had detected. When two such wrapped fields overlap, they either reinforce and pull the wires together or oppose and push them apart, depending on which way the currents run. The wires are not really reaching out and touching; they are leaning on each other through their invisible fields, the way two people can push each other around a swimming pool without touching, just by shoving the water between them.

What set Ampère apart from every investigator who had come before was not merely that he saw the effect, but that he refused to stop at seeing it. Ørsted had given the world a vivid phenomenon; Ampère gave it a *law*. He measured how the force between two wires grew when he increased the current, how it weakened as he moved the wires apart, how it changed with their length and arrangement, and he distilled all of it into precise mathematical formulas. This was the decisive turn in our whole story, and it is worth dwelling on. A curiosity is something you can point at and say, "How strange." A law is something you can *use*: it lets you predict, before you build anything, exactly how strong a force a given current will produce, which means you can design a machine on paper and trust that it will work when you build it in metal. Every electric motor and generator and transformer that follows in this book exists because men like Ampère converted the twitch of a needle into equations that an engineer could calculate with. Mathematics is what turns a marvel into a technology.

The speed and completeness of Ampère's work in those first weeks still astonishes historians, and it repays a closer look, because it shows what separates a great scientist from a merely competent one. When the news from Copenhagen arrived, dozens of European physicists heard it, and many of them repeated Ørsted's experiment and confirmed the twitch. But confirming a result is not the same as extending it. Where the others saw an interesting fact to be filed away, Ampère saw a whole unexplored country and set off at a run to map it. Within days he had asked the deeper question — if a current acts like a magnet, do two currents act on each other? — designed the experiment to test it, built the apparatus, run it, and begun the mathematics. He presented findings to the Academy on successive Mondays through the autumn of 1820, each week carrying the theory further, a sustained burst of creativity rarely matched in the annals of physics. It is a useful reminder that discovery is not merely a matter of stumbling onto a fact; it is a matter of knowing, once you have the fact, which questions to ask next and how fast to chase them. Ørsted opened the door by luck and preparation; Ampère sprinted through it by sheer force of intellect.

Ampère's apparatus for detecting the force between wires had to overcome a real practical difficulty that is worth appreciating, because it shows how delicate this pioneering work was. The force between two ordinary wires carrying modest currents is *tiny* — nothing like the firm tug of two bar magnets — and it is easily swamped by stray effects, by the pull of the earth's own magnetism, by drafts and vibrations. To see it at all, Ampère had to build cleverly balanced arrangements in which one movable wire was free to swing toward or away from a fixed one, isolating the feeble electrodynamic force from everything else. That he could measure such a whisper of a force accurately enough to extract a mathematical law from it, using the crude instruments of 1820, is a testament to his experimental care as much as his theoretical genius. We tend to remember Ampère as a pure mathematician of electricity, but he was a resourceful experimenter too, coaxing precise numbers out of forces barely strong enough to stir a hanging wire.

Ampère also proposed a bold and, for its time, almost heretical idea to explain *why* his wires behaved like magnets. If a loop of current acts like a little magnet, he suggested, then perhaps an ordinary iron magnet is nothing more than a vast collection of tiny, permanent electric currents, circulating endlessly inside the metal, all lined up in the same direction. There was no way to prove this in 1820 — no one had any notion of what could be circulating inside solid iron — and many of his contemporaries thought it a wild fancy. Yet it was, in essence, correct. We now know that the magnetism of iron comes from the motion and alignment of its electrons, each a minuscule circulating charge. Ampère had guessed, a century before anyone could confirm it, that *all* magnetism is ultimately electrical in origin — that there is really only one force here, wearing two costumes. It was one of the most prophetic guesses in the history of physics.

The man who saw all this so clearly was, by every account, a wreck of nerves and grief. And here the story turns from triumph to something close to tragedy, because Ampère's personal life was a catalogue of sorrows almost operatic in its cruelty. He was born in 1775 near Lyon, a child prodigy who devoured his father's library and taught himself advanced mathematics before he was a teenager. Then, when he was still a young man, the French Revolution reached his family: his father, a local justice of the peace, was arrested during the Terror and guillotined in 1793. The boy was so shattered by the news that he fell into a stupor for more than a year, unable to think or study, staring blankly at the world. He eventually recovered, married, and found happiness — only to have his beloved wife, Julie, die of illness in 1803, a few years into their marriage, leaving him with a young son and a grief he never fully escaped. A disastrous second marriage followed. For the rest of his life he was known as a brilliant, absentminded, melancholy man, forever losing his belongings and his train of thought, once famously mistaking a passing carriage for a blackboard and beginning to chalk equations on its back as it rolled away. He asked that his own tombstone be carved with the words *Tandem felix* — "Happy at last" — a phrase that tells you almost everything about how unhappy the living man had been.

There is a poignant symmetry worth pausing on. This grief-haunted, distracted, perpetually unlucky man produced work of crystalline precision and order — laws so exact that they still appear, essentially unchanged, in every physics textbook nearly two centuries later. It is as though the rigor he could not find in his life he poured entirely into his science. The great physicist James Clerk Maxwell, who will close out this part of our story, later called Ampère "the Newton of electricity," and it was no idle compliment: just as Newton had reduced the sprawling motions of the heavens to a handful of clean equations, Ampère had reduced the messy sparking business of currents and magnets to mathematical order.

For all his precision, Ampère was working with a foggy picture of what electricity actually *was*. This is a fact modern readers find hard to absorb: the electron — the tiny negative particle that we now know is the thing actually flowing in a current — would not be discovered until 1897, more than sixty years after Ampère's death. He wrote his laws of current without knowing what current was made of, the way a person might perfectly describe the flow of a river while having no idea that water is made of molecules. That such accurate laws could be built on such an incomplete foundation is one of the recurring wonders of physics, and a useful corrective to the assumption that you must fully understand something before you can measure and master it. Very often, mastery comes first and understanding limps along behind.

His name is now woven so deeply into daily life that most people who use it have no idea it was ever a man. The **ampere** — universally shortened to the *amp* — is the standard scientific unit of electric current, the measure of how much charge flows past a point each second. Every time you glance at the rating on a phone charger and see it deliver "two amps," or read that a household circuit is protected by a "fifteen-amp breaker," or notice that a car battery can supply hundreds of "cold cranking amps" to start an engine on a winter morning, you are speaking Ampère's name aloud without knowing it. The amp is one of the seven base units of the entire international system of measurement, sitting alongside the meter, the kilogram, and the second — a distinction that places this melancholy, bereaved Frenchman among the small handful of people whose names define the fundamental vocabulary of the physical universe. When an electrician sizes a wire, when an engineer specs a fuse, when your circuit breaker trips because a space heater and a hair dryer drew too much at once, Ampère's law is quietly doing the arithmetic.

By the early 1820s, then, two great facts stood established. Ørsted had shown that a current creates magnetism, and Ampère had measured the forces between currents and folded them into law. But a crucial piece of the puzzle was still missing, and it was the most practical piece of all. Ampère could tell you how *hard* a current would push, but not what determined how *much* current would flow through a given wire in the first place — why a thin wire carried less than a thick one, why some materials passed electricity freely and others choked it off. That relationship, the plainest and most universally used law in all of electricity, was being worked out at that very moment by a poor and overlooked German schoolteacher, laboring in obscurity with homemade wires and a stubbornness that the world would take years to reward.

Chapter 11: Ohm's Simple Law

In a cramped classroom at the Jesuit Gymnasium in Cologne, sometime in the mid-1820s, a middle-aged mathematics teacher named Georg Simon Ohm was doing something his employers would have considered a waste of time. Instead of grading exercises, he was drawing out lengths of copper and other metals into wires of carefully different thicknesses and lengths, connecting them to a battery, and painstakingly measuring how much current flowed through each one. He had almost no money for equipment. He drew his own wires by hand. His most reliable instrument was not even electrical: to measure the strength of a current, he used the tiny twist it produced in a suspended magnetic needle — Ørsted's discovery, put to work as a measuring tool only a few years after it was made. Ohm was a nobody, a provincial schoolteacher with no laboratory, no assistants, and no reputation, chasing a question that the great men of Paris had not bothered to ask. And yet what he found in that classroom is, without exaggeration, the single most-used equation in all of electrical engineering — the first thing every electrician, every technician, every physics student on earth still learns.

The question Ohm was chasing was deceptively humble. Ørsted had shown that a current makes magnetism; Ampère had measured the forces between currents. But nobody had pinned down the most basic bookkeeping fact of all: for a given battery, what decides how much current actually flows? Why does a fat, short wire carry a torrent of electricity while a long, thin one carries only a trickle? Ohm suspected there was a clean numerical relationship hiding in the mess, and after years of careful measurement he found it. It is so simple that it can be written in three letters, and it is worth stating plainly right now, because the rest of this book leans on it constantly: the current flowing through a wire equals the electrical "push" driving it, divided by the wire's "resistance" to it. In the shorthand that engineers have used ever since, *V equals I times R* — voltage equals current multiplied by resistance. This is **Ohm's law**, and it governs everything from the filament in a light bulb to the power lines strung across a continent.

To make sense of that equation, we need to define its three characters, and the cleanest way to do it is with an analogy that has been helping people understand electricity for almost two centuries: think of electric current as water flowing through a pipe. In this picture, the **current** — measured, as we saw, in Ampère's amps — is simply the *amount* of water rushing past, the flow rate. The **voltage** is the water *pressure* — the force pushing the water along, the thing a pump or a tall tank provides; in electrical terms it is the push supplied by a battery or a generator, and it is measured in volts, after Volta and his pile. And the **resistance** is how narrow or clogged the pipe is — how much the pipe fights the flow. A wide, smooth, short pipe offers little resistance and lets a lot of water through; a long, narrow, gunked-up pipe offers high resistance and only lets a dribble past. Ohm's law simply says these three quantities lock together in fixed proportion: raise the pressure and more water flows; narrow the pipe and less flows. Push harder, get more current; resist harder, get less. That is the entire idea, and its power lies in its plainness.

A second analogy may make it stick even better, because water pipes can feel abstract. Picture a crowd of people trying to move down a corridor. The number of people passing a doorway each second is the current. How eagerly they are pushing from behind — how motivated the crowd is to get through — is the voltage. And the width of the corridor, whether it is an open hall or a tight squeeze full of furniture, is the resistance. Widen the corridor and the same crowd flows through faster; clutter it and the flow chokes down to a shuffle. **Resistance**, in the electrical world, is exactly this: the opposition a material offers to the passage of current, the friction that electricity must overcome as it moves. Some materials — copper, silver, gold — offer so little resistance that current pours through them, and we call them conductors. Others — glass, rubber, dry wood — offer so much that almost nothing gets through, and we call them insulators. The rubbery coating on every wire in your home is there precisely to be a corridor no electricity can enter.

There is a lovely and important twist buried in the concept of resistance, and it is the reason light bulbs glow. When current is forced through a material that resists it, that resistance produces heat — the electrical equivalent of the warmth your hands feel when you rub them briskly together against friction. Push enough current through a thin enough wire of high enough resistance, and it grows hot enough to glow white and throw off light. This is exactly how the incandescent bulbs that will dominate the middle of our story produce their light, and it is why Thomas Edison's great breakthrough, decades later, will hinge on finding a filament with just the right resistance. Ohm's humble schoolroom law, in other words, contains within it the working principle of the electric lamp. The same effect warms the coils of a toaster, the element of an electric kettle, and the glowing bar of a space heater — all of them Ohm's law turned deliberately into heat.

Now comes the part of Ohm's story that ought to make every reader a little indignant, because it is a stark lesson in how badly the world can treat a person who is simply right.

When Ohm published his law in 1827, in a book with the forbidding title *The Galvanic Circuit Investigated Mathematically*, it was met not with acclaim but with indifference sliding into contempt. Some German critics dismissed it as a "web of naked fancies." The prevailing philosophy in German universities at the time was suspicious of reducing nature to cold equations; one influential minister of education reportedly declared that a physicist who preached such notions was unworthy to teach science at all. Ohm's work was ignored or ridiculed, and the blow to his career was real and lasting. Disheartened, he resigned his teaching post and spent years in a kind of professional wilderness, scraping by, his great discovery gathering dust. The most useful law in electrical engineering was, at its birth, treated as an embarrassment.

Vindication came, but slowly, and it came from abroad before it came at home — a recurring pattern in these stories. Foreign scientists, especially in Britain, gradually recognized the depth and correctness of what Ohm had done. In 1841, fourteen years after his book appeared, the Royal Society of London awarded him its Copley Medal, then the highest honor in world science, effectively telling the German establishment that the schoolteacher they had scorned was one of the finest minds of the age. Honors and a proper professorship finally followed in his own country, and Ohm ended his life, at last, as a respected man. But he had lost two decades to neglect, and the sense of wasted years hangs over his biography. It is a bracing corrective to a comforting myth many of us carry — the myth that good ideas win quickly, that truth has a kind of automatic buoyancy that floats it to the surface. Ohm's law is a permanent reminder that being right is often not enough; a discovery also needs the right audience, the right moment, and a fair share of luck, and when those are missing, even a truth as clean and useful as *V equals I times R* can languish unrecognized for a generation.

The reasons for Ohm's long neglect are worth understanding, because they reveal how much a discovery depends on the intellectual weather of its time. Part of the problem was philosophical, as we have seen — the dominant German romantic view of nature was hostile to reducing the living richness of natural phenomena to dry equations, and Ohm's clean mathematical law struck many of his countrymen as a kind of impoverishment, a draining away of nature's mystery. Part of it was social: Ohm was a mere schoolteacher, not a university professor, and in the rigid academic hierarchy of the German states, work from a lowly *Gymnasium* instructor commanded little automatic respect. And part of it was simple bad luck in presentation; Ohm's book was densely and abstractly written, opening with pages of forbidding theory before it reached the beautifully simple result, so that even sympathetic readers struggled to find the gold buried in it. A great idea, badly timed and badly packaged and delivered by a man of low status, can sink almost without a ripple — and Ohm's very nearly did. That his law is now the first thing every student of electricity learns, and his book a recognized classic, is a vindication that came agonizingly late to the living man.

There is a human detail in Ohm's story that lingers. He undertook his painstaking measurements not in a well-funded laboratory but in the margins of an exhausting teaching job, driven by a private hope that a real scientific discovery might finally win him the university post he craved and lift him out of the drudgery of schoolroom instruction. When his work was ignored and even ridiculed, that hope curdled into bitter disappointment, and for years he believed he had thrown away his best efforts on a world that did not care. There is something quietly heartbreaking in the image of the man who had found one of the most universally useful truths in all of science, convinced in his darker moments that his labor had come to nothing, waiting through his forties for a recognition that would not arrive until he was nearly old. His eventual vindication is one of the more satisfying reversals in this book, but it should not blind us to the long years of neglect that preceded it, nor to the many equally deserving discoverers whose vindication never came at all.

A few surprising facts round out the picture. The first is just how astonishingly *reliable* Ohm's law turned out to be: for the ordinary metals and everyday conditions of the electrical age, it holds with almost perfect accuracy across an enormous range, which is why a rule scratched out with homemade wires in the 1820s still sits, unchanged and unchallenged, at the foundation of a twenty-first-century industry that runs the planet. The second is that Ohm reached his result partly by consciously imitating an idea from an entirely different field: the French mathematician Joseph Fourier had recently described how heat flows through a metal bar — faster when the temperature difference is greater, slower when the bar is longer — and Ohm saw that electricity in a wire behaved by the very same logic, with voltage playing the role of temperature difference and resistance the role of the bar's length. He borrowed the shape of one science to crack open another, which is how a great deal of scientific progress quietly happens. The third fact is the one you now expect: Ohm's name became a unit. The **ohm** is the standard measure of electrical resistance, so that the same three quantities in his equation are today named for three different pioneers — the volt for Volta, the amp for Ampère, the ohm for Ohm — a small monument, hidden inside every engineering formula, to the international relay race that built this science.

You meet Ohm's law more often than you meet almost any other idea in this book, usually without noticing. Every time a device is designed so it draws the right amount of current and not so much that it melts, someone has used *V equals I times R*. The dimming of a lamp when you turn a knob, the sizing of the fuse that protects your house, the reason a phone charger delivers a safe trickle to a battery instead of a destructive flood, the careful matching of a speaker to an amplifier, the warmth of every heating element you own — all of it is Ohm's schoolroom arithmetic, running silently in the background of modern life. Of all the laws we have met, none is used more, by more people, more times a day.

With Ørsted's link, Ampère's force, and Ohm's law, the fundamental grammar of the electric current was now written down. But the deepest and most world-changing discovery of this whole era still lay ahead, and it would come not from a professor or an academician but from a poor, self-taught young man who had begun his working life not with wires and batteries but with paper and glue — a bookbinder's apprentice who would rise to become the greatest experimental scientist of the nineteenth century.

Chapter 12: The Bookbinder's Apprentice

In a bookbinder's shop in the crowded Blandford Street district of London, around the year 1810, a nineteen-year-old apprentice named Michael Faraday was doing something his master had never asked him to do. All day he sewed and glued and pressed the pages of other men's books into leather covers, learning a respectable trade. But in the evenings, and in stolen moments during the day, he read the books instead of merely binding them. One of them was a battered set of the *Encyclopaedia Britannica* that had come in for rebinding, and when he reached the entry on electricity, something in the young man caught fire. He read it over and over. He scraped together pennies to buy odds and ends, built himself a crude electrostatic machine and a small voltaic pile out of coins and scraps, and began repeating the experiments the article described on the workbench of the bindery. Here was a boy with almost no schooling, no money, no connections, and no conceivable path into the world of science — and he had decided, with a quiet stubbornness that would define his entire life, that he was going to become a natural philosopher anyway.

To understand how impossible that ambition was, you have to understand the England Faraday was born into. He came into the world in 1791, the third child of a poor blacksmith whose health was failing, in a family that sometimes did not have enough to eat; there is a famous account of the boy being given a single loaf of bread to last him an entire week. His family belonged to a small, devout Christian sect called the Sandemanians, whose faith prized humility, plain living, and service — values that stayed with Faraday to the end and shaped the modest, unassuming man he became. His formal education, by his own description, amounted to little more than "the rudiments of reading, writing, and arithmetic at a common day-school." In the rigidly stratified society of Georgian Britain, science was a gentleman's pursuit, the province of men with Latin, Greek, university degrees, and private incomes. A blacksmith's son who bound books for a living was supposed to bind books for a living, marry, and die in the same class he was born into. The wall between Faraday and the world of learning was not made of intelligence — he had that in abundance — but of class, and class walls in that England were very high and very thick.

What got him over the wall was a combination of relentless self-improvement and a piece of extraordinary luck, though as with Ørsted the luck came to a mind that had spent years preparing to seize it. Faraday had joined a self-education group called the City Philosophical Society, where working-class young men gathered in the evenings to hear lectures on science and better themselves. He took meticulous notes, taught himself to write clearly, and even worked to improve his own speech and reasoning. Then, in 1812, a customer of the bindery — noticing the young man's hunger — gave him tickets to attend a series of public lectures at the Royal Institution by the most famous scientist in Britain, the dashing chemist Sir Humphry Davy. The Royal Institution was a temple of fashionable science, a place where London's elite came to watch brilliant demonstrations, and Davy was its star: handsome, eloquent, celebrated across Europe for isolating new elements like sodium and potassium and for inventing the miner's safety lamp. Faraday sat in the audience, spellbound, and did something that reveals the whole cast of his character. He wrote up his notes from Davy's four lectures in beautiful detail, bound them himself into a neat little book — using the one skill his trade had given him — and sent the volume to Davy with a letter humbly asking for any position, however menial, that might let him escape trade and enter science.

It is one of the great hinge-moments in the history of the field, and it very nearly didn't happen. Davy was flattered but cautious; he reportedly advised the young man to stick with bookbinding, telling him that science was a harsh mistress who poorly rewarded those who served her. But fortune intervened in a small and almost comic way. Davy temporarily injured his eyes in a laboratory explosion involving a volatile chemical, and needed someone to write for him; he remembered the earnest young binder and took him on briefly. Then, in early 1813, a laboratory assistant at the Royal Institution was dismissed after a brawl, leaving a vacancy, and Davy offered the post to Faraday. The pay was low, the work included washing bottles and sweeping floors, and the position was near the bottom of the scientific world. Faraday took it without hesitation. At twenty-one, the bookbinder's apprentice had crossed the wall. He would spend the rest of his life inside those walls, and he would die more than half a century later as the most revered experimental scientist in the world, having risen to run the very institution he had entered as a bottle-washer.

The relationship between Davy and Faraday is one of the most fascinating and, in the end, most troubling in this whole history, and it turned on the same class divide that had nearly kept Faraday out altogether. At first it was a genuine mentorship, and a formative one. Later in 1813, Davy set out on a grand tour of Europe to meet the great scientists of the age — Ampère among them, in Paris — and he took Faraday along. But he took him along in an ambiguous role: officially as a scientific assistant, in practice often as a personal servant. Davy's wife, in particular, treated the young man as a valet, expecting him to ride outside the carriage with the servants and see to menial chores, and Faraday found the humiliation so acute that he several times considered abandoning the tour and going home. Yet the same journey put him in the same rooms as the finest minds in Europe and gave him, in effect, a traveling university education he could never have afforded. He swallowed the indignities and drank in the science. It is a bracing illustration of what talent from the wrong social class had to endure in that age simply to be allowed near the work it was born to do.

It is worth pausing on the small volume of bound lecture notes that Faraday sent to Davy, because that homemade book is one of the quiet turning points in the history of science, and it was possible only because of the very trade Faraday was trying to escape. A poorer, less skilled young man might have sent a sheaf of loose, ink-blotted pages that a busy celebrity would have glanced at and discarded. Faraday sent a beautifully organized, elegantly bound volume — a physical object that announced, before Davy read a word, that its maker was careful, diligent, and serious. The bookbindery had given Faraday not just a livelihood but a skill that, at the crucial moment, became his calling card. There is a lovely irony in it: the trade that symbolized everything he wanted to leave behind was the very thing that carried him out of it. It is a reminder that the humble skills we acquire in the jobs we do not want are rarely wasted, and that a life has a way of using everything it has gathered.

Faraday's habit of taking meticulous notes, formed in those self-improving years at the City Philosophical Society, became one of the defining features of his scientific life and deserves to be counted among his real contributions. Throughout his career he kept a laboratory diary of extraordinary thoroughness, recording every experiment — the failures as faithfully as the triumphs — with dates, diagrams, and numbered entries that eventually ran into the thousands.

This was not mere tidiness. It was a method. By writing down everything, Faraday could return years later to a half-finished idea, trace exactly what he had tried, and pick up the thread; and the discipline of recording forced a clarity of thought that his lack of mathematical training might otherwise have left him without. The poor boy who had taught himself to write clearly in a workingmen's evening society turned that hard-won skill into one of the most valuable scientific notebooks ever kept, a habit that later generations of researchers would consciously imitate. The bookbinder never really stopped making books; he simply began filling them with the discovery of how the world works.

A few surprising facts deserve their place. The first is that Faraday, for all the mathematical laws that would eventually bear his mark, was almost entirely innocent of higher mathematics. Where Ampère reasoned in equations, Faraday could barely handle algebra; he thought in pictures, in physical intuitions, in vivid mental images of what was actually happening in space around a wire or a magnet. It was one of the most remarkable cases in the history of science of a first-rate mind reaching first-rate results by a completely different route from his contemporaries — and, as we will see, his picture-thinking led him to a concept, the *field,* that the mathematicians would spend the next fifty years catching up to. The second surprising fact is that Faraday's deep religious faith and his humility were not incidental to his science but woven through it: he refused a knighthood, twice declined the presidency of the Royal Society, and turned down a lucrative request from the British government to help develop poison gas for the Crimean War, saying it was against his conscience. The third is that this man who reshaped physics and chemistry never took a university course in either; he remained, technically and proudly, self-taught to the end of his days.

It is worth correcting a misconception here, because the tidy version of Faraday's rise flatters our sense of fairness. We like to tell it as a pure meritocracy triumph — the poor boy whose sheer talent lifted him up, proof that ability will out. But Faraday's own life gives that story the lie. His talent was necessary but nowhere near sufficient. He needed a customer who happened to give him lecture tickets, a laboratory accident that happened to injure Davy's eyes, a fired assistant who happened to leave a vacancy, and a patron willing, however grudgingly, to reach across the class divide and pull him through. Countless equally gifted working-class children of that era got none of those breaks and vanished into obscurity, their gifts unrecorded because no encyclopedia crossed their workbench and no great man answered their letter. Faraday's ascent was a triumph of talent, yes — but also of luck, timing, and a rigid social order that let exactly one such boy through while shutting out thousands. The wonder is not that talent rises; it is how rarely, in such a world, it was permitted to.

You meet Faraday's beginnings, in a sense, every time someone insists that a brilliant scientist must come pre-stamped with elite credentials, and every time that assumption is proven wrong. The self-taught outsider who changes a field is a recurring figure in this book — we will meet him again in a partially deaf newsboy from Ohio and a priest's son from a mountain village in the Balkans — and Faraday is his patron saint. His portrait still hangs in laboratories around the world not only for what he discovered but for what he was: living proof that the door to the deepest questions in nature is not, in the end, guarded by birth or diploma. And there is a further legacy, quieter but no less real. Faraday never forgot where he had come from, and he became one of the great popularizers of science, founding the Royal Institution's Christmas Lectures for young people — a tradition that continues to this day — precisely so that ordinary children, the sort of children he had been, might catch the same spark from a public lecture that had once leapt into him from a rebound encyclopedia. He believed knowledge should be shared, not hoarded, and he spent his fame lowering the very walls he had struggled over, holding the door open behind him for whoever might come next. In an age of exclusive academies and gentlemen's clubs, that generosity was its own kind of revolution.

Faraday had gotten himself into the laboratory. What he would do there, over the next two decades, would surpass anything his teacher Davy achieved and would lay the direct foundation for the electrified world. It began in 1821, with a small, whirling, almost toy like device — a wire, a magnet, and a cup of mercury — that made a metal thread spin endlessly round and round, and became the first electric motor the world had ever seen.

Chapter 13: Faraday's Motor

On the third of September, 1821, in the basement laboratory of the Royal Institution in London, Michael Faraday bent over a small glass vessel and watched a piece of wire do something no wire had ever done before. He had filled the vessel with mercury — the liquid metal, silvery and heavy, that conducts electricity like a pool of liquid wire. Standing upright in the center of the mercury was a bar magnet, fixed in place. Dangling down from above, its lower tip just touching the surface of the mercury, hung a loose wire, free to swing. When Faraday sent an electric current down through the wire and out through the mercury, the wire did not merely twitch, as Ørsted's compass needle had twitched. It began to *move* — and not back and forth, but around and around, sweeping in a steady circle about the magnet, endlessly, for as long as the current flowed. A thread of metal was walking in circles around a magnet, driven by nothing but electricity. Faraday had built the first device in history to turn electrical energy continuously into motion. He had built, in the most literal sense, the first electric motor.

To feel the weight of this, you have to appreciate what "continuous motion" meant. Ørsted's current had nudged a needle to a new resting position and left it there — one push, one twitch, and then stillness. Ampère's wires had leaned together or sprung apart and then hung still. Every electrical effect anyone had produced so far was a single, static shove. What Faraday achieved was fundamentally different in kind: not a one-time push but *sustained rotation*, motion that renews itself and keeps going, the wire chasing forever around the magnet and never arriving. That is the difference between a doorbell that rings once and an engine that runs. And an engine — a device that will spin a wheel or turn a shaft as long as you feed it energy — is the thing that does the actual physical work of civilization: pumping, grinding, hauling, spinning, lifting. In that whirling wire lay the ancestor of every electric motor that has ever run, from the vast machines that drive subway trains to the coin-sized motor that makes your phone buzz.

The secret to the endless circling was the strange, sideways nature of the magnetic force that we first met with Ørsted's needle. Recall that the magnetic field wraps around a current-carrying wire in circles, like the grain of wood curling around a knot. Faraday grasped, more clearly than anyone before him, what this circling geometry implied: if the magnetic force pushes the wire *crosswise* rather than toward or away from the magnet, then the natural motion it produces is not a straight-line jerk but a *rotation*. The wire is perpetually pushed sideways, and sideways-forever means around-in-a-circle. Picture a tetherball on a rope: if you keep striking it in the same sideways direction, it will not fly toward you or away from you but will circle the pole endlessly. Or picture a toy boat in a round pond with a jet on its side that always fires crosswise — it will not cross the pond but will spiral around and around the center. Faraday's genius was to arrange a magnet, a wire, and a pool of mercury so that this eternal sideways push could express itself as smooth, continuous circular motion. It looks almost like a magic trick, a bit of metal walking in circles on its own, but it is pure geometry: a crosswise force, given room, becomes a spin.

It is worth being careful here about what Faraday's 1821

device was and was not, because the popular story tends to blur it. This was not yet the practical electric motor of later industry — the powerful, efficient machines that would eventually drive factories and trains looked very different and lay decades in the future. Faraday's apparatus, which he called an *electromagnetic rotation* device, was a demonstration, a proof of principle, a beautiful laboratory toy. What made it historic was not its power but its principle: it showed, for the first time and beyond any doubt, that electricity could be converted into continuous mechanical motion. Every real motor that followed is an elaboration of that single proven idea. To say Faraday invented "the electric motor" is true in the way that saying the Wright brothers invented "the airplane" is true — the first version could barely do anything useful, but it established that the thing was possible at all, and everything else was engineering.

And now the story darkens, because Faraday's triumph came wrapped in a betrayal — or at least in an accusation of one — that would poison his relationship with the mentor who had lifted him out of the bookbindery. To understand the quarrel, you have to know that Faraday had not been working entirely alone. Earlier in 1821, another British scientist, William Hyde Wollaston, had proposed an idea for making a current-carrying wire rotate on its own axis, and had discussed it in the laboratory with none other than Humphry Davy. Wollaston's idea was different from what Faraday actually achieved — Wollaston imagined a wire spinning about its *own* axis; Faraday produced a wire revolving *around a magnet* — and Faraday worked out the successful effect independently. But when Faraday rushed his discovery into print in a scientific journal, and did so under his own name without first consulting Wollaston or fully crediting the prior discussions, it looked to some observers as though the young assistant had quietly appropriated an idea that had been floating around his betters' conversations and claimed it for himself.

The accusation stung, and it was Davy, painfully, who did

the most to keep it alive. The relationship between the two men had been cooling for some time; the celebrated master seems to have grown uneasy watching his one-time bottle-washer blossom into a scientist whose gifts might soon eclipse his own. In 1823, when Faraday was proposed for membership in the Royal Society — the highest recognition British science could bestow — Davy, as president of the Society, opposed him, and pressed him to withdraw the nomination over the Wollaston affair. Faraday, according to the story, quietly replied that he had not put his own name forward and could not withdraw what he had not submitted; and when he asked Davy who *would* withdraw it, Davy is said to have answered, "I will." Faraday was elected anyway, by an overwhelming vote, with only Davy against him. Wollaston himself, to his credit, eventually accepted that Faraday had acted honorably and the two became friendly. But the wound between Faraday and Davy never fully healed. There is a bitter and possibly apocryphal remark attributed to Davy near the end of his life: asked to name his greatest discovery, he is supposed to have answered, "My greatest discovery was Michael Faraday" — a line that can be read as generous pride or as rueful jealousy, and which historians still argue over.

To understand why Faraday's little device mattered so much, it helps to see what the world had instead of electric motors in 1821, because the contrast reveals the size of the gap he began to close. Every machine that did continuous work in that age was driven either by muscle — human or animal, turning cranks and treadmills and capstans — by falling water spinning a wheel, by the wind filling a sail or turning a windmill's arms, or by the newest and most powerful prime mover of all, the steam engine, which burned coal to boil water and drove its piston back and forth. These sources shared a common limitation: they were bulky, fixed, and hard to divide. You could not put a waterfall in a workshop or a windmill in a kitchen, and even a steam engine was a large, hot, dangerous, coal-hungry beast that made sense only at a substantial scale. What the electric motor eventually offered — though not yet, not with Faraday's toy — was something none of these could provide: mechanical power that could be sent through a thin wire to any location, divided into portions large or small, switched on and off in an instant, and run clean and quiet in the corner of a room. Faraday's whirling wire was the first faint promise of power on tap, delivered anywhere, in any amount. That promise would take sixty years to be fulfilled, but it began in a dish of mercury.

It is also worth being clear about the direction of the great reversal Faraday would eventually complete, because his 1821 motor was only the first of the two halves. In this device, he put electricity *in* — from a battery — and got motion *out*. The motor is a converter that runs in that direction: electrical energy becomes mechanical energy, current becomes spin. Ten years later, as we will see, Faraday would discover how to run the same conversion backward, putting motion in and getting electricity out, which is the generator. The two machines are mirror images, and together they form the beating heart of the electrical age: generators at the power plants turning motion into current, motors in homes and factories turning that current back into motion. Everything in between — the wires, the transformers, the whole vast grid — exists merely to carry the current from the one to the other. Faraday built the first of the pair in 1821, almost as a plaything, with no notion that he and his successors were laying the cornerstone of a world.

A few facts about this episode are worth carrying forward, because they complicate the saintly image of Faraday that later generations built. The first is that Faraday, for all his famous humility and Christian scruple, was in this instance genuinely careless with credit — he moved too fast, published too eagerly, and did not adequately acknowledge the conversations that had preceded his work, and it caused real and lasting hurt. Even the gentlest of the great scientists was capable of the ordinary human hunger to be first. The second is that the priority dispute, ugly as it was, is a permanent feature of scientific life rather than an aberration: we will see the very same poison — who thought of it first, who deserves the credit, who stole from whom — flare up again and again in this book, most spectacularly in the war between Edison and Tesla. The third, and most poignant, is the reversal of roles: Davy, the glittering star who had reached down to rescue a poor young man, ended by trying to hold that same man back, apparently unable to bear being surpassed by his own protégé. It is a very old human story, older than electricity, and it played out in a basement laboratory in London between two of the finest minds of the age.

You meet Faraday's 1821 discovery every single day, more times than you could possibly count, though you never see a wire circling in a dish of mercury. The principle he proved — that electricity plus magnetism yields continuous rotation — is the beating heart of the electric motor, and electric motors are so thoroughly woven into modern life that a typical home contains dozens of them and a typical person is rarely more than a few feet from one. They spin the drum of your washing machine and the blades of your blender, drive the compressor in your refrigerator and the fan in your laptop, raise and lower your car windows, run the pump in your dishwasher, propel electric cars and elevators and power tools and the little vibrating alert in your pocket. Roughly half of all the electricity generated in the world is consumed by electric motors turning something. Every one of them is a great-grandchild of the whirling wire Faraday set spinning in 1821. He had shown the world how to turn current into motion — and he had done it with a wire, a magnet, and a cup of mercury, apparatus so simple that its very simplicity is part of the wonder. The mightiest machines of the modern age trace their ancestry to a device you could have assembled on a kitchen table, and that lineage is worth remembering the next time an elevator lifts you skyward or an electric train pulls silently out of a station: somewhere at the root of it is a loose thread of metal, walking in patient circles around a magnet in a London basement.

The reverse question, though, was still wide open, and it was the more profound of the two. Faraday had made electricity produce motion. Could motion, run backward, be made to produce electricity? Could you take a magnet, move it, and conjure a current out of thin air — a current with no battery behind it at all? For ten long years that question nagged at him, and when he finally cracked it, in 1831, he unlocked not a laboratory curiosity but the single principle on which the entire electrical power industry would come to rest.

Chapter 14: Induction: The Idea That Built the Grid

On the twenty-ninth of August, 1831, in his basement laboratory at the Royal Institution, Michael Faraday wound two separate coils of copper wire around opposite sides of a thick iron ring, taking care that the two coils never touched each other. One coil he connected to a battery, with a switch he could open and close. The other coil he connected to a galvanometer — a sensitive instrument whose needle would swing if even the faintest current passed through it. His idea was that the current in the first coil, magnetizing the iron ring, might somehow force a current to flow in the second, entirely separate coil. He closed the switch. Current surged into the first coil. And the galvanometer needle attached to the *second* coil gave a sudden kick — then fell back to zero, even though the first coil was still carrying a steady current. Puzzled, Faraday opened the switch, cutting the first current off. The needle kicked again, this time in the opposite direction, and again fell back to zero. He tried it over and over, and the pattern held with perfect stubbornness: the second coil produced a current only in the *instants* when the first current was changing — switching on, switching off — and produced nothing at all in between, while the first current held steady. In that maddening, flickering needle lay the discovery that would electrify the world.

What Faraday had stumbled into is the single most important principle in this entire book, so it is worth stating with great care, because everything — every power plant, every transmission line, every transformer, the entire electrical grid — rests upon it. He called the effect **electromagnetic induction**, and the heart of it is one crucial word: *change*. A steady, unchanging magnetic field does nothing to a nearby wire. But a *changing* magnetic field — one that grows or shrinks or moves — reaches into a nearby wire and pushes its charges into motion, conjuring a current out of what had been an inert loop of copper. This was the answer to the question that had nagged Faraday for ten years, ever since his little motor. He had made electricity produce motion; now he had made *magnetism produce electricity*. And the deep, non-obvious lesson buried in his flickering needle was that magnetism alone is not enough. It is the *changing* of magnetism that does the work. A magnet sitting still next to a wire is useless; a magnet in motion is a generator.

Faraday proved this even more vividly a few months later with an experiment so simple that any schoolchild can repeat it and so profound that it underlies the whole power industry. He took a coil of wire, connected it to a galvanometer, and simply pushed an ordinary bar magnet into the coil. As the magnet moved in, the needle swung — a current was flowing. When he held the magnet still inside the coil, the needle dropped back to zero, no matter how strong the magnet was. When he pulled the magnet back out, the needle swung the other way. Motion in, current one way; motion out, current the other way; no motion, no current. Wave a magnet near a coil of wire and you make electricity; hold it still and you make nothing. That homely little demonstration — a magnet and a coil, cheap enough for any classroom — is, in principle, exactly what is happening inside every generator in every power station on earth at this very moment. The coils and magnets are vastly larger and the motion is a blur of high-speed spinning, but the physics is identical: a magnet moving relative to a coil of wire produces a current.

Why should motion matter so much? Here two everyday analogies help. Think first of a sail. Still air — a dead calm — pushes a sail not at all, however much air surrounds it. It is *moving* air, wind, that fills the sail and drives the boat. What matters is not the presence of air but its movement. Magnetism works on a coil of wire in much the same way: a still magnetic field, however strong, drives no current; it is the *movement* or change of the field that pushes the charges along, like wind filling a sail. A second analogy: imagine you are floating in a swimming pool, and someone slides a large board through the water toward you. As the board moves, it drags the water and pushes you along with it. But once the board stops, even if it is right next to you, the water goes still and you drift to a halt. The moving board is the moving magnet; the water it drags is the current it induces; and the moment the motion stops, so does everything else. In both pictures, the lesson is the same one Faraday's flickering needle taught: it is change, motion, that transfers the energy. Steadiness transfers nothing.

Faraday did not stop at proving the principle; he built the first machine to exploit it, and in doing so he built the ancestor of every generator that has ever run. Later in 1831 he mounted a copper disc so that it could spin between the poles of a large magnet, with sliding contacts touching the disc's center and its rim. When he turned a crank and set the disc spinning, a steady current flowed out through the contacts — a current with no battery anywhere in the apparatus, produced purely by the mechanical work of his hand turning the crank. This device, now called the **Faraday disc** or the disc dynamo, was the first electric **generator**: a machine that converts mechanical motion into electricity. And this is the profound reversal at the center of our whole story. Faraday's 1821 motor turned electricity into motion. His 1831 generator turned motion into electricity. Together they are the two halves of a single door, swinging both ways, and through that door flows essentially all the electrical power that runs modern civilization. Understand these two machines and you understand, at its root, how the entire electrified world works.

Consider what a generator really means, because it is the answer to a question you may never have thought to ask: where does electricity actually *come from*? It does not come from nowhere. A generator produces electricity only when something turns it, and that something is always some other form of energy being converted. In a coal or gas power plant, burning fuel boils water into steam, and the rushing steam spins a turbine that turns the generator's coils past its magnets. In a hydroelectric dam, falling water spins the turbine. In a nuclear plant, the heat of splitting atoms boils the water that makes the steam. In a wind farm, the wind itself turns the blades. In every case — every single case but one — the recipe is the same: find some way to *turn a shaft*, use that shaft to move magnets and coils past each other, and Faraday's induction does the rest, converting the motion into current. The only major exception is the solar panel, which makes electricity directly from light without any spinning at all. But close to everything else — the overwhelming majority of the electricity humming through the wires of the world — is Faraday's disc dynamo, grown to titanic scale and spun by fire or falling water. Every time you flip a switch, somewhere a magnet is moving past a coil because Michael Faraday discovered, in 1831, that motion makes current.

There is a second, equally vital child of induction, and it is the one that will decide the great war at the center of this book: the **transformer**. Recall Faraday's very first experiment, the iron ring with two coils, where switching the first current on and off made the needle in the second coil kick. That arrangement — two coils sharing a magnetic link, so that a changing current in one induces a current in the other — is a transformer, and it has a near-magical property. By using a different number of wire turns in the two coils, you can make the induced voltage in the second coil higher or lower than the voltage in the first. Feed in electricity at a modest voltage and you can draw it out at a towering one, or the reverse. This is the device that lets electricity be pushed up to enormous voltages for its long journey across the countryside and then brought safely back down to gentle levels before it enters your home — and, crucially, it works *only* with the changing, back-and-forth current we call alternating current, and not at all with the steady current Edison would champion. When Tesla and Westinghouse do battle with Edison decades from now, the transformer will be their decisive weapon, and the transformer is nothing but Faraday's iron ring of 1831, scaled up and put to work.

It is worth dwelling for a moment on the ten-year gap between Faraday's motor and his discovery of induction, because it corrects a comfortable assumption about how discovery works. From our vantage point, the idea seems almost obvious: if electricity can make motion, surely motion can make electricity, and Faraday, having built the motor in 1821, ought to have found induction the following week. He did not. He wrestled with the problem, on and off, for a full decade, and so did other first-rate scientists across Europe, several of whom came close and missed. The reason is that the crucial ingredient — *change* — is deeply counterintuitive. Everyone was trying to get a current out of a *steady* magnetic field, reasoning that a strong magnet held near a wire ought to produce a steady current, the way a strong battery produces a steady current. But a steady field produces nothing at all; it is only the *changing* field that works, and change is a slippery thing to think about and a slippery thing to build an experiment around. Nature had hidden the secret in plain sight, behind a single word that ran against every intuition, and it took even Faraday ten years to see it. Great discoveries often look obvious only in the rearview mirror.

There is a further subtlety in Faraday's disc dynamo that is worth appreciating, because it made the machine genuinely new. His earlier induction experiments produced only brief flickers of current — a kick when the magnet moved in, another when it moved out, and nothing in between. That is not much use if you want a reliable supply of electricity; you cannot run a city on flickers. The spinning copper disc solved this by arranging for the change to be *continuous*: as the disc turns, every part of it is always moving through the magnetic field, so the induction never stops, and the current flows steadily for as long as the crank is turned. Faraday had converted the intermittent kick of a moving magnet into a smooth, unbroken stream of electricity — the first steady current ever produced by mechanical motion rather than by chemistry in a battery. In doing so he had invented not just a proof of principle but the working ancestor of a machine that would one day light continents.

A misconception worth correcting here is the tidy notion that Faraday "invented electricity" or "invented the generator" in the way an inventor patents a gadget. He did neither, in that sense. Faraday discovered a *principle of nature* — that changing magnetism induces current — and he built crude demonstrations of it, but he never patented anything, never sought to profit from it, and never built a practical, powerful generator; the useful dynamos of industry were the work of later engineers building on his principle. Faraday himself seems to have cared little for the applications. There is a famous, almost certainly embellished anecdote in which a politician — sometimes said to be William Gladstone — visits his laboratory, watches him fuss with his coils and magnets, and asks impatiently what possible use this electricity could ever be, to which Faraday supposedly replies, "Why, sir, there is every probability that you will soon be able to tax it." Whether or not he ever said it, the line captures a real truth: Faraday was chasing understanding, not profit, and he could not have foreseen the trillion-dollar industry his flickering needle would spawn. The man who made the modern electrical world possible died a relatively poor man by choice, having turned down the wealth and honors that his discoveries could easily have brought him.

You meet electromagnetic induction more intimately than almost any idea in this book, because it is, quite simply, where your electricity comes from — all of it, minus what your solar panels make. Every kilowatt-hour that has ever powered a light, a computer, a hospital, a city was, in all likelihood, generated by Faraday's principle: a coil and a magnet in motion. But induction hides in smaller places too. The wireless charging pad that powers your phone without a plug works by inducing a current in a coil inside the phone from a changing field in the pad — Faraday's two coils, exactly, a hundred and ninety years on. The card reader that beeps when you tap to pay, the induction cooktop that heats a pan while the glass stays cool, the metal detector at the airport, the electric guitar's pickup translating a vibrating string into sound — all of them are electromagnetic induction, Faraday's flickering needle, running quietly beneath the surface of ordinary life. No single discovery in these pages did more to build the world you live in.

Faraday had found the master principle. But across the Atlantic, at almost exactly the same time, an American schoolteacher had been building the most powerful electromagnets the world had ever seen and had brushed against the very same secret of induction — and had he published a little sooner, the honor might have been his.

Chapter 15: Joseph Henry and the American Current

In 1830, in a workshop at the Albany Academy in upstate New York, a tall, self-taught schoolteacher named Joseph Henry was carefully wrapping insulated wire around a horseshoe-shaped bar of iron, and he had solved a problem that no one in Europe had quite cracked. The trick was the insulation. Everyone knew by then that a coil of current-carrying wire wrapped around iron would turn the iron into a powerful magnet — an **electromagnet**, a magnet you could switch on and off at will simply by starting or stopping the current. But there was a limit to how much wire you could pile on, because if the bare wires touched, the current would take a shortcut across the coil instead of circling the iron, and the magnetism would collapse. Henry's homely, brilliant solution was to insulate the wire himself, wrapping it in strips of silk torn, according to the story, from his long-suffering wife's petticoats. Insulated this way, the turns could not short against one another, and he could wind coil upon coil, hundreds and then thousands of turns, packing far more magnetizing power onto a single iron core than anyone had managed before. The electromagnets he built were monsters for their day.

Just how monstrous is one of the genuine surprises of this history. Henry eventually constructed an electromagnet for Yale College that could lift more than a ton of iron — around two thousand pounds — from a device you could carry in your arms, powered by a modest battery. To a generation that thought of electricity as a thing that twitched frog legs and made sparks jump, the sight of a small horseshoe of iron hoisting a blacksmith's anvil, and then dropping it the instant the wire was disconnected, was very nearly miraculous. Here was electricity doing serious mechanical work, lifting real weight, and doing it under perfect control — on with a touch, off with a touch. The powerful electromagnet was not a toy. It was the component that would make the electric telegraph, the electric motor, the generator, and the electric bell all possible, because in every one of those machines the essential act is the same: use a current to create a controllable magnetic pull. Henry, more than anyone, turned the electromagnet from a curiosity into a practical, powerful tool.

But Henry's name endures for a subtler discovery, one that sits at the very center of alternating-current technology and that he came upon while playing with those same giant coils. He noticed something strange and, at first, annoying. When he switched off the current to one of his big electromagnets, he often got a vicious spark at the point where he broke the connection — a bright, snapping arc, sometimes strong enough to sting. The current, it seemed, did not want to stop. It fought to keep flowing for an instant after the circuit was opened, as if it had momentum. Henry realized that a coil of wire, carrying a current, resists any *change* to that current — it opposes being switched on, and once running, it opposes being switched off, lashing out with a spark in its effort to keep going. He had discovered what we now call **self-induction**: a coil, through its own magnetic field, induces a voltage in *itself* whenever its current changes, a voltage that always pushes back against the change. In effect, a coil gives an electric current a kind of inertia, a reluctance to speed up or slow down.

The cleanest way to picture self-induction is to think about the inertia of a heavy object. Imagine a loaded shopping cart. It takes real effort to get it rolling from a standstill, and once it is rolling, it takes real effort to stop it — shove it and it resists; try to halt it suddenly and it lurches forward, fighting you. A coil does exactly this to electric current. It resists the current starting, and when you try to cut the current off abruptly, the coil "lurches," dumping its stored magnetic energy into that snapping spark as it struggles to keep the current going. A second analogy is water in a long pipe: get a heavy column of water moving through a pipe and then slam a valve shut, and the water's momentum produces a violent bang — the plumber's "water hammer" — as the moving mass refuses to stop on command. Self-induction is electrical water hammer, and the property that measures how strongly a coil behaves this way is called **inductance**. It is one of the three or four fundamental electrical properties, sitting alongside resistance and capacitance, and it turns out to be indispensable: inductance is what makes transformers, radios, and the smooth running of alternating-current machinery possible. The very kickback Henry found irritating is now engineered into countless devices on purpose.

Here we arrive at one of the poignant near-misses of scientific history, the kind that keeps historians up at night. Joseph Henry, working in America, seems to have observed electromagnetic induction — the great principle of the previous chapter, that a changing magnetic field generates a current — at very nearly the same time as Michael Faraday, and quite possibly a little *before* him. But Henry was a schoolteacher with a punishing workload and no leisure for research; he taught long hours and could investigate only during the summer breaks. He made his observations, meant to write them up, and did not rush. Faraday, with a proper laboratory and time to work, made his own discovery in 1831 and, crucially, *published immediately*. Priority in science goes not to the first to observe but to the first to publish, and by that iron rule the discovery of induction belongs, rightly and permanently, to Faraday. Henry got there too, but a step slow to the printer, and it cost him the single greatest prize in the field. There is no evidence he was bitter about the induction question itself, and the two men later became warm friends; but it is hard not to feel the sting of it. A busy American teacher, wrapping wire in his wife's silk petticoats between classes, brushed his fingertips against the discovery that would electrify the planet, and lost it to a matter of months and a trip to the press.

Henry did, however, win clear priority for self-induction, and it is a measure of his stature that when the scientists of the world sat down, generations later, to assign names to the fundamental electrical units, they gave his to inductance. The **henry** is the standard unit of inductance, and it places this modest, overworked American schoolteacher in the same pantheon as Volta, Ampère, Ohm, and Faraday — the small company of researchers whose surnames became the permanent vocabulary of electricity. It is worth pausing on the fact that, in the roll call of those foundational units, four are European and one is American, and that one is Henry. In an age when serious science was overwhelmingly a European enterprise and the United States was regarded across the Atlantic as a provincial backwater of farmers and traders, Henry stood as living proof that first-rate physics could be done on American soil. He was, in a real sense, the founding figure of American physical science.

His practical influence ran even deeper than his fundamental discoveries, and it points straight toward the technologies at the heart of this book. Henry built an early electric motor. More importantly, he grasped that a current could be sent through a very long wire to work an electromagnet at the far end, and he strung a mile of wire around a room and used it to ring a bell remotely — a demonstration that contained, in embryo, the entire principle of the electric telegraph. Henry freely shared his ideas, and Samuel Morse, who would build the telegraph into a world-spanning commercial system and grow rich and famous from it, drew on Henry's work and his generous advice. In a controversy that would later turn bitter, Henry felt Morse gave him too little credit; Henry was a scientist who believed knowledge should be shared freely for the public good, while Morse was an entrepreneur who patented and profited. It is a small preview of the great tension running through this whole story — the open pursuit of knowledge versus the closed pursuit of gain, the scientist versus the businessman — a tension that will erupt spectacularly when Edison and Tesla meet.

The story of the silk petticoats deserves a second look, because behind the charming anecdote lies a genuine technical insight that mattered enormously. In Henry's day, insulated wire simply did not exist as a product you could buy; wire came bare, and if you wanted to wind a coil, you had to keep the turns from touching by some other means, usually by spacing them apart on a wooden frame, which strictly limited how many turns you could fit. Henry's decision to individually insulate the wire — first reportedly with silk cloth, later with more systematic wrappings — meant that adjacent turns could be packed tightly together, even layered on top of one another, without the current leaking across from one turn to the next. This is why his electromagnets so dramatically outperformed everyone else's: he could crowd thousands of turns of current-carrying wire around a single iron core, and since the magnetizing power grows with the number of turns, his magnets were monsters where others' were mice. It was not a flashy idea. It was, quite literally, a matter of wrapping thread around wire. But it is a perfect illustration of how a humble, practical refinement — insulation — can unlock a leap in performance that pure theory alone would never deliver. The whole later industry of electrical machinery would depend utterly on good insulated wire.

Henry's insight into how to design his magnets carried a further refinement that reveals his practical genius. He understood that the *shape* of the electrical source and the winding had to be matched to the job. For lifting great weights close at hand, he wound many turns of wire all connected so as to work together from a single large battery cell — what he called a "quantity" magnet, optimized for raw pulling power. For sending magnetism to the far end of a long wire, as in a telegraph, a different arrangement was needed, one suited to driving current through great length and resistance — an "intensity" magnet, wound and powered differently. This distinction, subtle as it sounds, was exactly the kind of understanding that separated a working engineer from a mere demonstrator, and it fed directly into the design of the telegraph systems that would soon web the continent. Henry grasped, earlier and more clearly than most, that electricity was not one-size-fits-all: the machine had to be shaped to the task.

A misconception worth correcting is the assumption, common in the popular retelling, that the foundations of the electrical age were laid entirely by Europeans — that America contributed nothing until the practical inventors like Edison arrived to commercialize what the Old World's geniuses had discovered. Henry is the standing refutation. Decades before Edison, an American had built the world's strongest electromagnets, independently discovered induction, pioneered self-induction, and sketched out the telegraph and the electric motor. And his influence did not end with his own experiments. In 1846 Henry was chosen as the first Secretary — the founding director — of the newly created Smithsonian Institution in Washington, and he spent the rest of his life building it into the great engine of American science it remains today, shaping the nation's entire scientific culture. He also organized the first national weather-observation network using the telegraph, a direct ancestor of the modern weather service. The American current in this story runs from Henry forward, and it is a stronger and older current than the popular Edison-centered version admits.

You meet Joseph Henry's work every time a relay clicks, every time a doorbell rings, every time a transformer hums on a utility pole, and every time an old-fashioned car engine fires its spark plugs — for the ignition coil that steps a car battery's humble twelve volts up to the tens of thousands of volts needed to make a spark is a direct application of the self-induction Henry discovered, using the coil's kickback to generate a high-voltage jolt. His electromagnets live on, grown to gigantic scale, in the machines that lift scrapped cars in junkyards and in the superconducting coils of medical MRI scanners. And the inductor — a coil whose whole purpose is the self-induction he found irritating — sits by the billions inside the electronics of the modern world, smoothing currents and tuning radios. The overworked schoolteacher's fingerprints are all over the machinery of daily life.

By the middle of the nineteenth century, then, the essential building blocks were in hand. Ørsted, Ampère, Ohm, Faraday, and Henry had established that electricity and magnetism were one, had measured the forces and the flows, and had discovered the motor, the generator, and the transformer in principle. What remained was to weave all of it — every scattered law and effect — into a single, unified theory, and that supreme act of synthesis fell to a quiet, devout Scottish mathematician whose four short equations would not only unite electricity and magnetism forever but would reveal, astonishingly, that light itself is an electric wave.

Chapter 16: Maxwell's Beautiful Equations

Sometime in the 1860s, in his study, a soft-spoken Scottish physicist named James Clerk Maxwell was turning over the tangled results of the previous half-century — Ørsted's twitching needle, Ampère's forces, Faraday's induction, Henry's coils — and doing what none of those men, for all their brilliance, had been able to do. He was reducing the whole sprawling, sparking, magnetic mess to mathematics. Faraday, who could barely handle algebra, had thought in vivid physical pictures: he imagined space around a magnet or a wire as filled with invisible lines of force, a texture of tension and pressure running through the emptiness, and he called this the *field*. Most mathematical physicists of the day regarded Faraday's field lines as a charming crutch for a man who could not do the equations. Maxwell did the opposite. He took Faraday's intuitive pictures seriously, more seriously than Faraday himself could, and set out to write them down in the rigorous language of mathematics. When he was finished, he had produced a handful of equations that captured every known electrical and magnetic phenomenon in one unified system — and then, almost as a byproduct, those equations told him something so staggering that he could scarcely believe it.

Before the payoff, we need the idea of the **field**, because it is Maxwell's foundation and one of the most important concepts in all of physics. A field is a way of describing how a force reaches across empty space. Instead of imagining that a magnet somehow grabs a distant nail directly, Faraday and Maxwell pictured the magnet as filling the space around it with an invisible condition — a magnetic field — so that the nail responds to the field right where the nail sits, not to the far-off magnet. Think of it like temperature in a room: near a fireplace the air is warm, and it grows cooler as you walk away, so that every point in the room has a temperature, an invisible quality filling the whole space. A magnetic field is like that, except that instead of warmth, every point carries a certain strength and direction of magnetic push. An electric field is the same kind of thing for electric charge: every point around a charged object carries a certain electric push, felt by any other charge that ventures there. These two invisible fields, electric and magnetic, filling all of space, are the true actors in Maxwell's story. The charges and magnets are just the sources that stir them up.

What Maxwell wrote down, in the form later generations tidied into **Maxwell's equations**, was a complete set of rules — four of them — governing how these two fields behave: how electric charges create electric fields, how magnetic fields form their endless loops, how a changing magnetic field creates an electric field (that is Faraday's induction, written in mathematics), and how electric currents and changing electric fields create magnetic fields (that is Ørsted and Ampère, likewise made exact). Four short statements, and between them they contained the entire behavior of electricity and magnetism, everything that every experimenter in this book had discovered, unified into a single seamless theory. It was one of the supreme achievements of the human mind, ranked by physicists alongside Newton's laws of motion — a moment when a vast, chaotic body of separate facts suddenly clicked into one elegant structure. Maxwell had shown, finally and completely, that electricity and magnetism are not two forces but one: **electromagnetism**, exactly as Ørsted's needle had first hinted forty years before, now proven with mathematical certainty.

But the equations had a secret, and it is the most beautiful moment in this entire history. Maxwell noticed that his four rules, taken together, allowed for a self-sustaining ripple. Suppose you jiggle an electric charge. Its motion creates a changing electric field. But by Maxwell's rules, a changing electric field creates a magnetic field. And a changing magnetic field, by Faraday's law, creates an electric field. So the changing electric field makes a magnetic field, which — because it too is changing — makes a new electric field a little farther out, which makes a new magnetic field farther still, each one endlessly regenerating the other as the disturbance races outward through empty space. Electric field begets magnetic field begets electric field, leapfrogging forever, a wave that carries itself along by its own bootstraps, needing no medium and no wire, traveling freely through the void. Maxwell had discovered, purely on paper, purely from his equations, that **electromagnetic waves** must exist — ripples of pure field, propagating through space at some definite speed.

And then came the thunderclap. Maxwell's equations did not just predict that such waves exist; they predicted exactly how fast they must travel, a speed fixed by two numbers that experimenters had already measured in the laboratory — one from experiments with electric charges, the other from experiments with magnets. Maxwell put those two humdrum electrical numbers into his formula, turned the crank, and out came a velocity. It was about 186,000 miles per second. He would have recognized that number instantly, because it was already famous: it was the measured speed of *light*. The coincidence was too precise to be an accident, and Maxwell drew the only possible conclusion, writing words that still raise the hairs on a physicist's neck — that light itself must *be* an electromagnetic wave, that the very illumination pouring through his study window was nothing but rippling electric and magnetic fields, the same fields Faraday had conjured with his coils and magnets, oscillating at an unimaginable speed. Two sciences that had seemed to have nothing to do with each other — electricity and optics, the study of currents and the study of light — were revealed to be one and the same. Light was electricity in flight.

It is hard to overstate how profound and unifying this was, so let an analogy carry it. Imagine that for centuries people had studied ocean waves and, quite separately, studied the ringing of bells, never dreaming the two were related — and then one day a mathematician proved that sound is simply waves in the air, that the bell and the ocean swell are the same phenomenon in different media. That is roughly the magnitude of what Maxwell did for light and electricity. He showed that the warmth of the sun, the color of a flower, the glow of a candle, and the current in a wire are all manifestations of one underlying thing: the electromagnetic field, trembling. And his theory implied something even grander. If visible light is an electromagnetic wave of a particular wavelength, then there ought to be other electromagnetic waves — longer and shorter — that our eyes simply cannot see. There should be an entire invisible spectrum of them, all traveling at the speed of light, all made of the same rippling fields, differing only in their wavelength. This was the prediction of what we now call the **electromagnetic spectrum**, and it opened the door to radio, microwaves, infrared, ultraviolet, X-rays, and gamma rays — all of them, every one, the same stuff as visible light, just tuned to a different pitch.

For years, Maxwell's electromagnetic waves remained a prediction on paper, an audacious deduction that many physicists admired but could not quite bring themselves to believe until someone actually produced such a wave in a laboratory and caught it. Maxwell himself did not live to see it; he died of cancer in 1879, at only forty-eight, the same year, as it happens, that Thomas Edison was perfecting his light bulb in New Jersey — a quiet passing of the torch from the age of theory to the age of application. The confirmation came in 1887, when a young German physicist named

Heinrich Hertz built an apparatus that generated invisible waves by making sparks jump across a gap, and detected them across the room with a small loop of wire in which the arriving waves induced a faint spark of their own. The waves behaved exactly as Maxwell's equations demanded — they traveled at the speed of light, they could be reflected and focused like light, they were light's invisible cousins. Hertz had made and caught the first **radio waves**, proving Maxwell right beyond any doubt. When admirers asked Hertz what practical use his discovery might have, he reportedly shrugged and said, "Nothing, I suppose" — one of the great wrong predictions in history, for within two decades those very waves would carry the first wireless messages across the Atlantic, and within a century they would carry every radio broadcast, television signal, cell-phone call, and Wi-Fi connection on earth. Hertz's name lives on as the unit of frequency, the **hertz**, which is why the dial of every radio and the specifications of every wireless device are marked in his honor.

It is worth knowing something of the man behind this supreme act of synthesis, because he stands in quiet contrast to the tormented and combative figures who fill much of this book. James Clerk Maxwell was born in Edinburgh in 1831 — the very year Faraday discovered induction — into a comfortable Scottish family, and he was, by every account, a person of rare gentleness, humor, and modesty, deeply religious in an undogmatic way, devoted to his wife, and beloved by nearly everyone who knew him. As a boy he was insatiably curious, forever asking "what's the go of it?" about every gadget and gimmick he encountered, and he published his first mathematical paper at the age of fourteen. His genius ranged far beyond electromagnetism: he did foundational work on the physics of gases, explained the nature of Saturn's rings decades before any spacecraft confirmed it, and produced the world's first true color photograph. Yet he wore his brilliance lightly and sought no fame, and when he died young, in 1879, he was less celebrated than lesser men. History has been slow to give Maxwell his due in the popular imagination, where his name is far less known than Edison's or Einstein's — an injustice, given that Albert Einstein himself kept a photograph of Maxwell on his study wall and declared that his own work stood on the foundations Maxwell had laid.

There is a surprising fact about the reception of Maxwell's equations that is easy to miss. When he first published his full theory, it was not embraced with immediate acclaim; it was regarded by many physicists as obscure, difficult, and philosophically suspect, cluttered with an elaborate mechanical scaffolding of imaginary gears and vortices that Maxwell had used to help himself visualize the fields. It took a later generation — including Heinrich Hertz, whose experiments confirmed the waves, and a British engineer named Oliver Heaviside, who stripped the theory down to the compact four-equation form we use today — to reveal the diamond hidden inside Maxwell's dense original presentation. In a real sense, the "Maxwell's equations" that adorn physics classrooms are Maxwell's physics in Heaviside's streamlined clothing. This is a recurring pattern worth noticing: a great theory is often born tangled and awkward, and only later polished by other hands into the clean, teachable form that makes it look, in hindsight, as though it had always been simple.

A misconception worth correcting is the popular notion that Maxwell "invented" something, in the way Edison invented the phonograph or Bell the telephone. Maxwell invented no device at all. He built no machine, held no relevant patent, and lit no lamp. What he did was purely theoretical — and it was, if anything, more consequential than any single invention, because it revealed the hidden rules that every future electrical technology would have to obey, and it predicted, decades in advance, whole categories of technology that did not yet exist. A second misconception is that such abstract theory is somehow impractical, the ivory-tower opposite of "real" engineering. The truth is precisely the reverse: Maxwell's four equations are among the most ruthlessly practical statements ever written. Every antenna, every radar dish, every fiber-optic cable, every electric motor and generator and transformer is designed, today, using Maxwell's equations. They are not a historical curiosity; they are working tools that a modern engineer uses every day. Abstract thought, it turns out, can be the most powerful practical force of all.

You meet Maxwell's discovery in the most literal way imaginable every waking second, because you are, right now, bathed in electromagnetic waves and utterly dependent on them. The light by which you read these words is one of Maxwell's waves striking your eye. The warmth of sunshine on your skin is another. But the invisible ones are the miracle of daily life: every text message and phone call reaches you as a Maxwell wave; every Wi-Fi signal filling your home, every GPS position pinpointing your car, every radio station, every television channel, every Bluetooth connection between your phone and your headphones, the microwaves that heat your dinner, the X-rays that image a broken bone, the infrared beam from your television remote — all of them are electromagnetic waves, all of them travel at the speed of light, and all of them exist in the engineered form they do because a quiet Scotsman proved in the 1860s that light and electricity are one, and that space can carry rippling fields from one place to another with no wire between. The wireless world runs on Maxwell.

With Maxwell's equations, the great scientific age of electricity reached its capstone. Everything essential about the nature of electricity and magnetism was now understood, written down, and mathematically complete. The remaining task was no longer to discover what electricity *was*, but to decide what humanity would *do* with it — how to make it cheaply, tame it safely, and pour it into the daily life of millions. That task belonged not to philosophers and mathematicians but to inventors, engineers, and businessmen, and the first and most famous of them was even now growing up along the railroads of the American Midwest: a partially deaf, restless, endlessly curious boy who would turn the science of the field into the machinery of the modern world.

Part III

The Wizard of Menlo Park

From a half-deaf boy selling newspapers on a Michigan railroad to the man who switched on a city — the rise of Thomas Edison and the invention of the electric light, the power station, and the modern research laboratory itself.

Chapter 17: A Deaf Boy on the Railroad

On a cold morning in the summer of 1862, a fifteen-year-old boy stood on the platform of the Mount Clemens station in Michigan, watching a small child toddle out onto the railroad tracks toward the path of a rolling boxcar. The boy on the platform was tall for his age, shabby, ink-stained, with a large head and quick gray eyes. He sold newspapers and candy on the train that ran between Port Huron and Detroit, and he had a hundred things to do that morning, but he did not stop to think about any of them. He dropped his papers and ran. He snatched the child off the rails a breath before the wheels reached the spot, and the two of them tumbled together onto the gravel, scraped and gasping, alive. The child's father, the station agent, was a man named James MacKenzie, and he had nothing to give in reward but the one thing he happened to possess: he knew how to work a telegraph, and he offered to teach the boy. The boy was Thomas Alva Edison, and that lesson would set the course of his entire life.

It is a good story, and like many of the best stories about Edison it is true in outline and burnished in the retelling — a distinction worth keeping in mind through everything that follows, because Edison lived long, told his own tale often, and was not above improving it. But the bones of it are solid. A telegraph operator taught a grateful newsboy the code, and that code carried the boy out of a hard rural boyhood and into the great electrical adventure of the nineteenth century. To understand how far that carried him, we have to go back to where he started, because almost nothing about the beginning suggested the end.

Thomas Edison was born on February 11, 1847, in Milan, Ohio, a small canal town that was, for a brief moment, one of the busiest grain-shipping ports in the world. He was the seventh and last child of Samuel and Nancy Edison. By the time Thomas — always called "Al," short for his middle name, Alva — was seven, the railroad had bypassed Milan, the canal traffic had dried up, and the family moved north to Port Huron, Michigan, a lumber town on the St. Clair River where Lake Huron pours south toward Detroit. It was there that the boy grew up, and there that the two facts most often repeated about his childhood took hold: that he was a poor student, and that he was going deaf.

Both facts deserve a careful look, because both are usually told wrong. Take the schooling first. The popular version holds that Edison had almost no formal education — and in the narrowest sense that is accurate. His total time in an actual schoolhouse amounted to only about twelve weeks. According to the story he told for the rest of his life, a schoolmaster named Reverend Engle called him "addled," meaning muddled or slow, and his furious mother marched to the school, pulled him out, and declared she would teach him herself. But the deeper truth is not that Edison went uneducated. It is that he was educated at home, intensively, by a mother who had herself been a schoolteacher. Nancy Edison set her restless son loose on books most adults would find daunting — Gibbon's *Decline and Fall of the Roman Empire*, histories, a school reader on natural philosophy that described simple experiments a boy could try. Edison later said, with the flat certainty he brought to most pronouncements, "My mother was the making of me." He did not mean she coddled him. He meant she took a child the system had discarded and refused to let him be discarded.

Here is the first surprising fact to carry forward: the man who would come to embody, more than any other single human being, the triumph of the American classroom-to-laboratory ideal barely set foot in a classroom. Edison's education was self-directed, chaotic, hands-on, and driven almost entirely by his own furious curiosity. He read what interested him and ignored what did not. He was, in the modern sense, a spectacularly successful dropout — and that pattern, learn-by-doing rather than learn-by-sitting, would define not just his life but the invention factory he later built.

Now the deafness, which matters more than any other single fact about his character. Edison began to lose his hearing as a boy and grew progressively deafer through his life, until in adulthood he could hear little more than a loud shout and, by his own account, no birdsong at all. What caused it is genuinely uncertain, and here the honest historian has to flag a contested point. Edison himself told at least two versions. In one, a train conductor boxed his ears or yanked him aboard a moving train by the ears — supposedly the moment "something snapped inside." In another telling, it was scarlet fever and a series of untreated middle-ear infections in childhood. The medical evidence points toward the infections; the ear-yanking story is almost certainly the kind of vivid, blame-assigning tale that memory likes to manufacture. The likeliest truth is the duller one: a childhood illness left him partly deaf, and it got worse with age.

What is *not* in doubt is how Edison chose to understand his own deafness. He insisted, again and again, that it was not an affliction but an advantage. It spared him idle chatter and let him concentrate with a ferocity that hearing people could not match. Bent over a workbench, he was sealed inside a bubble of silence that the world could not easily interrupt. Later, when he worked on the phonograph and the telephone, he would literally bite down on the wooden cabinet of a machine to feel its vibrations through the bones of his skull — hearing, in a sense, through his teeth. The point worth holding onto is this: a trait that most people would call a disability, Edison reframed as the source of his focus. Whether or not that reframing was strictly fair, it tells you everything about the man's relentless will to turn any hand he was dealt into a winning one.

Poverty and a big family meant that Al went to work young. At twelve or thirteen he took a job as a "news butcher" — a boy who worked the passenger cars of the Grand Trunk Railway, walking the aisles between Port Huron and Detroit selling newspapers, magazines, candy, peanuts, and cigars to travelers. It was a good beat for an ambitious child, and Edison worked it like a small enterprise. He hired other boys to help. He set up newsstands at stations. During the Civil War, he discovered a principle that every media mogul since has understood in his bones: bad news sells papers. When the Battle of Shiloh filled the wires with staggering casualty figures in April 1862, Edison — so the story goes — persuaded a telegraph operator to wire the headlines ahead to each station, bought far more papers than usual, and jacked up the price as the train rolled from stop to stop past crowds desperate for word of the dead. He made a small fortune in a single day. It was his first lesson in a truth he would return to: information has value, and the person who controls its delivery controls the profit.

But the detail from those railroad years that matters most for our story is the one that sounds the least likely. The Grand Trunk gave the boy the run of the baggage car during the train's long midday layover in Detroit, and Edison filled a corner of it with something no news butcher had any business owning: a chemistry laboratory. He had become obsessed with a school text on natural philosophy and with the experiments in it, and he could not afford a proper lab, so he built one in a moving train. He lined a section of the baggage car with shelves, filled them with bottles of chemicals bought with his newspaper profits, and spent the layover hours mixing reactions while the car sat in the Detroit yard. He even acquired a small secondhand printing press and, from that same baggage car, printed and sold his own newspaper, the *Grand Trunk Herald* — very likely the first newspaper ever composed and printed aboard a moving train. A teenage boy running a chemistry lab and a print shop out of a railway baggage car is such a strange and specific image that it is worth pausing on, because it is the whole future Edison in miniature: the fusion of tinkering, chemistry, publishing, and hustle, all crammed into whatever space the world would let him have.

The laboratory came to a spectacular and probably deserved end. A stick of phosphorus — a substance that can ignite on contact with air — is said to have jolted off a shelf when the train lurched, and a small fire broke out in the baggage car. The conductor, furious, threw the boy, his chemicals, and his printing press off the train at the next stop. (In some versions this is also the moment the conductor cuffed his ears, folding the fire and the deafness into a single dramatic scene — another reason to be suspicious of the ear-boxing tale.) The precise details are murky. What is clear is that Edison's makeshift laboratory ended in flames and expulsion, which is about the most Edison way for it to have ended.

And so we arrive back at the platform at Mount Clemens, and the toddler on the tracks, and the grateful station agent with a telegraph key. To grasp why that reward mattered, you have to understand what the telegraph *was* in 1862. It was the internet of its century — the first technology in human history to separate the speed of a message from the speed of a messenger. Before it, news traveled no faster than a horse, a ship, or a running man. A word from New York took days or weeks to reach the frontier. The telegraph shattered that limit. It sent messages as pulses of electric current down a wire, coded into the short and long taps we call dots and dashes, so that a few words could cross a continent in the time it takes to read this sentence. To a curious boy who had already glimpsed the value of information — who had made a fortune off war headlines and printed a paper in a baggage car — learning to command that current was like being handed the keys to the fastest vehicle on earth.

It is worth dwelling for a moment on how unlikely all of this was, because the shape of Edison's childhood defies almost every assumption we make about how extraordinary people are produced. He was the last of seven children in a family of modest means, in a town most Americans could not find on a map, in an age with no scholarships, no aptitude tests, no institutions scanning the population for hidden talent. He had a bare twelve weeks of formal schooling and a physical disability that walled him off from ordinary conversation. By every measure the world used then — and most it uses now — he should have vanished into the anonymous mass of nineteenth-century laborers. The historian is tempted to say he succeeded *because* of these obstacles, but that romanticizes too much. The truer statement is that he succeeded despite them, by an accident of temperament: he happened to be born with a curiosity so violent and a will so unbending that no amount of poverty or deafness or dismissal could smother it. Nancy Edison recognized that spark and fed it; the railroad gave it room to run; the telegraph gave it a direction. But the spark itself was simply there from the start, unearned and inexplicable, the way genius usually is.

There is a second lesson buried in the newsboy years, and it is about the peculiar American character of Edison's rise. He did not come up through a university or an academy or an apprenticeship to a master. He came up through commerce — through hustling papers, cornering the market on war headlines, printing his own newspaper, running boys of his own. From boyhood he understood technology and money as two halves of a single thing, inseparable. The chemistry lab and the printing press shared the same baggage car; the telegraph was both a marvel of physics and the fastest way to get a jump on the competition. This fusion — invention married to enterprise, the laboratory wedded to the ledger — would become the defining feature of Edison the man, and it was forged not in a classroom but in the aisles of a moving train, by a poor boy who had learned very early that a good idea was worth nothing until somebody paid for it.

So Edison learned the code. He practiced obsessively, MacKenzie coaching him, until his hand and ear (or, given his deafness, largely his hand and eye) could send and receive Morse with the fluent speed of a professional. By sixteen he had left the newspaper aisles behind and taken his first job as a telegraph operator. It was itinerant, ill-paid, nocturnal work that would carry him from town to town across the American Midwest over the next several years — but it put him, at last, with his own two hands on the live wire of electrical technology. Everything he had scavenged from the world so far — the home schooling, the chemistry, the hustle, the deafness that walled him into his own concentration — was about to be poured into a single trade.

The boy who had been called addled, who had barely gone to school and could barely hear, had found the one thing that would let all of it pay off: a wire humming with electric current, waiting for a mind quick enough to master it.

Chapter 18: The Telegraph Years

At two in the morning in a railroad telegraph office somewhere in the American Midwest, sometime in the mid-1860s, a young operator sat alone in a pool of lamplight, a live wire humming above his workbench, and quietly cheated his employer. His name was Thomas Edison, he was supposed to be awake and alert through the long night shift, and his job included proving he was awake by tapping a signal — the number six — into the wire every hour on the hour so the dispatcher down the line knew the night man had not fallen asleep. Edison wanted to sleep, or read, or tinker. So he built a small device out of a clock and a notched wheel that would, all by itself, tap out a perfect "six" onto the wire every hour, exactly on time. His superiors received their reassuring signal all night long. The operator sending it was often unconscious. When the trick was eventually discovered, Edison was in trouble — but the episode captures, better than any polished anecdote, the exact shape of his genius. Told to do a tedious job, his first instinct was never to endure it. It was to build a machine that would do it for him.

The telegraph years — roughly 1863 to 1869, from age sixteen to twenty-two — were Edison's apprenticeship, though not to any single master. He was what the trade called a "boomer" or "tramp" operator, drifting from job to job across the war-torn and postwar United States: Stratford Junction in Canada, then Indianapolis, Cincinnati, Nashville, Memphis, Louisville, Boston. He was good — very fast at receiving, though his sending was sometimes sloppy — and good operators were in demand, so he could quit or get fired in one town and be hired in the next within days. He preferred the night shift, partly because it paid a little more, partly because it left the long, quiet hours to himself, and partly, one suspects, because his deafness made the clatter of a busy daytime office a special kind of torment while the stillness of night suited his walled-in concentration.

To follow what he did in those years, you need a working picture of the machine he served, because Edison's first inventions were all attempts to make it do more. So let us define the telegraph plainly. A telegraph is, at its heart, the simplest electrical device imaginable: a battery, a long wire, and a switch at each end. Press the switch — the "key" — and you complete the circuit, sending a pulse of electric current all the way down the wire. At the far end, that pulse energizes an electromagnet, a coil of wire that becomes a magnet only while current flows through it. The electromagnet tugs a little iron lever, which makes a click. Release the key, the current stops, the magnet lets go, and the lever clicks back. That is the whole apparatus. Everything else — spanning continents, crossing oceans, building fortunes — is just this on-off click, repeated fast, arranged into a code. Samuel Morse's code assigned to each letter a pattern of short clicks (dots) and long ones (dashes), so that a skilled operator, listening to the rhythm of a chattering "sounder," could read language straight out of the noise the way you read these words off a page.

Think of the single telegraph wire as a narrow one-lane road, and you will understand the problem that consumed the young Edison and the entire industry. A road that carries only one car at a time in one direction is a bottleneck. Telegraph companies had strung enormous distances of copper wire at enormous cost, and each precious wire could carry only a single message at a time. Every operator who wanted to send had to wait for the line to clear. If you could somehow make one wire carry two messages at once, you would double the value of every mile of line already in the ground without stringing a single new strand of copper. That was the holy grail of telegraphy in the 1860s and 1870s, and it was worth staggering sums of money.

Edison threw himself at it. His early inventions were nearly all telegraph improvements, and the story of how he got there begins with a fish glue and an accident of chemistry — no, that is a different tale; the telegraph breakthroughs came from patient tinkering during those long night shifts and the workshops he haunted between them. First he helped perfect the **duplex**, a scheme for sending two messages down one wire simultaneously, one in each direction. Then he reached for the real prize: the **quadruplex**. As its name promises, the quadruplex telegraph sent *four* messages over a single wire at once — two in each direction. Understanding exactly how it worked requires more electrical detail than we need here, but the trick, in essence, was to make the two directions of transmission respond to two different *qualities* of the electric signal. One set of messages was encoded by changing the *strength* of the current — how much flowed — while the other was encoded by changing its *direction* — which way it flowed. Receivers tuned to notice only strength ignored the direction changes, and receivers tuned to notice only direction ignored the strength changes, so four conversations could ride the same wire without tangling. It was as if Edison had discovered that his one-lane road could carry four cars at once as long as two of them were distinguished by color and two by the direction they faced.

The quadruplex made Edison's name, and it made him a decent sum of money, and the circumstances of its sale taught him a lesson about business that was almost as valuable as the invention itself. In 1874 the device was so obviously worth a fortune to Western Union — the telegraph giant that dominated American communication the way a single company almost never dominates anything today — that Edison found himself, for the first time, holding real leverage. The quadruplex became a pawn in a corporate war between Western Union and a rival financier, Jay Gould, who was trying to break Western Union's grip. Edison, characteristically, played both sides and got tangled in litigation, but he came away with roughly $10,000 for the quadruplex rights — a great deal of money in 1874, enough to fund his ambitions and confirm to him that a single clever machine, sold to the right desperate buyer, could change a man's fortunes overnight. He was learning that invention was not merely a matter of cleverness with wires. It was a matter of knowing what the market would pay for, and of controlling the thing it wanted.

That lesson had been branded into him a few years earlier by his very first patented invention, which was also his first great commercial failure — and it is one of the most instructive failures in the whole history of technology. In 1869, Edison received U.S. Patent No. 90,646 for an **electric vote recorder**, a machine meant to let legislators register a yes or no vote instantly from their seats, tallying the result in seconds instead of the tedious quarter-hour of a spoken roll call. It was ingenious. It was fast. It worked. And it was a total commercial disaster, because Edison had solved a problem nobody with money wanted solved. When he demonstrated it to a committee of legislators, he was told, in effect, to take it away. The slowness of the roll-call vote was not a bug to the men who ran legislatures; it was a *feature*. The long minutes of an open, drawn-out vote were precisely when members lobbied one another, changed minds, and cut deals. A machine that recorded every vote instantly would abolish the very horse-trading that made the legislature run. The politicians did not want efficiency. They wanted the delay.

Edison took the rebuke to heart and drew from it a rule he

claimed to have followed ever after: never again waste time inventing something unless he was sure people would want to buy it. "Anything that won't sell," he said, "I don't want to invent. Its sale is proof of utility, and utility is success." This is worth dwelling on because it corrects a romantic misconception about Edison that clings to him even now — the image of the pure inventor, the tinkerer in love with the machine for its own sake. That was not the man. From the vote recorder onward, Edison was a *commercial* inventor, arguably the first great one, who aimed his enormous ingenuity deliberately and coldly at things the market was already hungry for. The telegraph, the stock ticker, the light bulb — he chased them not because they were the most beautiful problems but because they were the most *valuable* ones. His failures taught him this as surely as his successes.

The money from telegraphy let Edison do the thing he most wanted: stop working for other people. In the early 1870s he set himself up in Newark, New Jersey, first improving stock tickers — the machines that printed streams of stock prices onto ribbons of paper tape for Wall Street offices — and then running a bustling manufacturing shop that at its peak employed dozens of skilled machinists building telegraph and ticker equipment on contract. The Newark years turned Edison from a lone tramp operator into a boss, a manager of talent, a man who could take a rough idea in the morning and have his machinists producing a working model by evening. Here he gathered around him some of the men who would follow him for decades — most importantly a precise, unflappable English machinist named Charles Batchelor and a superb Swiss clockmaker named John Kruesi, both of whom we will meet again as central players in the drama to come.

There is one more strand of the telegraph years worth pulling, because it foreshadows the whole rest of Edison's life: this was where he learned to fail productively, and where he first tasted the courtroom. The quadruplex litigation was only the beginning. Edison's inventions, from the very start, were contested, copied, and fought over, and he learned early that a patent was not a trophy but a battlefield — a legal claim only as good as your willingness and ability to defend it. He learned to document obsessively, to file early, to keep witnesses. He also learned, in these itinerant years, the value of *improving* an existing technology rather than always inventing from nothing: nearly all his early triumphs were refinements of the telegraph, taking a machine the world already depended on and making it do more. This instinct — to find a large, established market and then invent the thing that market did not yet know it needed — was the exact instinct that would later carry him from the telegraph to the electric light. He did not invent lighting any more than he invented the telegraph. He found industries in motion and vaulted ahead of them.

And it was in these years, too, that the human tools of his later empire came together. The Newark shop was a rehearsal for Menlo Park: a place where Edison learned that he could multiply himself through other men's hands, that a boss who could keep a room full of skilled machinists fed with ideas could produce far more than any lone tinkerer. Batchelor and Kruesi were the first of the loyal crew who would build his miracles. The management, the delegation, the culture of long nights and shared purpose — all of it was prototyped in Newark on stock tickers and telegraph relays before it was perfected in New Jersey on phonographs and light bulbs. By the time Edison left Newark, he was no longer merely an inventor. He was becoming an organizer of invention, and that, far more than any single machine, was the thing that would change the world.

Where do you meet the telegraph years today? Every time your phone splits a single fiber-optic line into thousands of simultaneous calls and downloads, it is running an unimaginably sophisticated descendant of Edison's quadruplex — the same essential trick of squeezing many messages onto one channel at once, a technique engineers now call multiplexing. And every startup that builds first and asks "will anyone buy this?" only afterward is repeating, usually without knowing it, the mistake of the vote recorder. Edison learned it once and never forgot.

By the mid-1870s the pattern of the man was complete. He had the electrical mastery, the commercial instinct, the crew of loyal machinists, and enough money to gamble. What he wanted next was a place — a workshop unlike any that had ever existed, built for one purpose only: to invent things on purpose, on schedule, and for profit. In 1876 he found a quiet crossroads in New Jersey and set out to build it.

Chapter 19: The Invention Factory

In the spring of 1876, a two-story wooden building went up on a low rise beside a railroad line at a nearly empty crossroads in New Jersey, twenty-five miles southwest of New York City. From the outside it looked like a barn, or a modest rural meeting hall — clapboard siding, a pitched roof, a row of tall windows down each side to let in the daylight. There was almost nothing around it: a scattering of houses, a boarding hall, a hamlet so obscure that most people had never heard its name. The name was Menlo Park. Inside that plain wooden box, over the next six years, a thirty-year-old man and a small crew of assistants would invent the phonograph, perfect the electric light, and — most consequentially of all, though it is the least celebrated of the three — invent the very idea of inventing itself as an organized, industrial process. When Edison later said the ambition of the place was "to have a minor invention every ten days and a big thing every six months," he was not boasting so much as writing a production schedule. Menlo Park was a factory, and its product was invention.

To feel how radical that was, you have to picture how invention had worked before. The great electrical discoveries we have already followed — Franklin with his kite, Volta with his stack of metal disks, Faraday with his coils and magnets

— were the fruit of gifted individuals, often working alone or with a single assistant, following curiosity where it led, on no schedule and toward no product. Even Edison's own earlier work in Newark had been improvised, wedged into a commercial machine shop between paying jobs. The lone genius in his workshop was the model the whole culture carried in its head, and it is largely the model our culture still carries: the eureka moment, the solitary mind, the flash of insight. Edison's deepest and most disruptive idea was to replace that romantic picture with something that looked much more like a business. He would assemble a permanent team of specialists — machinists, chemists, mathematicians, glassblowers, draftsmen — house them under one roof with the best tools money could buy, aim them at commercially promising problems, and grind out inventions the way his old Newark shop had ground out stock tickers: reliably, repeatedly, on demand.

This is the single most important thing to understand about Thomas Edison, and it is the thing most often left out of the storybook version. His greatest invention was not the light bulb. It was the **industrial research laboratory** — the institution. Menlo Park was the world's first facility built and staffed for the sole purpose of turning out new technology as a continuous product. Every corporate research and development lab that came after it — every Bell Labs and IBM Watson Center and Silicon Valley skunkworks, every pharmaceutical research campus and university innovation lab — descends, directly or indirectly, from that clapboard building in New Jersey. When people say Edison "invented the modern world," they usually mean the light. They should mean this. He invented the machine that invents things, and it has been running ever since.

The building itself was arranged with a purpose that reflected the philosophy. The ground floor held the machine shop, powered by a steam engine, where ideas were turned into metal and wood by skilled hands. The upper floor was the laboratory proper: a long, open room lined floor to ceiling with shelves of chemical bottles — thousands of them, holding samples of very nearly every substance Edison could lay hands on, because his method depended on being able to *try things*. There were batteries, instruments, coils of wire, a small library. Later Edison added a dedicated library building and a glassblowing shed and a carpentry shop. The point of the layout was speed: a man with an idea in the morning could sketch it, have it machined downstairs, test it upstairs against any material on the shelves, and know by nightfall whether it was worth pursuing. The distance from thought to working model, which had been measured in weeks or months, Edison compressed to hours.

But a building is only lumber and glass. What made Menlo Park work was the men inside it, and Edison chose them with more care than he chose almost anything. Three deserve to be named, because the popular image of Edison as a solitary wizard erases the collaborators without whom he would have accomplished a fraction of what he did — and correcting that erasure is one of the honest debts this book owes. **Charles Batchelor**, the precise English machinist who had followed Edison from Newark, was his right hand and closest confidant, a man of such fine manual skill that he could execute the most delicate work in the lab and such steady judgment that Edison trusted him with a share of the profits. **John Kruesi**, the Swiss-trained master clockmaker, ran the machine shop and could translate Edison's rough sketches into flawless mechanisms — it was Kruesi who would build the first working phonograph from a drawing Edison handed him, and who reportedly muttered "Gott im Himmel" when the thing actually spoke. And **Francis Upton**, a young, university-trained mathematician and physicist hired in 1878 — nicknamed "Culture" by the rougher men of the shop — brought the theoretical rigor that Edison, for all his genius, lacked. Upton could calculate where Edison could only guess, and in the coming battle over the light bulb, that ability to *compute* the answer instead of merely stumbling onto it would prove decisive.

The relationship between Edison and Upton points to a truth about the lab that is worth stating plainly, because it cuts against a stubborn myth. It is often said that Edison was a pure empiricist who despised theory and mathematics, a man who found the light-bulb filament by blind, brute-force trial and error while sneering at book learning. There is a grain of truth in it — Edison did prize experiment over speculation, and he loved to needle college-trained men. But the fuller picture is that he *hired* the theory he lacked. Bringing Upton to Menlo Park was a deliberate acknowledgment that intuition and tinkering needed mathematics as a partner. The famous brute-force searches were real, but they were guided and narrowed by calculation more often than the legend admits. Edison's method was not anti-intellectual; it was pragmatically *combined* — hands and equations, shop floor and blackboard, working side by side. That combination was itself part of what he invented.

The culture of the place was as unusual as its purpose. Edison worked appalling hours and expected the same of his men, but he ran the lab less like a factory floor than like an all-male, all-night fraternity of tinkerers. The crew — the "muckers," they called themselves, and they called Edison "the Old Man" though he was barely older than some of them — worked through the night when a problem gripped them, then broke at midnight for food hauled in from town, singing songs around a pipe organ Edison had installed in the lab, trading dirty jokes and tall tales, sometimes sleeping on the workbenches for a few hours before starting again. Edison himself was famous for catnapping anywhere, curled on a bench or under a table with a stack of books for a pillow, then waking to work another twenty hours. There were pranks — a memorable one involved wiring a washstand or a metal object to a shocking induction coil and watching a visitor jump. The atmosphere was intense, informal, competitive, and joyful in the specific way that hard collective work on a fascinating problem can be joyful. It bred a loyalty that money alone could not have bought, and it was every bit as much a piece of Edison's invention as the building or the payroll.

Underneath the fun was a relentless philosophy, and Edison compressed it into the most famous thing he ever said: "Genius is one percent inspiration and ninety-nine percent perspiration." It is worth reading that line not as false modesty but as a literal description of how Menlo Park operated. The idea — the one percent — was cheap and common; everyone had ideas. What was rare was the willingness to test a thousand variations, to fail and record the failure and try again, night after night, until the thing actually worked. Menlo Park was an engine for converting sweat into product. Its shelves of ten thousand chemicals, its round-the-clock shifts, its meticulous notebooks recording every failed experiment, were all machinery for grinding through the ninety-nine percent that lesser inventors gave up on. This is why, when the great problem of the incandescent lamp finally landed on those workbenches, the lab could do what no lone genius in a private study could: test material after material after material, thousands of them, without losing heart.

Those notebooks deserve a word of their own, because they were among the quiet secrets of the operation and they correct another popular fantasy about how invention happens. We imagine discovery as a series of triumphant leaps, but the daily reality at Menlo Park was obsessive record-keeping. Edison insisted that every experiment — every failure, every dead end, every material that flared and died — be written down in detail, dated, and signed. Over the life of the laboratory his men filled thousands of notebooks with the granular minutiae of what did not work. This was partly good science: a failure carefully recorded is a question you never have to ask twice, and a path you never have to walk down again. But it was also, characteristically, good business. Those meticulous, dated, witnessed records were legal ammunition, the documentary proof of exactly when Edison had done exactly what — evidence he would need again and again in the patent courts, where fortunes turned on who could prove they had gotten there first. The notebook was a scientific instrument and a legal weapon at once, and it is a small emblem of how thoroughly Edison fused the pursuit of knowledge with the pursuit of profit.

It is worth being honest, too, about the darker side of the invention factory, because the same system that produced miracles also produced a pattern that would shadow Edison's reputation for the rest of his life and beyond. When a Menlo Park invention succeeded, the world credited one man: Edison. The muckers who had labored through the nights, the Uptons who supplied the mathematics, the Batchelors and Kruesis whose hands built the working models — their names were largely swallowed by the legend of the lone Wizard. Edison held the patents, took the fame, and made the fortune, while his collaborators, however well or poorly compensated, faded into the background of a story told as his alone. This is not unique to Edison; it is the near-universal fate of the team behind a famous name. But it is the reason we have to keep insisting, against the tide of the myth, that Menlo Park was a collective achievement. The genius of the place was precisely that it was *not* one man. Edison's real invention was a way of harnessing many talented people to a single purpose — and one of the costs of that invention was that the many were remembered as the one.

Where do you meet Menlo Park today? You meet it every time a company announces a breakthrough from its "R&D division," every time a drug emerges from a corporate research campus, every time a phone or a chip or a vaccine reaches you as the output of a large, funded team of specialists working to a schedule. The lone-genius story still sells — we love our Franklins and our Teslas — but the actual machinery of modern innovation runs on Edison's model, not the myth. He turned discovery from an accident into an industry.

By 1877 the invention factory was up and humming, its crew assembled, its method proven, its founder hungry for the "big thing every six months" he had promised. The first of those big things arrived at the end of that year, and it did something no machine had ever done before. It talked.

Chapter 20: Mary Had a Little Lamb

One night in early December 1877, in the upstairs laboratory at Menlo Park, John Kruesi finished building a small hand-cranked machine from a sketch Thomas Edison had handed him, and he had no idea what it was supposed to do. It was an odd little thing: a brass cylinder about the size of a large soup can, wrapped in a sheet of tinfoil, mounted on a threaded shaft with a crank at one end, and flanked by two small diaphragms — thin metal disks — each tipped with a needle, one on each side of the cylinder. Kruesi, the master machinist who had built a hundred ingenious mechanisms for Edison, thought it a waste of good workmanship. When he asked what it was for, the answer was so absurd that, according to the lab's lore, he laughed or scoffed: the machine, Edison said, was going to talk. Then Edison leaned close over one of the diaphragms, began to turn the crank at a steady pace, and shouted into the mouthpiece the first thing that came into his head — a children's nursery rhyme he had known since boyhood:

Mary had a little lamb, its fleece was white as snow.

He cranked the cylinder back to the start, moved the needle to the second diaphragm, and turned the crank again. And out of the machine, thin and scratchy and unmistakable, came a human voice reciting the rhyme back to him. It was Edison's own voice. The men in the room went silent, then loud. Kruesi, the story goes, blurted out "Gott im Himmel" — God in Heaven. Edison himself later admitted that he was "never so taken aback in my life." He had expected the thing to work in principle; he had not been prepared for the raw shock of actually hearing a machine speak. No one had. In the whole span of human existence up to that December night, sound had been the most fleeting thing there was. A word spoken was gone the instant it was uttered, surviving only in memory. Now a scrap of tin had caught a voice and held it, and given it back.

To understand why this was one of the most astonishing inventions in history — arguably more purely surprising than the light bulb — you have to grasp how genuinely unexpected it was, even to experts. Recorded sound had no real predecessors. The telegraph and telephone transmitted; they carried a signal from one place to another and let it vanish. The idea of *storing* sound, of freezing it in matter so it could be replayed a minute or a year or a century later, was so foreign that when Edison's assistants first understood what he was proposing, several of them simply did not believe it could be done. There had been a Frenchman, Édouard-Léon Scott de Martinville, who two decades earlier had built a device called the phonautograph that traced sound waves as squiggles on smoked paper — but his machine could only *draw* sound, never play it back; it was a seismograph for the voice, not a memory of it. Edison's leap was the second half: not just recording the wave, but reversing the process to make the wave sing again. That reversibility was the miracle.

The principle behind it is, once you see it, beautifully simple — so simple that people have since built working versions out of a paper cone, a sewing needle, and a plastic cup. Sound is nothing but vibration in the air: when you speak, your vocal cords set the air trembling in a rapid pattern of tiny pressure waves, and those waves travel out and rattle your listener's eardrum. Edison's machine did to a diaphragm what your voice does to an eardrum. Shout into the mouthpiece, and the pressure waves of your voice make the thin metal diaphragm vibrate in exactly the pattern of your speech. The needle attached to that diaphragm rides on the soft tinfoil wrapped around the turning cylinder, and as the diaphragm shakes, the needle bites deeper and shallower into the foil, cutting a wavering groove — a physical trace of the sound, a groove whose every wiggle is a frozen copy of a vibration in the air. That is recording. To play it back, you simply run the needle through the groove again: the wiggles in the foil now jostle the needle, the needle shakes the diaphragm, the diaphragm pushes the air, and the air carries the voice, faintly, back to your ear. The machine remembers a sound by writing its shape into metal and reads it back by feeling that shape again. Think of it as footprints in wet sand that could somehow re-create the walker, or a rut worn into a road that plays back the wheel that cut it.

Here is a surprising fact worth carrying forward: the phonograph, unlike almost every other great invention in this book, worked on the very first serious try. There was no long agony of failure, no thousand materials tested, no years of refinement before the first success. Edison sketched the idea, Kruesi built it in a matter of hours, and it spoke that same night. In a career defined by the ninety-nine percent perspiration of grinding trial and error, the phonograph was the rarest of things — a nearly pure flash of the one percent, an idea so right that it leapt from drawing to working machine in a single day. Edison later called it his favorite invention, his "baby," and it is easy to see why. It was the one that came to him whole.

A second surprise: the phonograph was, in a sense, an accident of a completely different project. Edison had not set out to record music or preserve human voices. He had been working on ways to improve the telegraph and the telephone — specifically, on a device to record incoming telegraph messages as indentations on paper tape so they could be replayed and re-sent at a slower speed. While experimenting with a needle riding over the moving tape, he noticed that when the tape ran fast, the indentations made a faint musical or vocal sound — a light, birdlike noise. That stray observation, the kind most people would have ignored, set his mind running: if a moving indented surface could make sound, could a controlled indentation *store* a particular sound and give it back? The phonograph was born not from a plan to capture music but from a telegraph man's stray notice of a hum in a machine. It is a small monument to the value of paying attention to the thing you were not looking for.

The world's reaction was not admiration so much as disbelief bordering on fear. When Edison and his men carried the tinfoil phonograph out into public in the winter and spring of 1878 — demonstrating it to the editors of *Scientific American* in New York, to crowds, eventually to President Rutherford B. Hayes, who reportedly had Edison woken at the White House at nearly midnight so his wife could hear it — the reaction was often frank astonishment that shaded into suspicion. Some listeners assumed a trick, a hidden ventriloquist. Clergymen worried aloud whether it was proper, or even natural, for a machine to counterfeit the God-given human voice. The device seemed to violate a rule everyone had lived under since the dawn of the species: that the dead are silent and the spoken word does not return. A machine that could store a voice implied, unsettlingly, that a person's voice might outlive the person — as indeed Edison's recitation of *Mary had a little lamb* has now outlived him by nearly a century. People had simply never had to think about such a thing before, and the newness of it landed with a jolt we can barely imagine, so thoroughly do we now swim in recorded sound.

There is a poignant coda to the invention that is too often left out, and it involves Edison's deafness. The man who gave the world recorded sound could barely hear it. As the phonograph developed over the following years, Edison tested and tuned his machines by biting down on the wooden cabinet and feeling the music travel up through the bones of his jaw into his inner ear — literally listening with his teeth. He judged the fidelity of recordings this way, and he was said to have a discriminating and demanding ear (or jaw) for musical quality, rejecting performances that a hearing man would have passed. There is something almost unbearably fitting in it: the deaf inventor of the machine that remembers sound, pressing his teeth to the wood to hear the voices he had taught the world to keep. His affliction, which he had always insisted was a gift of concentration, had led him — by a path no one could have predicted — to the one invention he loved above all the others.

It was the phonograph, more than any earlier invention, that made Edison world-famous and won him the nickname that would follow him for the rest of his life: the **Wizard of Menlo Park**. The word "wizard" is telling. Reporters reached for it because the machine seemed like magic, and its inventor — the young man in the obscure New Jersey hamlet who could apparently conjure impossible things out of tin and brass — seemed like a magician. The press descended on Menlo Park; the sleepy crossroads became a place of pilgrimage; and Edison, who understood publicity as well as he understood machinery, played the role of wizard to the hilt. The nickname was a gift to his ambitions. When, a year later, he announced he would tackle the electric light, the public was already primed to believe that the Wizard could pull it off, because he had already made metal speak.

One correction is due here, because the phonograph's fame has flattened its history. Edison invented the phonograph, and that credit is genuinely his — but the *practical, commercial* phonograph, the one that filled parlors and eventually became the record player, owes as much to others who came after. Edison's original tinfoil could be played only a few times before the delicate foil tore, and the sound was faint and short-lived; it was a marvel, not a product. It took a decade of work — much of it by rivals like Alexander Graham Bell's laboratory, which replaced the fragile foil with wax-coated cylinders — before recorded sound became something you could actually own and enjoy at home. Edison himself returned to improve it. The pattern here is one we will see again and again with him: he made the first thing that worked, and the world made the first thing that lasted. The distinction matters, and it will matter enormously when we come to the light.

Where do you meet that December night today? Everywhere. Every voice memo on your phone, every song streamed to your ear, every podcast, every film's recorded dialogue, every voicemail from someone now gone — all of it is the direct descendant of a needle scratching *Mary had a little lamb* into a sheet of tinfoil in 1877. The entire recorded-sound industry, the whole vast architecture by which the human voice is now stored, copied, and sent across the world, begins with that one crank and that one nursery rhyme. Edison had taught matter to remember.

And yet the phonograph, for all its wonder, was a marvel more than a necessity. The world was charmed by a talking machine, but it did not *need* one. The thing the world needed — the thing millions of people were waiting for, the thing that would make Edison not merely famous but rich beyond reckoning and pull him into the great war of his life — was still burning, dimly and dangerously, in every gas lamp and candle on earth. Edison meant to replace all of them with light drawn from a wire. And in that ambition he was very far from alone.

Chapter 21: The Race for the Lamp

In 1809, nearly seventy years before Edison turned his attention to the problem, an English chemist named Humphry Davy stood in the basement of the Royal Institution in London and made the most brilliant artificial light any human being had ever seen. He had assembled an enormous battery — a wall of some two thousand voltaic cells, the largest in the world, filling the room with racks of metal plates and acid — and he connected it to two sticks of charcoal, bringing their tips almost, but not quite, together. When he drew the charcoal points a hair's breadth apart, the electric current leapt the gap in a searing, sustained arc of blue-white flame, dazzling enough that observers had to shield their eyes. It was the first electric light. It was also, as a practical matter, useless — and the seventy-year gap between Davy's blinding arc and Edison's steady bulb is the subject of this chapter, because to understand what Edison actually accomplished, you must first understand the long parade of people who tried and failed before him.

Davy's device was called the **arc lamp**, and the principle he demonstrated in 1809 eventually did light the world's streets — decades later, once dynamos could supply the ferocious current it demanded. An arc lamp makes light by forcing electricity to jump across a small gap between two carbon rods; the current is so violent that it rips the air into a glowing plasma and heats the carbon tips to incandescence, throwing off a harsh, brilliant glare. By the 1870s arc lamps were being installed above city streets, public squares, and railway stations across Europe and America. The most successful American version was made by Charles Brush of Cleveland, whose arc lights turned night into a strange blue day over the avenues of a dozen cities. But anyone who has stood under a modern stadium floodlight knows instantly why the arc lamp could never come indoors. It was blindingly, punishingly bright — a single arc lamp put out the light of hundreds or thousands of candles, far too much for a parlor or a bedroom. It hissed and sputtered and flickered. It gave off noxious fumes and a faint danger of fire. And the carbon rods burned away as they worked, so the lamp needed constant tending and frequent replacement. Arc light was magnificent for a boulevard and impossible for a home.

What people wanted indoors was something quite different, and here we must draw a distinction that sits at the very heart of this chapter, because confusing the two kinds of electric light has muddled the story for a century. There are two fundamentally different ways to make light from electricity. The first is the arc — force current to jump a gap and glow. The second is **incandescence**, and it is the one Edison and his rivals were all chasing. To make something *incandescent* is simply to make it so hot that it glows. Anything, heated enough, will shine: a poker left in a fire glows dull red, then orange, then yellow-white as it gets hotter; the burner of an electric stove reddens; the sun itself is merely incandescent matter. An incandescent lamp works by passing electric current through a thin thread of material — the *filament* — that resists the flow of electricity. That resistance turns the electrical energy into heat, exactly as friction turns motion into heat when you rub your palms together, and the filament grows hot enough to glow. The dream was a light you could dim and brighten, divide into small gentle sources, and scatter through a house like so many tame little suns. The dream, in a word, was the light bulb.

The trouble — the single, maddening, decades-long trouble — is captured in a paradox. To glow brightly, the filament must get very hot. But almost anything hot enough to glow brightly will, in the open air, immediately catch fire and burn up. Heat a thread of carbon or metal until it shines, and the oxygen in the air attacks it and it is consumed in seconds, like the filament of a match. This is the problem of **burnout**, and it defeated inventor after inventor for the better part of forty years. There were only two possible escapes from it, and both were fiendishly hard. One was to find a filament material that could glow hot without disintegrating. The other was to remove the air itself — to seal the filament inside a glass bulb and pump out the oxygen, so there was nothing left to feed the fire. In practice the answer would require *both*, and a vacuum good enough to matter was, for most of the nineteenth century, beyond the reach of available pumps.

The list of those who tried and fell short before Edison is long, and honoring it is the whole point, because the popular belief that Edison simply "invented the light bulb" out of nothing is one of the most persistent falsehoods in the history of technology. He did not. He entered a crowded race with a decades-old prize, and men had been chasing it since before he was born. As early as the 1840s, an Englishman named Frederick de Moleyns and an American named John Starr held patents on incandescent lamps. A Belgian-French inventor, and a Russian named **Alexander Lodygin**, built incandescent lamps in the early 1870s — Lodygin's carbon-rod lamps briefly lit a St. Petersburg street and a naval yard. Another approach entirely came from the Russian engineer **Pavel Yablochkov**, whose "**Jablochkoff candle**" of 1876 — a clever arc lamp with two parallel carbon rods separated by a strip of insulating clay that burned away as the carbons did — briefly became a sensation, lighting the grand avenues of Paris and earning the nickname "the Russian light." For a moment in the late 1870s it seemed the Jablochkoff candle might be the future. But it was still an arc light at heart, still too bright and too short-lived for the home, and it flickered out of history as quickly as it had flared up.

The most important of Edison's predecessors, and the one whose name every fair account must set beside his, was an English chemist and physicist named **Joseph Swan**. Swan had been working on the incandescent lamp on and off since the 1850s, long before Edison gave it a thought. He understood the core ideas — a carbon filament, a glass bulb emptied of air — and by the late 1870s, aided by dramatic improvements in vacuum-pump technology, he was building carbon-filament lamps that genuinely worked. Swan gave a public demonstration of a working incandescent bulb in Newcastle in early 1879, and he was, by any honest reckoning, at least neck and neck with Edison and arguably a step ahead in the pure question of who first made a carbon thread glow inside an evacuated globe. In Britain, Swan is remembered as the true inventor of the light bulb, and the claim is not foolish. We will return to Swan, because the collision between his priority and Edison's would eventually be settled not in a courtroom but in a merger — the two men's companies joined into a firm called Ediswan rather than fight a war neither could win. For now, hold onto the essential fact: when Edison announced he would conquer the electric light, he was not opening virgin territory. He was joining a decades-long scramble in which serious, capable men had already planted flags.

Standing over the whole race, vast and entrenched and seemingly unassailable, was the incumbent it all aimed to overthrow: **gas lighting**. This is the piece modern readers most often forget, and without it the drama makes no sense. By the 1870s the developed cities of the world ran on gas. Coal was cooked in giant municipal gasworks to produce a flammable "town gas," which was piped through networks of underground mains beneath the streets and up into homes, shops, theaters, and factories, where it burned in open jets and, later, in mantles to give light. Gas lighting was a mature, capital-rich, deeply established industry with its pipes literally embedded in the ground beneath every major city and its shares held in every respectable investment portfolio. It had transformed urban life, extending the working and social day far into the night. But it had grave faults that everyone lived with because there was no alternative: the open flames were a constant fire hazard; the gas was poisonous and, if a jet was left open unlit, could asphyxiate a sleeping household; it fouled the air, blackened ceilings, consumed a room's oxygen, and threw off heat and soot. Gaslight was better than candles and oil lamps, but it was dirty and dangerous, and everyone knew it. A clean, cool, safe, flameless light — light from a wire — would be worth an empire to whoever could deliver it. That prize, measured in the size of the gas industry it would replace, is what drew Edison in.

He entered the race in the autumn of 1878, and he entered it in the most characteristic way possible: by loudly announcing that he had essentially already solved a problem he had barely begun to work on. After a visit to see William Wallace's arc-lighting dynamos in Connecticut, Edison came home electrified with confidence and told reporters that he had the incandescent light nearly in hand and would have homes and streets lit within weeks. It was wildly premature — he had no working lamp and no idea yet what the filament should be made of — but the announcement did something shrewd. It rattled the gas companies so badly that gas stocks tumbled on the London and New York exchanges at the mere news that the Wizard of Menlo Park had turned his attention to light. It also, crucially, drew investors. On the strength of little more than his reputation and his nerve, Edison secured the backing of a group of financiers — including interests connected to the banker J. P. Morgan — who formed the Edison Electric Light Company to fund the work. Edison had sold the light before he had invented it, exactly as the vote-recorder failure had taught him never to invent anything the market would not buy. Here the market was ravenous, and he made sure everyone knew he was coming.

Before we follow Edison into the laboratory, it is worth naming plainly the misconception this chapter exists to demolish, because it is perhaps the single most widespread falsehood in the popular history of technology: the belief that Thomas Edison invented the light bulb. He did not, and no serious historian claims he did. The idea of the incandescent lamp was decades old when he took it up. Working incandescent bulbs had already been built, demonstrated, and even patented by others — Swan foremost among them — before Edison produced his own. What Edison did was different, and in some ways more impressive than pure invention: he took an idea that many capable people had partially realized and made it *practical, durable, and commercially viable at scale*, and — as the coming chapters will show — he embedded it in a complete system that could actually light a city. The distinction matters not to diminish Edison but to see him accurately. He was not a lone genius conjuring light from nothing. He was the greatest closer in the history of invention, the man who took a race already crowded with runners and found the last, decisive burst of speed that carried one design across the line and into every home on earth. Understanding that is understanding what kind of genius he actually possessed.

Where do you meet this long race today? The distinction it turned on is still glowing in your appliances. The harsh arc that defeated the parlor lives on in the brilliant xenon and metal-halide lamps of stadiums, film sets, and old movie projectors. The gentle incandescence Edison chased lit nearly every home on earth for the following century, and though the simple filament bulb is now being retired in favor of the cooler, thriftier LED, the word "incandescent" still names the warm glow we instinctively call homey. And the very phrase "a light-bulb moment" for a flash of insight is a monument to a race most people do not know was a race at all.

Edison had staked his reputation and other men's money

on a boast. Now he had to make good on it — and to do that, he and his muckers at Menlo Park had to solve the one problem that had beaten everyone before them. They had to find the filament.

Chapter 22: The Filament

Late in the evening of October 22, 1879, a small knot of exhausted men gathered around a workbench in the upstairs laboratory at Menlo Park to watch a thread of burnt cotton glow. Charles Batchelor had spent hours that day performing one of the most delicate operations the lab had ever attempted: taking an ordinary piece of cotton sewing thread, coiling it, packing it in a mold, and baking it in a furnace until it was carbonized — reduced to a fragile, brittle filament of nearly pure carbon, so delicate that it crumbled at a touch and broke again and again before he managed to seal an intact one inside a glass bulb. The lab's improved vacuum pump drew the air out of the globe until almost nothing remained inside. Then they connected the current. The carbon thread warmed, reddened, brightened, and settled into a steady, warm, golden glow — not the searing glare of an arc, but a soft and gentle light, the kind you could read by, the kind you could live with. And then they waited. The great question was never whether a filament would light. It was whether it would *last*. Hour after hour the little lamp burned. It burned past midnight and into the next day. It burned for something on the order of thirteen or fourteen hours before the filament finally gave out — an eternity compared to anything the lab had achieved before, and enough, at last, to prove the principle. Edison had his lamp.

That night is usually where the storybook stops, but the thirteen-hour bulb was the culmination of more than a year of grinding failure, and the shape of that failure is the real story — because it is the clearest illustration in Edison's whole career of what he meant by "ninety-nine percent perspiration." To reach the burnt cotton thread, the muckers of Menlo Park had spent well over a year testing materials with a doggedness that borders on the absurd. The central problem, as we have seen, was burnout: find something that would glow white-hot without destroying itself. Edison and his men attacked it from both directions at once — improving the vacuum inside the bulb until it was among the best achievable anywhere, and searching, endlessly, for the right thing to put inside.

They began, logically, with metals. Metals conduct electricity well and can be drawn into fine wire, and Edison's early bet was on platinum, a metal with an extraordinarily high melting point. He and Francis Upton labored over platinum filaments for months, even devising clever regulating circuits to shut off the current before the wire melted. But platinum was maddening — it was ruinously expensive, a fortune in every lamp, and it never glowed as brightly or as reliably as they needed. Other metals melted or oxidized. And so, sometime in 1879, Edison made a conceptual turn that mattered as much as any single material he tried. He abandoned metals and went back to carbon.

Before we follow him there, we have to pause on the single most important idea in this entire chapter, because it is the piece the legend almost always omits and the piece that actually explains why Edison won. The popular story is all about the *material* of the filament — the romantic image of Edison testing ten thousand substances until one happened to work. But the deeper breakthrough was not a material at all. It was a decision about electrical design, and it was worked out with mathematics, much of it by the university-trained Upton. The insight was this: the filament should be made **thin and highly resistant to electricity**, rather than thick and freely conducting.

To see why that mattered, we need to define **resistance** plainly. Electrical resistance is a material's reluctance to let current flow through it — the electrical equivalent of friction, or of a narrow pipe choking the flow of water. A thick copper wire has very low resistance; current pours through it easily, like water through a wide pipe. A thin carbon thread has high resistance; the same current has to force its way through, and that struggle is exactly what generates heat and light. Now here is the economic heart of it. There is an iron law of electrical circuits, part of the family of relationships we met with Ohm, that says the power lost as heat in a wire depends on the *square* of the current flowing through it. Double the current, and you quadruple the heat wasted in the wires carrying it. This meant that a lamp designed to run on a large current would demand thick, heavy, hugely expensive copper conductors to feed it without melting the supply wires — and copper was the single largest cost in any lighting system. But a **high-resistance** lamp needed only a *small* current to reach the same brightness, because it wrung the light out of a modest flow by making that flow work hard. A small current meant thin, cheap copper wires could serve a whole district. In one stroke, the high-resistance filament transformed the light bulb from a scientific curiosity into an economically possible product. Edison did not just want a bulb that glowed. He wanted a bulb that a city could *afford* to wire, and the physics of resistance is what made that difference.

This is the correction that most needs making: Edison's genius in the light bulb was not that he found a magic material by luck. It was that he understood — and calculated — what *properties* the filament had to have to make the whole surrounding system cheap enough to sell. He was reverse-engineering the filament from the economics of copper. That is why he beat Swan and the rest, several of whom built lamps that glowed perfectly well but were designed around low resistance and would have bankrupted anyone who tried to wire a neighborhood with them. The bulb was a system-design problem disguised as a materials problem, and Edison, almost alone, saw it that way.

With the high-resistance principle fixed, the search narrowed to finding a thin carbon filament that would endure — and *this* is where the famous brute-force testing comes in, and where it belongs, not as the whole story but as its final chapter. Carbonized cotton thread proved the concept on that October night. But cotton was fragile and inconsistent, and Edison, ever the pragmatist, wanted a filament that could be manufactured by the thousand with reliable results. So began one of the most exhaustive material searches in the history of invention. The muckers carbonized and tested a bewildering catalog of substances: cardboard, tar-coated paper, wood shavings, celluloid, boxwood, coconut shell and fiber, flax, and — in a burst of Edison showmanship — hairs plucked from the beards of the men in the lab, with a small betting pool over whose whiskers would burn longest. Legend, encouraged by Edison himself, put the number of materials tested in the thousands, and while the exact count is impossible to verify and the figure grew in the retelling — a fact worth flagging — the underlying reality is not in doubt. The Menlo Park notebooks record a relentless, systematic march through material after material, each carbonized, sealed, lit, and timed.

The winner, for a time, was one of the humbler substances on earth, and the story of how Edison found it has the flavor of legend but appears to be roughly true. The tale goes that Edison, turning something over in his hands one day, idly picked at the frayed edge of a bamboo fan or a strip of split bamboo and wondered about its long, fibrous structure. Carbonized **bamboo** turned out to make a superb filament — tough, long-lasting, burning for hundreds of hours where the cotton thread had lasted fourteen. Edison became so convinced of bamboo's virtues that he dispatched men to the far corners of the earth — to Japan, to South America, to the jungles of Southeast Asia — on expensive expeditions to hunt down the ideal species of bamboo for lamp filaments. One of his scouts reportedly combed the interior of Japan and found a variety of bamboo that outperformed all the rest, and for years Edison's lamps glowed by the carbonized fibers of Japanese bamboo. That a man solving one of the great technical problems of the industrial age ended up sending agents into tropical forests to gather grass is one of the delightful surprises of the whole affair, and a reminder of how unpredictable the path from problem to solution really is.

All of this culminated in one of the great public spectacles of the nineteenth century. On New Year's Eve, December 31, 1879, Edison threw open the grounds of Menlo Park to the public for a mass demonstration of his lighting system. He had strung dozens of the new incandescent lamps through the laboratory, the office, the outbuildings, and along the wires overhead and down the lane to the railway station. Special trains carried thousands of curious visitors out from the cities to the little New Jersey crossroads on a cold winter night — and there, spread across the dark countryside, they saw a constellation of soft, steady, flameless electric lights glowing in the December dark, warm and silent and utterly unlike the harsh gas jets and blinding arcs they knew. People walked from lamp to lamp in wonder. Reporters filed breathless dispatches. Some skeptics, suspecting a trick, are said to have peered behind the lamps looking for hidden wires or concealed gas. It was pure theater, staged by a master showman who understood that an invention seen by ten thousand awestruck people on New Year's Eve was worth more than any patent. The Wizard had made good, in public, on the boast he had made a year before, and the news raced around the world.

There is a human texture to the filament search that the tidy summary flattens, and it is worth restoring, because it shows what "ninety-nine percent perspiration" actually felt like from the inside. The work was miserable. Carbonizing a filament meant hours of exacting, repetitive labor over furnaces, and the carbonized threads were so fragile that most of them shattered before they could even be mounted — Batchelor reportedly ruined dozens of promising filaments simply trying to attach them to their wires, each failure sending him back to start again. Sealing a filament into a bulb and pumping it down to a high vacuum could take the better part of a day, only for the lamp to fail in minutes. The men worked in shifts around the clock through the autumn of 1879, red-eyed and short-tempered, testing, timing, recording, discarding. When the burnt cotton thread finally burned through the night of October 22, the exhaustion in the room was as real as the triumph. This is the reality behind every clean sentence in a textbook that says "Edison discovered the carbon filament in 1879." Discovery, at Menlo Park, was mostly a matter of enduring — of being the last team still standing when everyone else had given up.

And there is a final correction to make, because the bamboo triumph was not the end of the story, and the popular telling usually stops too soon. Carbonized bamboo made Edison's lamps commercially viable for the better part of a decade, but it was never the perfect answer, and the search never truly stopped. By the late 1880s Edison's own chemists had moved to "squirted" cellulose filaments — a treated plant fiber forced through a die into a uniform thread — which were more consistent than any natural fiber. And within a generation, carbon itself was abandoned entirely in favor of drawn metallic tungsten, which glowed brighter and lasted far longer. So the filament Edison is famous for was, in the end, a way station, not a destination. What endured was not the particular material but the method — the willingness to test everything, and the deeper insight about high resistance that no change of material ever overturned. Edison found *a* filament that worked. His more lasting gift was the reasoning that told him what any good filament had to be.

Where do you meet the filament today? For over a century,

in nearly every room you ever entered. The carbon thread of 1879 gave way to a tungsten wire in the early twentieth century — brighter, longer-lived, the coiled metal filament that glowed in ordinary incandescent bulbs until our own era. When you switch on an old-style bulb, or admire the deliberately visible glowing loop of an "Edison bulb" sold now for its nostalgic warmth, you are looking at the direct heir of Batchelor's baked cotton thread. And the phrase we reach for when we praise stubborn, unglamorous effort — that success is mostly perspiration — was minted, quite literally, in the year-long slog to find that thread.

But a glowing filament in a single bulb, however many hours it burned, lit nothing but a laboratory. To light a city, Edison would need far more than a lamp. He would need everything that had to stand behind it — and building all of it at once was the real work still to come.

Chapter 23: Building a System

Imagine, for a moment, that you are handed a working light bulb in the year 1880 — a perfect one, that glows steadily for hundreds of hours — and told to light your home with it. You quickly discover that you own the single most useless miracle in history. There is no wire in your walls to carry current to it. There is no generator anywhere to make that current, and no way to send it to your street if there were. There is no socket to hold the bulb, no switch to turn it on and off, no fuse to stop it from setting your house ablaze if something goes wrong, and no meter to tell the company how much to charge you. The bulb, alone, is nothing. This is the fact that Thomas Edison understood more deeply than any of his rivals, and it is the true reason he, and not Joseph Swan or Alexander Lodygin or any of the others who made a filament glow, is remembered as the man who lit the world. Edison's genius was not the bulb. It was the realization that a bulb is meaningless without a whole invisible system standing behind it — and that he would have to invent, more or less simultaneously, every single part of that system.

This is the most important and least appreciated chapter in the entire Edison story, so let us state its thesis without hedging. The light bulb was perhaps the least difficult and least original of Edison's contributions to electric lighting. The truly staggering achievement — the thing no one else even fully attempted — was to conceive of electric light as a *complete integrated system* and then to design and build every component of that system from scratch, all at once, so that they worked together. Between the New Year's Eve demonstration of 1879 and the opening of his first power station in 1882, Edison and his growing army of workers did not perfect a lamp. They invented an industry, part by part, and each part was its own act of invention.

Edison had a brilliant guiding model for how to do this, and it reveals how his mind worked. He did not invent the electric-lighting system from pure imagination. He copied its architecture, almost point for point, from the industry he intended to destroy: the gas industry. Gas lighting already had a complete distribution system, and it worked. A central gasworks manufactured gas; underground mains carried it beneath the streets; smaller pipes branched off to individual buildings; a meter at each building measured how much gas the customer used so the company could bill for it; and valves let the user turn each jet on and off. Edison looked at this mature, proven, profitable arrangement and essentially said: I will build exactly this, but with electricity flowing through copper instead of gas flowing through pipe. The central gasworks becomes a central generating station. The gas mains become electrical conductors. The meter becomes an electric meter. The valve becomes a switch. By modeling his electrical system on the gas system, Edison gave himself a complete blueprint for every piece he needed to invent, and he gave his future customers a service that felt reassuringly familiar. He was not asking them to imagine something wholly new. He was offering them a better gas.

So consider the parts, because the sheer breadth of what had to be invented is the point. First, the lamp needed something to hold it and something to control it. Edison's people devised the **screw socket** — the spiraling threaded base still called an "Edison screw" and stamped with an "E" on the bulbs in your closet at this very moment — so a bulb could be twisted securely into place and replaced when it burned out. They devised practical **switches** to open and close the circuit, letting an ordinary person turn a light on and off with a fingertip, safely, without sparks or danger. These sound trivial until you realize that no standardized version of them existed and every one had to be designed, tested, and made manufacturable.

Then came the matter of safety, which was not trivial at all, because Edison was proposing to run enough electric current through the walls of people's homes to make metal glow, and electric current in a fault can start a fire in seconds. Edison's answer was one of the quiet masterstrokes of the whole system: the **safety fuse**. A fuse is a deliberately weak link — a short piece of thin wire or metal strip, built into the circuit, that is designed to be the first thing to fail. If something goes wrong and too much current surges through the wires — a short circuit, an overload — the fuse wire heats up and melts before the household wiring does, breaking the circuit and shutting off the flow before a fire can start. Think of it as a sentry posted at the door, instructed to fall on its sword at the first sign of trouble so the rest of the house is spared. The fuse (and its modern reincarnation, the circuit breaker in your home's electrical panel) is the reason a wiring fault trips a switch instead of burning your house down, and it descends directly from Edison's insistence that a lighting system had to be *safe* enough for ordinary families to trust.

Next, and most fundamental of all, there had to be a source of power, because a light bulb without a generator is a lamp without oil. Edison needed a machine to convert the mechanical energy of a steam engine into electric current on a massive scale — a **dynamo**, or generator. Generators existed, built on Faraday's principle of induction, but the dynamos of the 1870s were inefficient, wasting much of the engine's power as heat. Worse, the received wisdom among electrical engineers held that a generator worked best when its internal resistance matched the resistance of whatever it was powering — a doctrine that, had Edison followed it, would have thrown away half his power. Edison and Upton defied the experts and designed a generator with very *low* internal resistance, so that it delivered the great majority of its energy to the lamps rather than wasting it inside itself. The result was a series of increasingly large and efficient dynamos, culminating in a giant weighing some twenty-seven tons that the men nicknamed "**Jumbo**," after a famous circus elephant of the day. Jumbo could power over a thousand lamps, and it was the beating heart Edison's central stations would need.

Finally — and this is the piece that reveals Edison the businessman as clearly as any — there was the problem of getting *paid*. A system that gives light is a charity unless you can measure and bill each customer for what they use, exactly as the gas company did. So Edison invented an **electric meter**, an instrument to measure how much electrical energy each household consumed. His early version was ingenious and a little strange: it used electricity to slowly deposit metal from a chemical solution onto a plate, so that the amount of metal built up over a month was proportional to the electricity used, and a company man would come read it. The device was clumsy and would be much improved by others, but the principle — the metering of electric energy so it could be sold as a commodity by the unit — was essential, and it is the ancestor of the meter spinning or blinking on the side of your house today. Without the meter, there is no electric-utility *business*; there is only a very expensive gift.

Now stand back and look at what binds all these pieces together, because there is one more design decision that made the whole system possible, and it is worth understanding. Edison wired his lamps in **parallel**, not in series, and the difference is the difference between a usable system and a useless one. To picture it: if you string lights in **series**, they sit one after another along a single loop of wire, like a line of buckets passing water hand to hand — the same current must pass through every lamp in turn. This has a fatal flaw for a home or a city: if one bulb burns out, it breaks the single loop and *every* lamp on the circuit goes dark at once, and you cannot turn one light off without killing them all. (Anyone who has wrestled with an old strand of Christmas lights where one dead bulb blacks out the whole string has suffered a series circuit firsthand.) In a **parallel** system, by contrast, each lamp sits on its own little branch bridging two main wires, like the rungs of a ladder between two long rails. Every lamp gets the full voltage independently, and each can be switched on or off, or burn out, without affecting any of the others. This is exactly how you want the lights in a building to behave — each one independent — and it is how every circuit in your home is wired to this day. Designing the whole system around parallel wiring, and around a steady, standardized voltage that every lamp and appliance could share, was the organizing principle that let all the other pieces cohere.

Here is the surprising truth that all of this adds up to, and the misconception it corrects. We say "Edison invented the light bulb," and it makes him sound like a man who had one bright idea. But between 1878 and 1882 Edison and his team filed for and received an astonishing cascade of patents — not on one thing but on dozens: on lamps, sockets, switches, fuses, meters, dynamos, conductors, junction boxes, regulators, and the distribution scheme itself. He was not inventing an object. He was inventing an *ecosystem*, and doing it as a coordinated engineering campaign of a kind the world had never seen. The integration mattered more than any single part, because a chain of miracles is worthless if the links do not connect. Swan made a fine bulb. Edison made a fine bulb *and* everything a bulb needs to be worth owning, and that is why the world remembers one name and not the other.

It is worth pausing on why this systems thinking was so rare, because it is the quality that most distinguishes Edison from the brilliant tinkerers who lost to him. Most inventors fall in love with a device. They see the elegant object — the bulb, the motor, the gadget — and pour their genius into perfecting it, assuming the world around it will somehow arrange itself to make the object useful. Edison's mind ran the other way. He seems to have been almost incapable of thinking about a component in isolation; he could not look at a lamp without immediately seeing the socket it needed, the wire that would feed it, the generator behind the wire, the meter that would bill for it, and the customer who would pay. He thought in whole networks, in complete circuits from the coal pile to the glowing filament and back to the cash register. This is why his notebooks from these years are filled not just with sketches of bulbs but with calculations of copper costs, diagrams of distribution networks, and estimates of how many lamps a district could support. He was designing an infrastructure, and he knew it. The lamp was merely the part the customer could see.

This habit of mind has a name in the modern world — systems engineering — and Edison is often cited, rightly, as one of its founding practitioners, even though the term did not yet exist. The discipline of designing all the interlocking parts of a complex technology together, so that each is optimized for its role in the whole rather than perfected in isolation, is now fundamental to everything from spacecraft to computer networks to the power grid itself. Edison arrived at it not from theory but from hard commercial logic: he understood that customers do not buy components, they buy *outcomes* — in this case, reliable, affordable, safe light at the flick of a switch — and that delivering an outcome requires mastering every link in the chain. The rival who built a better bulb but ignored the system was building a better part of a machine that did not exist. Edison built the machine.

Where do you meet this system today? You are surrounded by it, unchanged in its essentials. The screw base of your bulbs still bears Edison's name and his "E." The parallel wiring in your walls still lets you switch off the bedroom light without darkening the kitchen. The fuse or breaker in your electrical panel still stands sentry against fire. The meter on the outside wall still measures what you owe, just as his did. When you flip a switch and a single lamp comes on while everything else stays as it was, you are operating, part for part, the system a deaf man and his muckers designed in a New Jersey laboratory a century and a half ago.

Edison now had every piece he needed, proven and patented, sitting in the workshops of Menlo Park. What remained was to assemble them at full scale, in the most demanding and most public place on earth, and prove that a system that worked on a New Jersey lawn could light the beating financial heart of America. He set his sights on lower Manhattan.

Chapter 24: Pearl Street

At three o'clock in the afternoon on September 4, 1882, Thomas Edison stood in the Wall Street offices of the banker J. P. Morgan, dressed uncharacteristically well for the occasion, and watched a clock. A few blocks away, in a converted building at 255–257 Pearl Street in lower Manhattan, his chief electrician stood beside the massive Jumbo dynamos with a hand on a switch, waiting for the signal. When the appointed moment came, the switch was thrown, current surged out through miles of copper laid beneath the streets, and in offices and shops across a square-mile district of downtown New York, hundreds of incandescent lamps came softly, steadily alight in the middle of a sunny afternoon — so that, in the daylight, almost no one noticed. It was the most understated world-changing event imaginable. There was no thunderclap, no blinding flash, just a scatter of small warm lights glowing pointlessly in the sunshine, waiting for a dusk that would, from that day forward, never again be entirely dark. The age of electric power had begun, and it began so quietly that most of the city missed it.

The Pearl Street Station was the payoff of everything we have followed — the filament, the dynamo, the socket and switch and fuse and meter, the whole integrated system — assembled at last into the thing Edison had been building toward all along: the world's first commercial **central power station**. The phrase deserves unpacking, because the word "central" carries the entire revolution. Up to this point, if you wanted electric light, you needed your own generator on your own premises — an arrangement suited only to a factory, a hotel, or a rich man's mansion. What Edison built at Pearl Street was something categorically new: a single generating plant that made electricity in one place and sold it, through a network of wires, to many separate customers scattered across a district, each of whom simply paid for what they used and needed no machinery of their own. This is the idea of the **electric utility** — the model by which nearly all the world's power has been supplied ever since. Pearl Street was its birthplace. When you pay a monthly bill to a power company you have never visited, for electricity made at a plant you will never see, you are a customer of the institution that was born at 255 Pearl Street in 1882.

Getting there had been a grueling, and largely invisible, ordeal — and the invisible part is the part worth telling, because it shows how much of the electrical revolution happened underground. Edison had chosen, deliberately and at great expense, to bury his distribution wires beneath the streets rather than string them on poles overhead. This was partly for safety and reliability and partly for appearances — he wanted his system to look like a permanent public utility, not a temporary experiment. But burying the conductors meant digging up the streets of one of the busiest cities on earth. Through 1881 and 1882, Edison's crews tore open the pavement of lower Manhattan and laid roughly fifteen miles of underground copper conductors, insulated and packed into iron pipes, threading them beneath the cobblestones and around the existing tangle of gas, water, and sewer lines. Edison himself, by several accounts, worked in the trenches at night alongside his men, personally supervising the laying of the mains, because a fault in a buried cable would be a nightmare to find and fix once the streets were closed again. It was slow, filthy, expensive work, and it is a useful corrective to the image of electrification as a matter of brilliant sparks and glowing bulbs. Much of it was men with shovels, in the dark, in the mud, laying pipe.

There was a reason Edison chose lower Manhattan's **First District**, the area around Wall Street and the financial houses, and it was a shrewd one. He was not merely lighting a neighborhood; he was staging a demonstration for the people who could make him rich. By lighting the offices of bankers, brokers, newspapers, and the financiers of Wall Street — including the office of J. P. Morgan himself, whose home Edison had earlier wired as a private showcase — Edison ensured that the men with the capital to fund a national rollout would see his light glowing on their own desks every evening. He was lighting the wallets of America. Morgan, the most powerful banker in the country, had backed Edison's lighting venture and would remain entangled with the electrical business, for better and worse, for the rest of Edison's career. The choice of district was pure Edison: a technical decision and a marketing decision fused into one.

Now the technical heart of it, because Pearl Street ran on a specific electrical recipe that would matter enormously in the war to come. The station generated **direct current** — abbreviated **DC** — at a nominal **110 volts**. These two terms deserve plain definitions, because the entire drama of the next fifteen years turns on them. *Direct current* means electricity that flows steadily in one constant direction through the wire, like water flowing down a river always toward the sea — a smooth, one-way stream. (Its rival, which we will meet soon, is *alternating current*, which sloshes rapidly back and forth.) *Voltage* is the electrical "pressure" that pushes the current along, much as water pressure pushes water through a pipe; higher voltage shoves current harder and farther. Edison settled on 110 volts as a compromise: high enough to drive current usefully through the wires and make the lamps glow, but low enough to be reasonably safe for people to have in their homes and offices — a voltage that would sting but was unlikely to kill. That figure is why, to this day, the ordinary wall outlets of North America deliver electricity at roughly 110 to 120 volts. The number in your walls is Edison's number, chosen in 1882 and never fully abandoned.

The station's debut was modest by the standards of what it would become, and it is worth being precise about the scale, because the smallness of the beginning makes the speed of the growth all the more remarkable. When Pearl Street opened on that September afternoon, it served on the order of **eighty-five customers**, powering roughly **four hundred lamps**. Within a couple of years those numbers had multiplied many times over, into the thousands of lamps and hundreds of customers, and Edison's model was being copied in cities across the country and the world. But the seed was those first four hundred bulbs — a handful of lights in a single district of a single city, from which the entire electrified world would grow.

The reception, once dusk finally fell and the lights could actually be seen, was rapturous. *The New York Times*, whose own offices sat within the First District and were lit that first evening by Edison's lamps, reported the experience with something close to wonder — describing how the electric light was soft and steady and easy on the eyes, how it did not flicker or smoke or foul the air as gaslight did, how a man could work by it or read fine print without strain. The lights, the paper marveled, looked almost like daylight, "soft, mellow, and grateful to the eye," and it "seemed almost like writing by daylight." Coming from journalists who had spent their whole careers squinting through the fumes and flicker of gas jets, this was extraordinary praise, and it captured what made the moment historic. This was not a brighter version of an old thing. It was a fundamentally better and cleaner thing, and the people who experienced it first understood immediately that the world had changed.

Here are three facts about Pearl Street worth carrying forward, each of which surprises people who think they know the story. The first: the station was not, in its early years, a financial triumph. It was a technological miracle and a business struggle. The copper was ruinously expensive, the plant was costly to run, coal had to be shoveled by hand to raise the steam, and for its first years the operation lost money. Edison had proved the system worked long before he proved it could turn a reliable profit — a gap that other men, notably a young clerk of his named Samuel Insull, would later have to close. The second: the original Pearl Street Station did not last. In 1890 it was gutted by a fire, one of its Jumbo dynamos surviving as a relic; the physical birthplace of the electrical age burned, even as the idea it embodied spread unstoppably. The third, and most consequential: everything about Pearl Street — the buried copper, the 110 volts, the very economics of the central station — was built around direct current, and direct current had a hidden, crippling limitation that Edison either did not yet grasp or refused to face. The system that lit Wall Street so triumphantly in 1882 could not, by the unbending laws of physics, reach much beyond the district it stood in.

It is worth lingering on what it actually took to keep those first four hundred lamps burning, because the romance of the switch-throwing moment hides an unglamorous truth about early electric power: it was brutally physical work. Behind the soft, clean, flameless glow in the offices of Wall Street stood a hot, filthy, deafening room full of men and coal. The Jumbo dynamos were driven by steam engines, and the steam had to be raised by burning coal, and the coal had to be shoveled — by hand, ton after ton, by stokers laboring in the heat before the boilers, so that gentlemen upstairs could read their newspapers by a light that seemed to come from nowhere. The cleanliness of electric light at the point of use was, in a sense, an illusion purchased by concentrating all the dirt and danger in one place, out of sight. This is a pattern worth carrying through the rest of the book, because it never really goes away: the electricity that arrives so cleanly at the wall has always been made, somewhere out of view, by burning or damming or splitting something, and the tidiness of the outlet conceals the furnace behind it.

Pearl Street was the first demonstration of that bargain, and we are still living inside it today.

There is also a lesson in Pearl Street about the difference between an invention and a business, and it is one Edison learned the hard way. He had the finest lighting technology in the world, and for years he could not make it pay, because the economics of a central station are cruelly unforgiving in the early going: you must sink enormous capital into copper and plant *before* the first customer signs on, and then you must sell enough electricity, reliably, to cover that mountain of fixed cost. A plant running at a fraction of its capacity is a plant losing money, no matter how brilliant its engineering. The men who ultimately made the electric utility profitable were not primarily inventors but organizers and financiers — men who understood load, and rates, and the art of keeping the generators busy around the clock. Edison built the machine that made the power. It would take others to build the business that sold it. That gap between the miracle and the ledger is one of the quiet themes of the entire electrical age, and it opened for the first time on Pearl Street.

Where do you meet Pearl Street today? Every time you flip a switch and draw power from a distant plant you neither own nor tend, you are living inside the model it created — the central station selling electricity as a service to strangers. The roughly 120 volts at your North American outlets is a direct inheritance from Edison's 110. And the "First District" idea — build the network where the money and the influence are, and let success spread outward from there — remains the template by which new technologies, from streetlights to broadband to charging stations, are still rolled out.

Edison had lit the financial capital of America and shown the world what a central power station could do. Now he had to sell it — city by city, customer by customer, against a gas industry fighting for its life. And in that great campaign of expansion he would stake his entire fortune and reputation on the one current his masterpiece at Pearl Street ran on, never fully reckoning with the fatal flaw buried inside it.

Chapter 25: Selling the Light

In the years right after Pearl Street opened, a curious ritual played out in city after American city. A well-dressed representative of the Edison Electric Light Company would call on a factory owner, a hotelier, or a department-store magnate, and would offer to install electric lighting on the premises. And very often, before he could finish his pitch, a second man — the agent of the local gas company — would already be in the outer office, or would arrive the next morning, warning the same customer that this newfangled electric light was dangerous, unproven, ruinously expensive, and very likely a passing fad. The two men were foot soldiers in a commercial war, and for Thomas Edison the years from 1882 through the mid-1880s were consumed by it: the grinding, unglamorous, city-by-city work of *selling the light* against an entrenched and frightened incumbent. Inventing electric lighting had been the hard part scientifically. Making the world actually buy it — and buy it in the particular form Edison had staked everything on — was the hard part commercially, and it would prove, in the end, to contain the seeds of his undoing.

The enemy, as we have seen, was gas. And it is worth pausing to appreciate how formidable an enemy it was, because hindsight makes electric light look inevitable, and in the

1880s it was anything but. The gas industry was rich, established, politically connected, and fighting for its existence. It had decades of infrastructure in the ground, contracts with every municipality, and a product that, whatever its faults, worked and was familiar. Gas companies slashed their prices to undercut the newcomer. They spread doubt about electricity's safety — and here they had genuine ammunition, because early electric wiring, badly installed by inexperienced crews, did sometimes cause fires and shocks. They lobbied city governments to deny Edison the permits he needed to tear up streets and string wires. Edison, who had rattled gas stocks with a mere announcement back in 1878, now had to grind out his victory building by building, contract by contract, against an opponent that would not go quietly. The romance of invention gave way to the trench warfare of business.

Edison fought this war on two fronts at once, and understanding the two fronts is essential to understanding what came next. The first front was the **central station** model we saw at Pearl Street — a generating plant serving a whole district of separate customers over a shared network of wires. This was Edison's grand vision, the utility model, but it was slow and enormously capital-intensive: you had to dig up the streets, lay miles of costly copper, build a plant, and wait years to sign up enough customers to recoup the investment. Cities were reluctant, capital was scarce, and central stations spread only gradually.

So Edison leaned heavily on a second, faster front: the **isolated plant**. An isolated plant was a small, self-contained Edison generating system installed on a single customer's own premises — a dynamo in the basement of a hotel, a mill, a theater, a wealthy home, an office building, or an ocean liner — powering only that one establishment's lights. It required no city permits, no torn-up streets, no network; Edison's company simply sold the customer a complete lighting plant and walked away. Isolated plants were quick to install and quick to pay for themselves, and in the first years after Pearl Street they vastly outnumbered central-station customers. They spread Edison lighting into hundreds of buildings across the country and around the world far faster than the slow, expensive central stations could, and they brought in badly needed revenue. J. P. Morgan's own Wall Street mansion had been an early isolated plant; so were factories, steamships, and grand hotels from New York to Europe. The isolated plant was Edison's workhorse, and it was profitable in a way the central stations, for years, were not.

But notice what every one of these installations — central station and isolated plant alike — had in common. Every single one of them ran on **direct current**. The dynamos generated DC; the lamps were designed for DC at 110 volts; the meters, the wiring, the whole system Edison had patented and manufactured was a DC system, root and branch. And as Edison's business grew — as he built the manufacturing companies to make the dynamos and lamps and cables, as he licensed his system across the nation and overseas, as his central stations multiplied and his isolated plants proliferated — every dollar of it was invested in direct current. Edison did not merely *prefer* DC. He was becoming, financially and institutionally and psychologically, the incarnation of DC. His factories were tooled for it. His patents protected it. His reputation was staked on it. His fortune was riding on it. To question direct current was, for Edison, to question the whole edifice he had built.

This is the moment to correct a misconception that flattens the drama, because it is often said that Edison "backed the wrong technology," as though he had blundered into DC out of ignorance. He did not. In 1882, direct current was, for the job it was doing, genuinely the right choice. It was well understood, it was safe at the low voltages he used, it could run motors directly, it could be stored in batteries as a backup — a thing alternating current could not do. For lighting a dense district like lower Manhattan, where customers were packed close around the station, DC worked beautifully. Edison's commitment to it was not stupidity; it was a rational bet on the best available technology of the moment. The tragedy — and it is a genuine tragedy, in the old sense of a great man undone by the very qualities that made him great — is that the technology of the moment does not stay the technology of the moment, and Edison's greatest strengths would become, in the face of a changing world, his greatest weaknesses.

For here we arrive at the psychological core of the coming war, and it is worth stating plainly because it explains behavior that would otherwise look inexplicable. Edison's stupendous success with the DC system created in him an overcommitment to it that went far beyond reason. He had invented it, fought for it, sunk his fortune into it, and been proved right by it. Having been so spectacularly right, he could not entertain the possibility of being wrong. The same relentless, stubborn will that had let him test ten thousand filaments and dig ditches through Manhattan at midnight — the refusal to quit, the certainty that he could bend the world to his purpose — curdled, when applied to direct current, into a blindness he would never overcome. The man's defining strength was his tenacity. His defining flaw was that his tenacity had no reverse gear. Once Edison committed to something, no evidence, no argument, no demonstration could pry him loose. And he had committed, totally, to direct current.

There is a revealing human detail here that captures the depth of the commitment. When, in the years to come, the superiority of alternating current for long-distance distribution became a matter of open engineering debate, some of Edison's own most brilliant associates — men who could read the mathematics — quietly concluded that AC was the future and urged him to embrace it. Edison would not hear it. He is reported to have said, in effect, that alternating current was a waste of time and would soon be abandoned, and he treated advocacy of AC within his own organization almost as a kind of disloyalty. One of the great ironies of the whole story is that some of Edison's ablest people saw the truth clearly and could not move him — a man who had spent his life being smarter and more persistent than everyone around him, now using that same persistence to hold shut a door the future was trying to open. The empire he had built was, in the most literal sense, a DC empire, and its emperor had resolved that it would remain one.

It is worth pausing to appreciate just how much there was to defend, because the scale of Edison's commercial machine explains the ferocity with which he would soon fight for it. By the mid-1880s "Edison" was no longer one man in a laboratory; it was a sprawling constellation of companies, each a pillar of the DC edifice. There was the Edison Electric Light Company, which held the patents and licensed the system. There was the Edison Machine Works, which built the great dynamos. There was the Edison Lamp Works, which manufactured the bulbs by the tens of thousands. There was the Edison Tube Company, which made the underground conductors, and Bergmann & Company, which turned out the sockets, switches, meters, and fixtures. Taken together they formed a vertically integrated industrial enterprise — Edison made everything, from the filament to the fuse to the generator — and every factory, every payroll, every patent, every contract in the whole apparatus assumed direct current. To switch to AC would not have meant changing a mind. It would have meant scrapping an industry and rebuilding it from the copper up. The sheer sunk weight of it all was a kind of gravity, and it held Edison in place.

And the customers came. Through the 1880s the isolated plants multiplied into the hundreds and then thousands, and the central stations, though slower, took root in city after city at home and abroad — London, Milan, Berlin, and beyond all lit, in their first electrical districts, by Edison's DC system. Every new installation deepened the investment and hardened the commitment, so that the very momentum of success made retreat more unthinkable with each passing year. A younger, smaller Edison might have been able to change course when the physics turned against him.

The Edison of the mid-1880s, standing atop a global business built entirely on one kind of current, could not — and would not.

Where do you meet this chapter today? You meet it every time an industry giant, having ridden one great technology to dominance, cannot bring itself to embrace the next one — the film company that dismissed digital, the phone maker that scorned the touchscreen, the encyclopedia that laughed at the internet. The pattern historians and business writers now call the "incumbent's dilemma," or the innovator's trap, has no cleaner nineteenth-century illustration than Thomas Edison and direct current. The very success of a technology, and the fortune and identity built upon it, can make the people who mastered it constitutionally unable to see past it. Edison, the supreme disruptor of the gas industry, was about to become the incumbent defending his own ground against a disruption he refused to believe in. And you meet this chapter, too, in a subtler way: the "isolated plant" — your own generator, serving only you — has quietly returned in the age of rooftop solar panels and home batteries, the old idea that you might make your own power on your own premises rather than buy it from a central station, alive again a century and a half later.

Edison had built a magnificent thing and bet everything on it. But buried in the physics of his beloved direct current was a limitation so fundamental that no amount of will could overcome it — a wall, written into the laws of nature themselves, that meant his system could never reach the farms, the towns, and the distances that a truly electrified world would demand. To understand the war that was coming, and why Edison was destined to lose it even as he refused to believe he could, we have to look squarely at that wall. We have to understand exactly why direct current could not travel.

Chapter 26: The Limits of Direct Current

Picture a map of an American city around 1885, with Edison's central stations marked on it as small dark circles — and around each circle, draw a ring about one mile across. Everything inside a ring could have Edison's electric light. Everything outside it could not. Not because Edison lacked ambition, or capital, or customers beyond the ring — the demand was everywhere — but because of a wall written into the laws of physics, a wall that direct current could not climb over no matter how much money or genius was thrown at it. Beyond roughly a mile from the generating station, Edison's electricity simply gave out, dwindling away in the wires until the lamps at the far end glowed dim and useless. To light the next mile, you had to build another entire power station. And another for the mile after that. This is the fatal flaw at the heart of Edison's empire — the reason his beautiful system could light dense city districts but could never, ever reach across the distances that a whole electrified nation would require. This chapter is about that wall: what it was, why it existed, and why it doomed direct current to lose the war that was coming.

The problem, stated simply, is that **direct current at low voltage cannot travel far without wasting most of its energy as heat in the wires.** To understand why — and it is genuinely worth understanding, because this single fact decided the entire War of the Currents — we need two ideas we have already met, and we need to put them together.

The first idea is that wires resist the flow of electricity. No wire is a perfect conductor; even good copper fights the current passing through it, just a little, all along its length. The longer the wire, the more total resistance the current must overcome — a mile of copper resists far more than a foot of it, the way a mile of gravel road slows a cart far more than a few yards of it. And here is the crucial consequence: whenever current forces its way through resistance, some of the electrical energy is converted into heat and lost. The wire warms up, and that warmth is energy that came out of your power and will never light a single lamp. It is simply gone, radiated uselessly into the air along the length of the line.

The second idea is the precise law governing how much energy is lost, and it is the hinge on which everything turns. The power wasted as heat in a wire is proportional to the **square of the current** flowing through it — a relationship engineers write as "I-squared-R," where I stands for the current and R for the wire's resistance. The "squared" is what makes this so brutal. It means the losses do not grow gently as the current rises; they explode. Double the current, and you do not double the wasted heat — you *quadruple* it. Triple the current, and the losses grow ninefold. High current in a long wire is catastrophic: the wire bleeds away energy at a ferocious, accelerating rate, and by the time the current has traveled any real distance, most of the power has been cooked off into useless warmth before it ever reaches the customer.

Now put the two ideas together with Edison's design choice, and the trap springs shut. Recall that Edison's system ran at a low voltage — 110 volts — chosen to be safe in homes. But there is an unbreakable relationship between voltage, current, and power: to deliver a given amount of power, if the voltage is *low*, the current must be *high*. It is exactly like delivering a fixed amount of water: through a low-pressure pipe you must send a large volume of water, while through a high-pressure pipe you can send the same amount of water as a thin, fast jet. Power is voltage multiplied by current, so low voltage forces high current, and high current — by the merciless I-squared-R law — means ruinous losses in any long wire. Edison's insistence on a safe, low voltage, which was exactly right for the lamp in your parlor, was exactly wrong for sending power across a distance. The very choice that made his system safe indoors made it helpless outdoors.

There was a way to fight back, but it only deepened the bind. If your losses come from high current in a resistive wire, you can reduce the resistance by using a *thicker* wire — a fatter copper conductor fights the current less, the way a wider pipe passes water more easily. And this is precisely what Edison had to do: his distribution mains were made of thick, heavy copper to keep the losses tolerable over even short distances. But copper was expensive — one of the single largest costs in the whole system — and the thickness required grew punishingly as the distance increased. To push DC power just a little farther, you needed dramatically more copper, until at even modest distances the cost of the copper conductors alone became financially impossible. You would have to bury a fortune in metal in the ground to light a customer two miles away. So Edison was caught in a vise with no escape: keep the wires cheap and the power dies within a mile; make the power travel and the copper bankrupts you. Either way, direct current could not economically leave the neighborhood of its birth.

The practical upshot was the ring on the map. An Edison DC central station could serve customers only within roughly a mile or so of the plant. To electrify a whole city, you would need a power station every mile or two — dozens of them, each with its own coal supply, its own engineers, its own thundering dynamos, dotting the map like fire stations. To electrify the countryside, the small towns, the farms, the vast distances between American communities, was simply out of the question. And here is a fact that should stop us, because it reveals how absolute the barrier was: the greatest sources of cheap electrical power in the world were waterfalls and rivers, where falling water could turn a generator for free — and the greatest of these, Niagara Falls, sat about twenty miles from the city of Buffalo. With direct current, that power was worthless to Buffalo. Twenty miles might as well have been twenty thousand. The energy of Niagara could not be brought to the people who needed it, because DC could not make the journey. The richest power source on the continent was locked away by the physics of Edison's current.

Now the reader can see the shape of the solution, even before its champions appear, because the trap has an obvious-looking key. The whole disaster came from high current. And current is high only because voltage is low. So: what if you could send the power at *very high* voltage — thousands of volts — so that the current, for the same power, became tiny? Tiny current means tiny losses, because the murderous I-squared term shrinks to almost nothing. At high voltage, power could race across hundreds of miles through a modest wire, losing almost nothing on the way. The physics practically begs for it. There is only one problem, and it is a fatal one for Edison: you cannot pipe thousands of volts into a person's home. High voltage that can travel a hundred miles will also kill instantly anyone who touches it and burn down any building it enters. To be *safe*, electricity must arrive at the home at a low voltage, like Edison's 110. But to *travel*, it must cross the countryside at a high voltage. You need it high for the journey and low for the destination — and there is no way, with direct current, to easily change a voltage from one level to another. That was the thing DC could not do. Direct current, once generated at a given voltage, was stubbornly stuck at roughly that voltage. There was no simple, efficient device that could take low-voltage DC, boost it to thousands of volts for the long haul, and then step it back down to a safe level at the far end.

But such a device existed — for alternating current. This is the crux of the entire War of the Currents, and we state it here so the coming chapters make sense: alternating current, the sloshing back-and-forth current that Edison dismissed, could be effortlessly transformed from one voltage to another by a simple, cheap, motionless device with no moving parts, called a **transformer**. Because AC constantly changes, it could be stepped *up* to enormous voltage at the power plant for the long journey — losing almost nothing on the way — and then stepped *down* to a safe, gentle voltage at the customer's door. The transformer was the key that fit the lock the I-squared-R law had built, and it was a key that worked only with alternating current. DC had no equivalent. This — not some vague notion that "AC is simply better" — is the precise, physical reason alternating current would win the war. It could do the one thing that mattered for distributing power across a country, and DC could not.

It is worth flagging a common misconception here, because the popular telling gets it backward. People often say that alternating current is simply the superior technology and Edison was a fool to resist it. That is too crude. For its purpose — safe local power, direct running of motors, storage in batteries — direct current was and remains excellent, and, as this book will show much later, high-voltage DC would stage a remarkable comeback for very-long-distance transmission in the twentieth century, once the technology to change *its* voltage was finally invented. The war was never about eternal, absolute superiority. It was about one specific, decisive question at one specific moment in history: how do you distribute electric power cheaply across the distances of a real country in the 1880s and 1890s? To *that* question, the transformer gave AC an answer, and DC had none. Edison lost not because his current was bad, but because he was trying to make it do the one job it physically could not do, and he refused to admit the job was impossible.

And that refusal is the human tragedy waiting at the end

of this chapter. Edison, the most empirical man alive, the man who worshipped what worked and mocked mere theory, was confronted with a limitation that his own admired mathematics could have shown him plainly, and he would not look at it. His blind spot was not ignorance; it was will. He had built his life and his fortune on direct current, and he could not concede that it had a ceiling. So when a quiet, methodical industrialist named George Westinghouse began building an alternating-current system that could leap the wall Edison could not climb — and when a strange, brilliant immigrant named Nikola Tesla supplied the missing piece that let AC run motors as well as lamps — Edison did not adapt. He went to war.

Where do you meet this chapter today? You meet it every single time you use electricity. The reason the power lines marching across the countryside on their tall steel towers carry warnings of tens or hundreds of thousands of volts, while the outlet in your wall delivers a safe 120, is the exact solution to the problem this chapter describes: send it high to make it travel, step it down to make it safe. The humming gray transformer on the utility pole outside your window, or in the green box on the corner, is performing at this moment the one trick direct current could not, the trick that decided the war. Every mile of high-voltage line and every transformer on earth is a monument to the wall Edison could not climb.

The wall was real, the key existed, and other men were already reaching for it. The stage was set for the bitterest fight of the electrical age — and to understand the men who would beat Edison at his own game, we must now leave New Jersey behind and travel to a mountain village in the Austrian Empire, where, nine years after Edison, a priest's son had been born with a mind that saw machines turning in the air.

Part IV

The Dreamer from Smiljan

From a priest's son with a photographic mind and visions he could not silence to the immigrant who arrived in New York with four cents and a letter to Edison — the making of Nikola Tesla and the invention that would decide the war: the rotating magnetic field.

Chapter 27: A Priest's Son and a Photographic Mind

At the stroke of midnight between the ninth and tenth of July, 1856, in the village of Smiljan, a summer thunderstorm broke over the limestone hills of the Lika region, in the mountainous military frontier of the Austrian Empire that is today part of Croatia. Rain hammered the roof of the parish house; lightning flickered against the windows. Inside, a woman named Đuka was giving birth to her fourth child, a son. The village midwife, unnerved by the storm, is said to have called the boy a child of the darkness. His mother, according to the story the family later told, answered without hesitation: no — he will be a child of the light. The infant was baptized in the Serbian Orthodox faith and given the name Nikola. Nikola Tesla.

We should be careful with that midnight-lightning scene, because it is exactly the kind of detail that grows more perfect with each retelling, and Tesla, who understood the value of a good story, did nothing in his later years to sand off its edges. Whether the storm raged at the precise moment of his birth we cannot verify. What is solidly documented is the date, July 10, 1856; the place, the tiny village of Smiljan;

and the family into which he was born — a household of the Serbian Orthodox Church, ethnically Serb, living in a region that was then a patchwork borderland of empires and faiths. This is the first fact worth fixing in place, because popular accounts routinely muddle it. Tesla was an ethnic Serb, born in a village that lies within the borders of modern Croatia, in what was then Austrian territory. All three of those things are true at once, and the tangle of them tells you something about the contested corner of the world that produced him.

His father, Milutin Tesla, was a priest of the Serbian Orthodox Church — a tall, learned, sharp-tongued man with a library of theology and poetry and a reputation for delivering sermons that could hold a congregation rapt. Milutin wanted his gifted son to follow him into the priesthood, and much of the drama of Tesla's youth turns on the boy's quiet, stubborn refusal of that destiny. But the father left his mark all the same. Milutin drilled his children in mental exercises that sound, to modern ears, almost like eccentric parlor games: he would make young Nikola guess his thoughts, solve mental arithmetic, repeat long passages back from memory, find the flaws in a piece of reasoning. Tesla later credited these drills with sharpening the extraordinary machinery of his mind. Whether or not that is the whole explanation, it is a lovely corrective to a common assumption — that a genius simply arrives fully formed. Tesla's mental discipline was, at least in part, trained into him at the dinner table.

And then there was his mother, who may have mattered even more. Đuka Tesla — born Georgina Mandić, the daughter and granddaughter of Orthodox clergy — could neither read nor write. In the standard telling of great inventors' lives this is the sort of detail meant to signal a humble, unremarkable origin. In Tesla's case it signals the opposite. Đuka was, by every account, a natural engineer. She built her own household tools and mechanical appliances — looms, churns, a mechanical eggbeater — and she could tie three knots with the fingers of one hand. She had committed thousands of lines of Serbian epic poetry to memory and could recite them by the hour, which is its own kind of intellectual feat and which passed directly to her son, who inherited a memory that seemed to have no floor. Here is a surprising fact worth pausing on: the man who would design the alternating-current system that powers the modern world learned his first lessons in invention from an illiterate mountain woman who made machines with her hands because she had no others to buy. Tesla said so himself, and said it with pride. "My mother," he wrote, "was an inventor of the first order." He believed his own gift descended from hers.

That inherited knack showed itself early, and often comically. As a small boy in Smiljan, Tesla was forever building things and forever nearly killing himself in the process. He constructed a tiny bladeless waterwheel, smooth as a disc, and set it spinning in a mountain brook — a childish toy that, he later insisted, planted the seed of a lifelong obsession. For around the same age he saw an engraving of Niagara Falls and announced to a startled uncle that one day he would go to America and capture the power of that colossal cataract in a great wheel. It is one of those boyhood boasts that biography usually files under charming exaggeration, except that Tesla did precisely that: some forty years later, his alternating-current machinery would harness Niagara, exactly as the child had promised. Another of his early engines ended less gloriously. Having captured sixteen May bugs — the fat, droning beetles of the Balkan summer — the boy glued them to a light wheel mounted on a spindle, reasoning that their frantic beating wings would drive it around. To his delight the contraption actually turned. The experiment ended only when a neighbor's boy, who had a taste for the insects, began cheerfully eating them one by one straight off the wheel, at which point the horrified young inventor abandoned bug-powered engineering forever.

The Tesla household held five children, and it is impossible to understand Nikola without understanding the shadow cast by the eldest, his brother Dane. Dane was, in the family's estimation, the truly brilliant one — quick, gifted, the golden son on whom the parents' hopes rested. When Nikola was about five years old, Dane died. The circumstances are murky and were told different ways over the years; the version Tesla carried was that Dane was thrown from a horse and died of his injuries, and that Nikola, present at the scene, felt himself somehow responsible. Whatever the exact truth, the death fell on the small boy like a permanent weather system. He grew up in the house of a dead prodigy, measured against a brother who could no longer disappoint anyone, whose gifts glowed brighter in memory than any living child's could. It is not hard to read Tesla's lifelong hunger to astonish — the compulsion to produce, again and again, some feat no one had ever seen — as an argument he never stopped having with a ghost. This is speculation, and should be labeled as such; the death itself is fact, and its weight on him is unmistakable in his own writings.

But the strangest and most consequential fact about the boy is the one hardest to convey to anyone who has not experienced it, because it sits so far outside ordinary experience. From early childhood, Tesla was tormented by intense involuntary visions. When a word was spoken to him, an image of the thing it named would appear before his eyes — not a memory, not imagination as most of us know it, but a vivid, luminous apparition that seemed to hang in the air, so real that he sometimes reached out to see whether his hand would pass through it. A word like "book" or "horse" or "sister" would summon its object in blazing detail. Flashes of light would fill the air around him, obscuring real objects and real people. He described being unable to tell, for a moment, whether what he saw was in the room or in his skull.

To understand how disorienting this was, borrow two everyday comparisons. Most people can, if asked, picture an apple — a faint, private, controllable image behind the eyes, easily dismissed. Now imagine that image were not faint and not private and not, above all, controllable: that it flared up unbidden, bright as a lamp, every time anyone said "apple," and would not go away when you wanted it gone. Or think of the way a catchy song can lodge in your head and play itself against your will for hours. Tesla's affliction was that, but visual, and constant, and triggered by ordinary speech. Modern researchers have reached back for names to describe it — a form of hyper-vivid mental imagery, sometimes discussed alongside synesthesia and eidetic or so-called photographic memory — but no tidy diagnosis captures it, and we should resist pretending otherwise. What the phenomenon was, precisely, we cannot say. That he suffered it, and that it shaped everything, we can.

Because here is the turn that makes the story extraordinary rather than merely sad. The same faculty that tormented the child became, in the man, the single most powerful tool an engineer has ever possessed. Tesla learned to summon and control the visions rather than merely endure them, and once he could control them he discovered he could build machines inside his own mind. He could conceive of a device, construct it in his imagination down to the last screw, set it running in his head, and let it operate for weeks. Then he would inspect the imagined machine for wear — checking, in his mind's eye, which parts had heated up or worn thin — and revise the design before a single piece of metal had been cut. "It is absolutely immaterial to me," he wrote, "whether I run my turbine in thought or test it in my shop. I even note if it is out of balance." He claimed, and there is good reason to believe him, that he could develop an invention completely in imagination and have it work correctly on the first physical trial. This is a genuinely different way of inventing from Edison's, and the contrast between the two methods is one of the deepest currents running through the rest of this book. Edison built and rebuilt and tested by exhausting brute trial; Tesla ran the trials in a theater behind his eyes.

The self-discipline this required was total, and it bordered on the pathological. As a young man Tesla drove himself with a monastic severity — reading through entire libraries, working eighteen-hour days, forcing his prodigious memory to swallow whole books. In his youth he nearly killed himself with overwork more than once; a bout of cholera in his teens laid him near death for months, and he used even that as leverage, extracting from his reluctant father a promise that if he recovered he could study engineering rather than enter the priesthood. He recovered. The bargain held. The boy who saw visions of horses and books when the words were spoken would go to study the invisible force that neither casts a shadow nor makes a sound.

It is worth correcting one more misconception here, gently, because it colors everything that comes later. The Tesla of internet legend is a kind of serene wizard, a being of pure intellect untroubled by the grubby human world. The real young Tesla was a nervous, driven, physically fragile boy, prey to phobias and compulsions that would deepen into the famous eccentricities of his old age — the horror of pearls and of human hair, the counting of steps, the obsession with the number three. His genius did not float free of his afflictions. It grew directly out of them, twisted together with them at the root, the gift and the torment two faces of the same overloaded nervous system. To romanticize the one while ignoring the other is to misunderstand the man completely.

His schooling only sharpened the picture of a mind that ran hot and strange. At the higher Real Gymnasium in Karlovac, Tesla devoured mathematics with such ease that his teachers suspected him of cheating; he could perform lightning calculations in his head, arriving at answers so quickly that the working seemed to be missing, because for him much of it was — the figures assembled themselves behind his eyes without visible effort. He read compulsively and mastered languages the way other people collect habits, eventually speaking or reading something like eight of them, from his native Serbian through German, French, and Italian to a bookish English he polished by memorizing whole passages of literature. Here is a final surprising fact to carry forward: the man who would spend his life among dynamos and transformers thought of himself, at least as much, as a poet and a philosopher. He wrote verse, he could recite Goethe and the Serbian epics for hours, and he approached engineering as a kind of art, a search for elegance and completeness rather than mere function. That aesthetic sense — the refusal to accept an ugly, sparking, wasteful machine when a beautiful one might exist — was not incidental to his genius. It was the engine of it.

You meet Tesla's peculiar way of thinking every time a modern engineer runs a design as a computer simulation — testing a bridge, an engine, a microchip in a virtual world before committing a cent to steel or silicon. We call it modeling, and we treat it as an obvious labor-saving idea. Tesla did it a century before the machines existed to do it for him, using only the fevered visual apparatus he had learned to master since childhood. He was, in a sense, a walking simulation engine, and the whole architecture of modern alternating current was assembled first inside that instrument.

The boy would soon leave the mountains for the wider empire, carrying his visions with him into the lecture halls where, for the first time, someone would tell him that the very machine he dreamed of building was impossible.

Chapter 28: Graz, Prague, and a Vision in a Park

In a lecture hall at the Imperial-Royal Technical College in Graz, in the Austrian province of Styria, sometime in 1878, a professor of physics named Jakob Pöschl was demonstrating a marvel of the age. On the bench before him stood a Gramme dynamo — a French-built electrical machine, gleaming with brass and copper, that could be run two ways. Feed it mechanical motion by turning its shaft, and it generated a direct electric current; feed it current instead, and it ran the trick backward, spinning as a motor. It was one of the finest electrical devices in the world, and at its heart sat a component that would become the villain of this chapter: the commutator. As Pöschl ran the machine, the commutator sparked and sputtered, throwing off tiny blue flashes and a smell of ozone, the brushes chattering against the spinning contacts. Among the students watching was a thin, intense young man of twenty-two, and something about that sparking made him deeply, physically uneasy.

To follow what happened next you need to understand the commutator, because Tesla's whole future turns on his hatred of it. A direct-current motor has a problem built into its nature. To keep the motor turning, the current running through its spinning coils has to reverse direction at exactly the right instant on every rotation — otherwise the magnetic forces that push the shaft around would, half a turn later, push it back the other way, and the motor would simply lock and shudder in place. The commutator is the mechanical trick that solves this: a split ring of copper segments on the motor's shaft, against which press a pair of carbon or copper brushes, flipping the current's direction at each half-turn like a person frantically swapping two battery cables back and forth many times a second. It works. But it works the way a clattering, sparking, self-destroying contraption works. The brushes wear down. The contacts pit and arc. The sparks waste energy and, in the presence of flammable dust or gas, threaten fire. To Tesla, watching the Gramme machine spit and flash, the commutator was not a clever solution. It was an ugly wound in an otherwise beautiful machine.

So he said so. He suggested to Pöschl that surely one ought to be able to build a motor with no commutator at all — a motor that ran on alternating current directly, without the whole sparking apparatus of brushes and split rings. Alternating current, or AC, is current that does not flow steadily in one direction the way a river runs downhill, but instead sloshes back and forth, reversing itself many times a second, like a tide rushing in and out of a harbor. It seemed to Tesla that a current which already reversed itself ought to be able to run a motor without any need for a commutator to do the reversing mechanically. Pöschl, a competent and conventional physicist, dismissed the idea in front of the whole class. He devoted an entire lecture to explaining why the young man was wrong. "Mr. Tesla may accomplish great things," he concluded, "but he certainly never will do this. It would be equivalent to converting a steadily pulling force, like that of gravity, into a rotary effort. It is a perpetual-motion scheme, an impossible idea." The professor was, in effect, telling him that the machine he could already half-see was a violation of physics.

And here is the first surprising fact of this chapter: the professor was not being stupid. By the physics as it was commonly understood in 1878, the objection had real force. If you simply feed alternating current into an ordinary motor coil, the coil is yanked one way, then immediately yanked back the other, and the net result is a violent buzzing rather than smooth rotation. The magnetic pull has no consistent direction to turn toward. What Tesla intuited, and could not yet articulate, was that the answer lay not in a single reversing pull but in making the magnetism itself rotate — in creating a field that swept around in a circle, so that the motor's rotor would chase it endlessly, the way a dog chases a lure it can never quite reach. He did not have that answer in Graz. He had only the certainty, immovable and slightly mad, that his professor was wrong. For a young student to be told by his teacher that his one great idea is a species of perpetual-motion fantasy, and to walk out of the room still believing in it, requires a kind of arrogance that is indistinguishable, at this distance, from faith.

The obsession took hold and would not let go. For the next several years, as his life scattered and darkened, the problem of the sparkless motor ran continuously in the background of Tesla's mind. His time at Graz ended badly — he had begun as a brilliant, prize-winning student and unraveled into a compulsive gambler, losing his allowance and his standing, eventually breaking with the school and, for a period, with his family. His father's death in 1879 deepened the disorder. He drifted, taught for a while, and eventually made his way to Prague, where in 1880 he attended lectures at the university, though he never took a degree there. He was, by any external measure, a young man in trouble: gifted, unmoored, chasing a machine that a professor had certified as impossible.

There is a revealing anecdote from the depth of the Graz gambling years. Tesla had gambled away not only his own allowance but money the family could ill spare, and his mother, rather than berating him, is said to have pressed a roll of bills into his hand and told him to go and enjoy himself — that the sooner he lost everything they owned, the better, since she was confident he would soon get over the vice on his own. The reverse psychology, if that is what it was, worked. Tesla later claimed he conquered the gambling passion by an act of will so complete that he never again felt its pull, tearing the habit out of himself the way he tore out every other appetite he judged beneath him. Whether the cure was truly that clean we cannot know — self-mastery was central to the legend Tesla built of himself, and he was not above improving a moral tale in the retelling. But the pattern is consistent with everything else about him: the conviction that the mind, properly disciplined, could simply command the body and the will, and that both would obey. It is worth holding this image of him — the young man who believed he could conquer any weakness by sheer force of concentration — because that same ferocious inwardness is what would soon let him build a working motor entirely inside his skull.

Then, in early 1882, in Budapest, the machine came to him. Tesla had taken a job with a telephone exchange in the Hungarian capital, and the strain of the period — poverty, overwork, the endless subterranean churning of the motor problem — had brought him to the edge of a complete nervous collapse. He described symptoms so extreme they read like the notes of a physician documenting a rare disorder: his senses became grotesquely, unbearably acute. He could hear a watch ticking three rooms away. A passing carriage sent shudders through his whole body. The light of the sun felt like blows. He could feel the ground trembling under distant footsteps. A doctor, baffled, declared his condition beyond the reach of medicine. Whatever was actually happening inside his overtaxed nervous system — and we should not pretend to diagnose it across a century and a half — Tesla emerged from it slowly, aided by a devoted friend, a former athlete named Anthony Szigeti who took him out walking to rebuild his strength.

It was on one of those walks, on a late afternoon in February 1882, that the most famous single moment in Tesla's life occurred. He and Szigeti were strolling through the City Park in Budapest as the sun went down, and Tesla, whose head was as full of poetry as of machinery, was reciting aloud — from memory, of course — a passage from Goethe's *Faust*. It was the sunset speech, in which Faust watches the sun sinking and longs for wings to follow it, to chase the dying light around the turning world. "The glow retreats, done is the day of toil," Tesla recited, "it yonder hastes, new fields of life exploring." And as he spoke the lines about the sun's westward flight, watching the actual sun redden and fall in front of him, the answer to the impossible motor arrived all at once, complete.

He saw it. Not a hint of it, not the beginning of it, but the whole thing — a motor with no commutator, in which two or more alternating currents, timed to peak at different moments, would produce a magnetic field that did not merely pulse but *rotated*, sweeping steadily around a circle, dragging the rotor along behind it in smooth, continuous motion. He saw the fields turning in his mind with the same blazing clarity that had tormented him as a child, and he understood in that instant that the vision was correct. So overcome was he that he seized a stick and drew a diagram in the dirt of the path, sketching the crossed currents and the whirling field for the bewildered Szigeti, trying to make his friend see what he could see. "Watch me reverse it," he is supposed to have said, showing how the imagined machine could be made to turn the other way merely by changing the currents. The diagram he drew in the dust that evening was, in essence, the diagram of the alternating-current world we still live in.

What is worth dwelling on is the sheer strangeness of how the insight arrived — not as the last step of a chain of reasoning, but all at once, whole, in a flash, at the least "scientific" moment imaginable. Tesla had spent years attacking the motor problem head-on, by direct assault, and had gotten nowhere. The answer came only when he had stopped consciously working on it, when his mind was full of poetry and sunset rather than currents and coils. This is a pattern psychologists of creativity would later study exhaustively: the way solutions to problems long chewed over tend to surface during rest, during a walk, in the drifting state between concentration and reverie. The conscious mind lays the groundwork through years of effort; the answer then bubbles up unbidden when the pressure is released. Tesla's Budapest vision is one of the most vivid examples in all of science of this quiet, involuntary process — the reason so many discoveries are made in bathtubs and on strolls rather than at desks. His years of frustrated labor were not wasted; they were exactly what made the sudden flash possible. The park merely supplied the moment of release.

We should be honest that this scene, too, comes to us mainly through Tesla's own much later memoir, written decades afterward with all the shaping instincts of a man who knew he was composing his own legend. The theatrical perfection of it — the poetry, the sunset, the stick in the sand — invites suspicion. But unlike the midnight lightning of his birth, the substance of this claim is corroborated by the technology itself. Whatever the exact texture of that February afternoon, the rotating magnetic field is a real and coherent idea, it is the correct solution to the problem Pöschl called impossible, and Tesla demonstrably possessed it years before he built a working motor or filed a patent. The insight was genuine even if the staging was burnished. That is a useful distinction to carry through the rest of Tesla's story, where the line between documented achievement and self-authored myth runs through nearly every chapter.

You meet the fruit of that Budapest sunset every single day, in more places than you could count. The rotating magnetic field Tesla saw in the park is the operating principle of the induction motor, and induction motors are the quiet, brushless, nearly maintenance-free workhorses inside your washing machine, your refrigerator compressor, your air conditioner, the pumps that move water through your city, the fans in ventilation shafts, the drills in factories, and the drive motors of most electric trains. They run for decades without the sparking commutator Tesla so despised, because he found the way to make the magnetism itself do the turning. The machine the professor swore could never exist is now so ordinary that it hums, unremarked, in the walls of nearly every building on earth.

But an idea drawn in the dust of a Budapest park is not yet an invention, and Tesla had neither the money, the materials, nor the freedom to build it. To do that he would first have to go to work for the one company in Europe carrying the name of the man who would become his great antagonist — the Continental Edison Company in Paris.

Chapter 29: Paris and the Continental Edison

In the spring of 1882, a young engineer stepped off the train into the noise and glare of Paris carrying a head full of a machine no one else could see. Nikola Tesla had come to the French capital to work for the Continental Edison Company, the European arm of the growing empire of Thomas Edison, and for a provincial from the mountains of the Austrian military frontier the city was an assault of pleasures and shocks. He was twenty-six, fluent in half a dozen languages, physically restored from his Budapest collapse, and — for almost the first time in his life — earning real money. He promptly threw much of it away. In his memoirs he describes a life of dawn swims in the Seine, laps of the city, work all day at the Edison works in the suburb of Ivry, and evenings squandered at billiards, at which he was formidable. Underneath the pleasant surface, the rotating field he had glimpsed in the Budapest park kept turning in the back of his mind, waiting for a chance to be built in metal instead of imagination.

His duties at Continental Edison were varied and demanding, and they turned him, almost by accident, into a specialist in failure. When an Edison lighting plant somewhere in Europe went wrong — and in those pioneering years the plants frequently went wrong — the company needed someone who could travel to the site, diagnose the fault, and set it right, often under the anxious eyes of a customer who had paid a great deal of money for light that would not stay lit. Tesla proved unusually good at this thankless work. He possessed a rare combination of deep theoretical understanding and genuine mechanical skill, so that he could reason his way to the hidden cause of a breakdown and then, sleeves rolled up, actually repair it with his own hands. Most engineers of the day were strong in one of these and weak in the other; Tesla was formidable in both. It was this reputation as a troubleshooter, more than any single invention, that first marked him out within the Edison organization as a man worth watching.

The Continental Edison Company existed to spread the Edison system across Europe — the incandescent bulbs, the dynamos, the whole integrated apparatus of direct-current lighting that had been perfected at Menlo Park and was now being sold city by city. Tesla was hired as an engineer to install and repair these systems, and it was here, working inside the Edison organization itself, that he first came into the orbit of the men who would carry him to America. Chief among them was Charles Batchelor, an English-born mechanic and one of Edison's closest and most trusted lieutenants — the man who had stood beside Edison through the invention of the phonograph and the light bulb and who now managed Edison's European interests. Batchelor was no mere administrator; he was a first-rate engineer with a discerning eye for ability, and he noticed the intense young Serb almost at once. That recognition would change Tesla's life. But before it could, Tesla had to prove himself in the field, and the field, in the winter of 1883, meant the German city of Strasbourg.

The Strasbourg job was a mess, and it is worth dwelling on because it shows the ordinary, unglamorous engineering labor out of which Tesla's legend grew. Continental Edison had installed a lighting plant at the new railway station in

Strasbourg — the region had recently passed to Germany after the Franco-Prussian War, which made every technical failure a diplomatic embarrassment as well. At the opening ceremony, with dignitaries present and, by some accounts, the German Emperor Wilhelm I himself in attendance, something went badly wrong: a short circuit in the wiring produced a violent flash and blew out a section of a wall. The whole installation was faulty, the German authorities were furious and refused to accept the plant, and the company needed someone to go and quietly fix the disaster without further scandal. They sent Tesla.

He spent much of 1883 in Strasbourg untangling the botched wiring, soothing offended officials, and rebuilding the plant so it would actually work. It was tedious, thankless troubleshooting — the electrical equivalent of being sent to a foreign city to repair someone else's plumbing — but it gave Tesla something priceless: unsupervised time, a workshop, and materials. In a mechanical shop across from the railway station, in his hours away from the repair work, he finally built the thing he had been carrying in his mind for years. Using the parts he could scrounge, he assembled the first crude physical model of his induction motor — a motor with no commutator, no brushes, no sparking contacts of any kind, driven by the rotating magnetic field he had seen in the Budapest sunset. And it ran.

Try to feel the weight of that moment. For five years Tesla had held an idea that a university professor had publicly declared impossible, an idea he could demonstrate only inside his own skull. Now, in a rented shop in a contested border city, he watched a disc of metal begin to spin, smoothly and silently, pulled around by nothing but invisible fields sweeping through the air — no wires touching it, no commutator flashing, no brushes chattering. The vision had crossed over into the world. He was, by his own account, overcome. He tried to interest a local acquaintance, a former mayor named Bauzin, in financing further development, but nothing came of it. The provinces of eastern France were not where a revolution in electrical power was going to be funded, and Tesla knew it.

It is worth pausing on how ordinary the setting of that first triumph was, because it says something true about how invention actually happens. There was no grand laboratory, no team of assistants, no patron leaning over his shoulder. There was a borrowed corner of a mechanical shop, a heap of scavenged parts, and a young man stealing hours from a repair contract to build a device that no one had asked for and no one yet wanted. The machine Tesla assembled in Strasbourg was crude — little more than a proof that the principle worked, a piece of metal that would rotate when the staggered currents were fed to it in the proper rhythm. But it was, so far as we know, the first time in human history that a motor had been driven purely by a rotating magnetic field, and it happened almost in secret, in the off-hours of a man who had been sent to fix somebody else's broken lights. The revolutions in technology rarely announce themselves. This one first turned in a rented workshop while the official business of the day was patching up a railway station.

He returned to Paris in 1883 with two things: a working proof of his motor, and the promise of a reward. This is the first appearance in Tesla's life of a pattern that would repeat itself with cruel regularity, so it deserves careful attention. Before he left for Strasbourg, Tesla had been promised a substantial bonus for successfully salvaging the installation — the figure usually given is twenty-five thousand francs, a genuine fortune for a young engineer, enough to change his circumstances entirely. When he came back, having done exactly what was asked, he went to collect. And he was passed from one manager to another in a small circle of polite evasion. Each administrator told him the bonus was authorized by someone else; each someone else referred him onward; the money receded before him like the horizon. He never received it. Tesla, proud and easily wounded, concluded that the promise had never been sincere, and the episode left a lasting bruise.

It is worth flagging here, because it bears directly on a more famous grievance still to come, that we know of this Paris bonus almost entirely from Tesla's own later account. There is no independent ledger confirming that twenty-five thousand francs was ever formally promised. What we can say is that the *shape* of the story — a diligent young engineer stiffed on a bonus by a chain of managers who each pass the buck — is entirely plausible for a large corporation of the era, and that it establishes a template. When Tesla later told an almost identical story about being denied a much larger sum in America, the Paris episode was already there in his memory as the pattern the American betrayal supposedly followed. A careful reader should hold both possibilities in mind: that Tesla was genuinely cheated more than once, and that a man cheated once may come to see the same betrayal everywhere.

What the Paris disappointment did, unambiguously, was help make up his mind to leave Europe. Batchelor had been watching him, and Batchelor understood two things: that Tesla was an engineer of rare quality, and that the young man's talents were being wasted patching up other people's installations in provincial railway stations. There was a bigger arena. Edison's operation in New York — the Edison Machine Works, the beating heart of the whole enterprise, where the dynamos themselves were designed and built — could use a mind like Tesla's. Batchelor urged him to go to America, and offered to smooth the way with a letter of introduction to the great man himself. Here a small but persistent misconception is worth correcting. The popular image is of Tesla arriving in New York as an anonymous stranger who talked his way in off the street. In fact he arrived as a known quantity, a trusted transfer within the Edison organization, carrying a personal recommendation from one of Edison's most valued associates. He was, in effect, being promoted from the European branch to headquarters.

There is one more surprising fact about the Paris years that undercuts the tidiest version of the Tesla legend, and it concerns the motor itself. Tesla is often portrayed as a lone visionary who conceived alternating-current power in total isolation, a prophet without antecedents. But Paris in the early 1880s was precisely the place where the possibilities of alternating current were being seriously explored by others. The transformer — the device that would make long-distance AC transmission practical — was being developed in these same years by the Frenchman Lucien Gaulard and his English partner John Dixon Gibbs, whose work Tesla could hardly have been unaware of in the electrical circles of the city. Tesla's specific contribution, the rotating-field motor, was genuinely his own and genuinely original; nobody else had it. But he was working within a broader European current of AC experimentation, not inventing the entire idea of alternating power from a clear blue sky. The legend of the solitary genius flattens a richer truth, in which Tesla's unique insight fit into, and completed, a larger effort that many hands were pushing forward at once.

The decision to cross the Atlantic was not made lightly, for it meant abandoning a great deal. Tesla was, in Europe, a rising engineer with a secure position, a network of respected colleagues, and the comforts of the cultured cities he loved. America was a rumor of opportunity and little else — a raw, distant country he had never seen, where he knew almost no one and where, by reputation, the manners were coarse and the streets were mud. But the logic was inescapable. His motor needed capital and industrial muscle to become real, and the largest concentration of electrical enterprise in the world was gathering around Edison in New York. If the future of electricity was being built anywhere on earth, it was being built there. And so Tesla resolved to follow the current back to its source, trading the settled disappointments of the Old World for the unknown promise of the New. Batchelor's letter would be his passport; the motor in his head would be his fortune, if only he could find someone willing to build it.

You meet the legacy of these unglamorous troubleshooting

years every time a field engineer is dispatched to a customer's site to fix an installation that looked perfect on paper and failed in reality. The work Tesla did in Strasbourg — diagnosing a fault in someone else's system, working under political pressure, salvaging a botched job while inventing on the side — is the daily texture of engineering as it is actually practiced, then and now, far from the tidy heroism of the textbook. The rotating-field motor was born not in a laboratory of pure thought but in the margins of a repair contract, built by a man who was supposed to be fixing something else.

Late in the spring of 1884, Nikola Tesla boarded a ship for the United States, carrying almost nothing of value except a satchel of poems and calculations, a head full of a motor the American Edison had never heard of, and a folded letter of introduction from Charles Batchelor to Thomas Edison — a letter whose contents, when Tesla finally presented it, would become one of the most quoted sentences in the entire history of invention.

Chapter 30: Arrival in America

On a June morning in 1884, a tall, gaunt young man in a threadbare suit walked down the gangplank of a steamer into the chaos of the New York waterfront and stood, for a moment, taking the measure of his new country. Nikola Tesla was twenty-eight years old. By his own famous accounting, he had in his pockets four cents in coin, a book of poems he had written, a few technical articles, some drawings for a flying machine, the calculations for his impossible motor, and a letter of introduction addressed to Thomas Edison. Everything else — his luggage, his money, most of his possessions — had been lost or stolen along the way, in a journey that had already thrown nearly every misfortune at him that a transatlantic passage could hold.

The trip had been a catalogue of small disasters, and Tesla told the story with a rueful relish for the rest of his life. He had nearly missed the boat train in Paris; his money and tickets had gone astray; he had talked his way onto the ship at the last moment. During the ocean crossing there had been a mutiny among some of the passengers or crew, and Tesla, characteristically, had waded into the middle of the fracas and gotten roughed up for his trouble. He landed at last in New York with, as he put it, his fortune reduced to those four cents and the contents of his head. It is one of the great arrival stories in American history, the immigrant genius stepping onto the docks with nothing but his mind — and precisely because it is so perfect, we should handle the details with some care.

The "four cents" is Tesla's own figure, from his own memoir, and it may be exactly true or it may be the rounding-down of an old man polishing a good tale. The essential situation, however, is not in doubt: Tesla arrived poor, disoriented, and dependent, a stranger in an enormous foreign city. What separated him from the millions of other penniless immigrants pouring through New York that decade was the folded letter he carried — and here we come to the single most quoted sentence in the story of these three men. The letter of introduction had been written by Charles Batchelor, Edison's trusted European manager, and according to the tradition it contained a line that Batchelor addressed directly to his chief. "I know two great men," it read, "one is you; the other is this young man."

That sentence has been repeated so often, in so many books and articles and documentaries, that it has taken on the solidity of scripture. It deserves both its fame and a note of caution. If Batchelor truly wrote it, it is a remarkable thing for one of Edison's closest associates to have said — an assertion, to Edison's face, that this unknown immigrant stood on the same plane as the most celebrated inventor in the world. It would mean that at least one person who knew both men recognized, from the very start, what Tesla was. But like so much of the Tesla record, the line survives largely through later retellings rather than the original document, and a wary reader should file it under "cherished tradition, probably rooted in something real" rather than "verified quotation." What is not in doubt is that Batchelor thought highly enough of Tesla to send him to Edison at all, and that Tesla walked into the Edison organization with a genuine endorsement from a genuine insider.

There is a small, human detail from the crossing that Tesla liked to tell, and it is worth repeating because it shows the mixture of vulnerability and stubborn dignity that defined him. Somewhere in the confusion of his departure, his money and much of his baggage were gone, and by the time the ship neared New York he had almost nothing left. Yet he had used the long days at sea, characteristically, not to despair but to work — reviewing calculations, brooding over his motor, even, by some accounts, helping the crew with mechanical repairs. He stepped off the boat exhausted and robbed, and rather than collapsing he went almost immediately to work, so eager was he to make himself useful. That combination — a man who arrived with four cents and threw himself into the nearest problem within hours of landing — tells you more about Tesla than any figure in his pockets. Money slid through his hands his whole life. Work never did.

New York in 1884 was itself a kind of shock to the system, and Tesla's first impressions of America were mixed with a foreigner's clear-eyed dismay. He had come from Paris, from Budapest, from the old European cities with their long boulevards and their sense of accumulated civilization, and the raw, roaring, half-finished quality of Manhattan appalled the fastidious side of his nature even as its energy thrilled the engineer in him. In a line he liked to repeat, he said that America, compared to the Europe he knew, was "a century behind Europe in civilization" — the streets filthier, the manners coarser, the whole place cruder than he had expected. And yet it was also, unmistakably, the future. The city crackled with construction and ambition and machinery. It was building itself upward and outward at a pace no European capital could match, and it was, above all, the one place on earth where a man with a revolutionary electrical idea might actually find the capital and the appetite to build it. Tesla's ambivalence — repelled by America's crudeness, seduced by its possibility — is the ambivalence of a great many immigrants before and since, and it never entirely left him.

Within a day or two of landing, Tesla presented himself and

his letter at the Edison operation in lower Manhattan and met the man himself. The encounter between the two future antagonists is one of the pivotal scenes in this book, and it is worth setting down plainly, because the contrast could hardly have been sharper. Edison, then thirty-seven, was at the height of his fame — the Wizard of Menlo Park, the conqueror of the incandescent lamp, a stocky, rumpled, half-deaf force of nature who chewed tobacco, worked around the clock, and cared nothing for appearances. He spoke in the blunt, practical, faintly mocking idiom of the American self-made man. Tesla, by contrast, was tall and immaculate even in his poverty, courtly in manner, fluent in the classics, given to quoting poetry in several languages, a man who bowed and observed the formalities of the Old World. The empiricist and the visionary, the tinkerer and the theorist, the man who trusted his hands and the man who trusted his mind — they stood facing each other in a New York machine shop, and neither, in that first moment, can have quite understood what he was looking at.

There is a well-worn anecdote about this first meeting that captures the gulf between the two men, though like most Tesla anecdotes it should be held at arm's length. Tesla, dressed with his usual immaculate care, is said to have arrived and been struck at once by Edison's utter indifference to appearances — the tobacco stains, the disheveled clothes, the blunt manner. Edison, for his part, reportedly sized up the courtly young European with amused skepticism, unimpressed by talk of grand theories and more interested in whether the man could actually fix a machine. Whatever passed between them in words, the deeper meeting was a collision of two civilizations: the Old World gentleman-scholar who reasoned from principle and the New World self-made tinkerer who trusted only results. Neither would ever fully respect the other's way of working, and the sparks of that incompatibility, struck at their very first encounter, would eventually set the whole electrical industry alight.

Tesla, naturally, tried immediately to interest Edison in his great idea: the alternating-current motor, the rotating field, the whole vision he had carried across the ocean. And Edison, just as naturally, was not interested. This is the crucial fact that shaped everything that followed, and it was not stupidity or blindness on Edison's part. Edison had staked his entire fortune, his reputation, and years of relentless work on direct current. His Pearl Street station, his patents, his dynamos, his whole integrated system ran on DC. He had every commercial and psychological reason in the world to regard alternating current not as a promising alternative but as a threat and a distraction — and, being the man he was, he said so bluntly. Alternating current, in Edison's settled view, was dangerous, impractical, and a dead end. He wanted nothing to do with the young man's motor. What he wanted was a good engineer to help him improve the machines he already had.

And so, instead of a partnership between the two greatest electrical minds of the age — the collaboration that, in a kinder version of history, might have spared everyone the war that was coming — Tesla was simply hired. He went to work for Edison as an employee, one skilled hand among many at the Edison Machine Works, put to the practical labor of fixing and improving direct-current dynamos while the alternating-current revolution folded up quietly inside his head for a while longer. It is one of the great near-misses of technological history: the man with the answer to Edison's deepest technical limitation stood in Edison's own workshop, offered it to him, and was told to go make the old system run better instead.

To grasp what Tesla had walked into, picture the state of American electricity in 1884. It was barely two years since Edison had switched on the Pearl Street station and lit a few square blocks of lower Manhattan with direct current — the first central power station in the country, and still very nearly the only one of its kind. Electric light was a marvel confined to a handful of wealthy districts and showpiece buildings; most of America still burned gas, oil, and candles after dark. The whole industry was an infant, and Edison was its undisputed father, his name practically synonymous with electric light itself. For an ambitious young engineer, there was no more thrilling place on earth to be, and no more powerful patron to serve. That is what makes the coming rupture so poignant. Tesla had reached the very center of the electrical world; he stood beside the one man with the resources and the reputation to build his motor; and he found the door to the future politely, firmly closed against the very idea that would open it.

There is a surprising fact worth pausing on here, one that cuts against the popular image of Tesla as a man who could never function in the workaday world. In these first American months he was, by all accounts, an outstanding employee — diligent, tireless, ingenious, exactly the kind of problem-solver Edison prized. He worked the long hours that were the culture of the place, sometimes around the clock, and he quickly earned the respect of the hard-bitten machinists around him. The friction between Tesla and Edison, when it finally came, was not the friction of a dreamer who couldn't hold a job. It was the deeper, more tragic friction of two men whose entire ways of thinking about the world were incompatible, thrown together over a technology on which they fundamentally disagreed.

You meet the shape of this story constantly in the modern world, wherever a brilliant newcomer arrives at an established company carrying an idea that threatens the very product the company has bet its future on. The employer, invested in what already works, cannot afford to see the disruptive thing clearly, and the innovator, unable to get a hearing, eventually walks out the door to build it elsewhere — sometimes into the arms of a rival who will use it to overturn the very empire that turned it away. It is the oldest pattern in the history of innovation, and Tesla and Edison enacted it in its purest form in a Manhattan workshop in 1884, before either of them had any idea that they were rehearsing a war.

For now, though, the young immigrant had a job, a wage,

and a foothold in the New World, and he bent himself to the task Edison had set: making the Wizard's dynamos better. What he could not yet know was how quickly that work would sour, and how a dispute over money — real or imagined, generous or cruel — would send him back out onto the streets of New York within a year.

Chapter 31: Working for the Wizard

Late one evening in the winter of 1884, on the passenger steamer *Oregon* moored at a New York pier, two of the ship's dynamos had failed, and the finest ocean liner of its day sat dark and crippled in the water, unable to sail. Its electric lighting plant — one of the showpiece installations of the Edison empire — had broken down completely, and the ordinary repairmen could not fix it. Edison, embarrassed and pressed for a solution, turned to the tall young immigrant he had recently hired and sent him down to the docks. Tesla went aboard that night with a crew of assistants, found the dynamos' armatures short-circuited and their commutators badly damaged, and worked straight through until dawn, tearing the machines down and rebuilding them by lamplight. By daybreak both dynamos were running and the *Oregon* was cleared to sail.

Walking back through the sleeping streets at five in the morning, Tesla ran into Edison and a couple of his managers, who were only then heading home after their own night's work. Edison, seeing his new engineer out at that hour, made a joke about the young man being out on the town. Tesla told him he was returning from repairing the *Oregon*. Edison, according to Tesla's account, said nothing for a moment, then walked off — but Tesla overheard him remark to a companion, with grudging admiration, that this was a "damned good man." The anecdote, like most that survive from this period, comes to us through Tesla, and may have gained a little polish in the telling. But it captures something true and important: in the workshop, on the practical merits, Tesla was superb, and Edison knew it.

To understand what happened next, you have to understand the place where these two men were briefly yoked together. The Edison Machine Works was the industrial heart of Edison's operation — a grimy, thunderous factory where the dynamos that generated the current for the whole system were designed, built, and repaired. A dynamo, in the plain terms this book insists on, is simply a machine that turns motion into electricity: you spin coils of wire past magnets (or magnets past coils), and the moving magnetic field pushes a current through the wire, exactly as Faraday had discovered half a century before. Run a steam engine to spin the dynamo's shaft, and out the other end comes the electricity to light a city. The dynamo is the beating heart of any power system, and in 1884 Edison's dynamos, though they worked, were inefficient, wasteful of energy, and prone to the very sparking and overheating that Tesla had hated since his student days in Graz. They were, to his eye, crude.

So Edison set him to improving them, and here the famous incompatibility of the two men first showed itself in the daily texture of the work. Edison's method — the method that had built the light bulb and the phonograph and made him the most famous inventor alive — was relentless, systematic trial and error. He believed in trying everything. Faced with a problem, he would test one material, then another, then a thousand others, working through the possibilities by sheer exhaustive labor, trusting his hands and his experiments over any theory. "I have not failed," runs his most famous saying, "I've just found ten thousand ways that won't work." He was suspicious of pure mathematics and of men who reasoned from first principles; he trusted what he could build and measure on the bench. When a young mathematician once tried to calculate the volume of a light bulb with equations, Edison, the legend goes, simply filled it with water and poured the water into a measuring glass. That was Edison: the empiricist, the tinkerer, the man who beat nature into submission by trying every door until one opened.

A famous story captures Edison's whole cast of mind, and it is worth telling even though it may have been sharpened over the years. Edison, so the tale goes, once handed a newly hired mathematical physicist an empty glass bulb of an odd, curving shape and asked him to calculate its internal volume. The young man set to work with rulers and calculus, covering pages with integrals as he tried to model the bulb's complicated curves, and after some hours had made little progress. Edison, passing by, grew impatient, took the bulb, filled it with water, and poured the water into a graduated measuring cylinder. There, he said — that is its volume. The anecdote is usually told to make the mathematician look foolish, but it cuts both ways, and Tesla would have seen it exactly the other way around. To Edison, the water trick was the triumph of practical common sense over useless theory. To a mind like Tesla's, it was a confession — proof that Edison did not truly understand the shape he was measuring, and could answer a question only by brute physical trial rather than by grasping the underlying form. Both men were, in their fashion, right. The difference between them was the difference between the whole coming age of engineering.

Tesla worked in almost exactly the opposite way, and the opposition ran all the way down to the bone. Where Edison built and tested, Tesla thought and saw. He designed his machines complete in his imagination, ran them in the theater behind his eyes, and regarded Edison's endless bench-trials as a colossal, undisciplined waste of effort. Years later Tesla put the criticism into words that were both admiring and cutting. Edison's method, he wrote, was so inefficient that "an immense ground had to be covered to get anything at all unless blind chance intervened, and, at first, I was almost a sorry witness of such doings, knowing that a little theory and calculation would have saved him ninety per cent of the labor." If Edison needed a needle in a haystack, Tesla said, he would search straw by straw with the diligence of a bee, when a moment's calculation would have told him where to look. It was the complaint of the theorist against the tinkerer, and there was real truth in it — as there was real truth in the tinkerer's suspicion that elegant theories often shatter on contact with the messy physical world.

This was not merely a difference in technique. It was a difference in temperament so total that the two men could hardly have been designed as more perfect opposites if a novelist had invented them. Edison was earthy, profane, indifferent to food and dress and hygiene, working in shirtsleeves stained with machine oil, napping on laboratory benches, chewing tobacco and firing off crude jokes. He cared about money and results and had a showman's genius for publicity. Tesla was ascetic and fastidious to the point of obsession — impeccably dressed even when broke, revolted by dirt, formal in his manners, a man who dined alone at precise hours and could not bear to shake certain hands. Edison distrusted anything he could not touch; Tesla lived most fully in a world he could only see. Edison was a businessman who invented; Tesla was an artist who happened to work in copper and iron. Thrown together in the same clangorous workshop, they were like two craftsmen from different centuries forced to share a bench.

And yet — this is the surprising part, the part the legend of their enmity tends to erase — Tesla genuinely admired Edison, and would go on admiring him, in a complicated way, for the rest of his life. In the same essays where he mocked Edison's wasteful methods, Tesla wrote with real reverence about the older man's superhuman capacity for work, his courage, his refusal to be beaten. "His success," Tesla wrote, "was due to sheer force of will and boundless energy." The younger man never denied Edison's greatness;

he only insisted that it was a greatness of a fundamentally different, and to his mind lesser, order than the greatness of pure theoretical insight. Their story is usually told as pure hostility, hero against villain, and that is a distortion worth correcting. What lay between them at the Machine Works in 1884 was not yet hatred. It was mutual, wary respect straining against a total mismatch of methods and minds. It is one of history's small ironies that the two are remembered as bitter enemies, when in the workshop of that year what bound them was closer to a frustrated, grudging esteem — each recognizing in the other a magnitude of talent he could neither match nor abide.

During these months Tesla did real, documented work redesigning Edison's dynamos — improving their efficiency, reworking the design of the machines and the systems that regulated them. He later claimed to have designed some two dozen improved types of standard machines, replacing Edison's shorter magnets with more efficient designs, and to have devised automatic regulators along the way. It is difficult, at this distance, to verify the precise scope of every claim, and Tesla was not a man to understate his own contributions. But the core of it is solid: he was a highly productive engineer whose improvements made Edison's direct-current machinery meaningfully better, and he did the work fast and well. This matters, because it sets up the bitterness to come. Tesla did not walk away from Edison as a failure who couldn't cut it. He walked away as a man who believed — rightly or wrongly — that he had delivered enormous value and been cheated of his reward.

There is one more misconception worth dismantling before we reach the break. In the mythologized version, Edison is a small-minded thief who could not recognize genius when it stood in front of him, and Tesla a pure soul betrayed. The reality is more human and more interesting. Edison ran a hard-driving commercial enterprise in a ruthless industry, on the constant edge of financial danger, in the middle of a period when he was fighting to keep his DC empire solvent against gathering competition. He did not have the luxury, or the temperament, to nurture a subordinate's radical alternative to the very technology his business depended on. Tesla, for his part, was proud, thin-skinned, and prone — as the Paris bonus episode had already shown — to reading betrayal into ambiguous dealings. When these two combustible temperaments collided over money, each was primed to see the worst in the other.

It is worth noting, too, how physically brutal the work itself was, because the friction between the two men played out against a backdrop of genuine industrial hardship. The Edison Machine Works was hot, loud, and dangerous, and the hours were merciless. Tesla later recalled laboring from half past ten in the morning until five o'clock the following morning, day after day, a schedule that even Edison's hardened foremen found remarkable; one of them is said to have remarked that he had never seen a man work so hard. Tesla drove himself past the point of exhaustion partly out of ambition, partly out of the monastic self-discipline that governed his whole life, and partly, one suspects, because the deep dissatisfaction gnawing at him — the knowledge that he was refining a technology he believed doomed — could be quieted only by relentless labor. A man who cannot build the machine he dreams of will sometimes bury the frustration in the machines he is permitted to build. For the better part of a year, that is what Tesla did, and did brilliantly, all the while nursing the certainty that the real future lay in the very direction his employer forbade him to look.

You meet the Tesla-Edison clash of methods in nearly every workplace where two brilliant people cannot stand the way the other one works — the intuitive designer against the data-driven engineer, the theorist who wants to model the whole system before building anything against the maker who wants to build a rough version today and fix it tomorrow. Modern technology firms are full of exactly this tension, and the honest lesson of Tesla and Edison is that both methods work and both have limits. Edison's exhaustive trial-and-error gave the world the practical light bulb; Tesla's imaginative theorizing gave it the alternating-current motor. Neither man could have done the other's work, and their tragedy is that they could not combine their gifts instead of colliding them.

The collision, when it came, would center on a promise — a bonus, a sum of money, a supposed pledge that Edison, in Tesla's telling, made and then laughed off as a joke. It is the most famous falling-out in the history of invention, and it is also, as the next chapter will show, one of the most misremembered.

Chapter 32: The Break

Sometime in the first half of 1885, in the offices of the Edison Machine Works in Manhattan, Nikola Tesla walked in to collect what he believed he was owed and walked out having quit. The exact scene we can only reconstruct through his own account, given decades later, but in that account it is one of the sharpest and most wounding moments of his life. He had done what was asked. He had redesigned Edison's dynamos, improved their efficiency, solved the problems set before him. And when he came to claim the reward he believed had been promised, he was met not with money but with a joke — and the joke, more than the lost money, was what he could never forgive.

Here is the story as Tesla told it. Early in his employment, he says, one of the managers — in most versions Edison himself — offered him a bonus of fifty thousand dollars if he could redesign the company's inefficient dynamos and devise a set of improvements. Fifty thousand dollars in 1885 was an enormous sum, the equivalent of well over a million in today's money, a figure that could have made Tesla independent for life. He threw himself at the problem, worked punishing hours for the better part of a year, and delivered the improvements. Then he went to Edison and asked for his fifty thousand dollars. And Edison, according to Tesla, laughed at him and said: "Tesla, you don't understand our American humor." The promise, Edison supposedly explained, had been a joke; there had never been any real intention to pay such a sum. When Tesla, stunned and insulted, asked at least for a raise — from eighteen dollars a week to twenty-five — and was refused, he resigned on the spot.

It is a devastating little story, and it has done more than almost any other single anecdote to fix the popular image of Edison as a cheat and Tesla as his victim. So we owe it a careful, honest examination, because the truth is considerably murkier than the meme, and this book has promised to separate documented fact from cherished legend wherever the two diverge. The first thing to say is that the incident cannot be independently verified. It rests almost entirely on Tesla's own recollection, published long afterward, and Tesla told the tale in more than one form over the years, with details that shifted. No corroborating document — no letter, no ledger, no contemporaneous account — establishes that Edison ever offered fifty thousand dollars for the dynamo work. We have one man's memory of a promise made in private, recounted at a distance of decades by someone with a lifelong sense of grievance.

It is worth being precise about how anecdotes like this one harden into fact, because the process is instructive and it recurs throughout Tesla's story. A striking claim is made once, late in life, by a single person with an obvious emotional stake. It is repeated by an admiring biographer, then by another who copies the first, then by a popular magazine, then by a documentary, then by a thousand websites — and at each step the hedges and qualifications fall away, the "reportedly" and "according to Tesla" quietly vanish, and what began as one man's decades-old memory arrives in the present as flat, confident history. The "you don't understand our American humor" line has traveled exactly this road. By now it is quoted as though a stenographer had recorded it in the room. In truth it rests on the least reliable kind of evidence there is: uncorroborated reminiscence, produced long after the fact, by an interested party. That does not make it false. It makes it unproven, which is a different and more honest thing to say.

The second thing to say is that the sum itself strains belief, and this is the crux of the skeptical case. Fifty thousand dollars was a staggering amount to dangle before a junior engineer earning eighteen dollars a week — more than the whole enterprise could plausibly have paid out as a bonus to a single employee for improving existing machines. Some historians have suggested that if any figure was mentioned at all, it was likely far smaller, and that Tesla, in memory, may have inflated a modest promise into a princely one, or misunderstood a casual, exaggerated remark as a firm commitment. Edison and his managers were known for a rough, teasing shop humor, and it is entirely possible that a throwaway line — something like "improve these and there's a fortune in it for you" — was meant by the speaker as ordinary workplace hyperbole and taken by the literal-minded, proud young foreigner as a binding contract. In that reading, no one exactly lied, but two men with very different senses of humor and very different assumptions about money talked past each other with lasting consequences.

The third thing to say, and the most important, is that the specific dollar figure almost doesn't matter, because the break was overdetermined. Even if there had never been any bonus, promised or joked, Tesla and Edison were on an unavoidable collision course, and the reason had nothing to do with fifty thousand dollars. The reason was alternating current. Tesla had come to America carrying the idea for the AC induction motor, the rotating field, the whole system that he believed would supersede everything Edison had built. Edison had staked his fortune and his name on direct current and regarded AC as a dangerous rival to be crushed, not developed. There was simply no future for Tesla inside the Edison organization. He could redesign Edison's dynamos forever and never be allowed to build the one machine he had crossed an ocean to build. The disagreement was not personal or financial at its root; it was structural, technological, total. A visionary of alternating current could not remain an employee of the great empire of direct current. Sooner or later, over some pretext or other, he had to leave.

Consider what it would actually have meant for Edison to embrace Tesla's motor. It would have meant conceding that alternating current — the technology Edison would soon be publicly branding as deadly and impractical — was in fact the future, and that his own direct-current system, on which he had staked his fortune, his patents, his hundreds of employees, and his very identity as the man who electrified America, was a dead end. No one at the height of such an empire willingly declares his own creation obsolete. Edison had every rational and every emotional incentive to reject AC, and reject it he did, not out of simple ignorance but out of interest. This is a crucial and often-missed point: Edison was not too stupid to see the merit of Tesla's idea. He was too invested to afford seeing it. Sometimes the most dangerous blindness is not the inability to understand a threat but the inability to accept one, and Edison, for all his towering gifts, could not bring himself to accept that the courtly young foreigner tinkering with his dynamos was holding the blueprint that would one day bury the whole direct-current world he had labored to build.

This is the deeper truth that the fifty-thousand-dollar story both dramatizes and obscures. The anecdote gives us a satisfying villain and a clean grievance — Edison the liar, Tesla the wronged — and human memory loves such stories, which is exactly why we should distrust their tidiness. The real break was slower, sadder, and more impersonal than a broken promise: two men whose entire understandings of electricity, of method, of the future itself were incompatible, who had been thrown together by an accident of employment, and who could not have stayed together no matter how the money had gone. The bonus, real or imagined, was the spark. The incompatibility was the powder.

There is also a cultural dimension to the misunderstanding that the meme entirely misses, and it repays a moment's thought. Tesla was a formal, literal-minded man from a world of ceremony and honor, where a gentleman's spoken word was a serious thing and a promise was very nearly a contract. He had been raised in a priest's household on the military frontier of an empire, drilled in courtesy and precision, and he took language at face value. Edison's Machine Works, by contrast, ran on a rough, teasing, wisecracking American shop humor in which extravagant remarks — "do this and there's a fortune in it for you" — were common coin, meant and understood by everyone present as encouragement rather than commitment. Drop a solemn, literal European into that atmosphere and misunderstandings become almost inevitable. What one man tossed off as a bit of motivational banter, the other filed away as a debt of honor. When the "fortune" failed to materialize, each could sincerely believe the other had behaved dishonorably: Tesla certain he had been cheated of a promise, Edison baffled that anyone would have taken a bit of shop talk so literally. Neither had to be a villain for both to feel wronged. That, in the end, may be the truest reading of the famous break — not a swindle, but a collision of two cultures that could not hear each other.

We should also be fair to Edison here, because fairness is in short supply in the popular version. Edison never, as far as the record shows, told this story at all; the entire account is Tesla's. It is possible Edison genuinely made and reneged on a promise, in which case the meme is right. It is possible he made a joking remark that Tesla misread. It is possible Tesla, decades later and nursing a long list of grievances against a world he felt had cheated him, reconstructed the episode into something sharper and more villainous than it had been. All three are consistent with the evidence, which is thin. An honest historian holds them in suspension rather than picking the most dramatic and presenting it as fact. What we can say with confidence is only this: Tesla left Edison's employ in 1885, believing he had been cheated, carrying a resentment that never fully healed, and convinced — correctly, as it turned out — that the older man had refused the greatest opportunity ever to walk through his door.

There is a surprising and often-forgotten fact that complicates the neat hero-villain picture even further. When Tesla left Edison, he did not go straight into the arms of a rival with his AC motor. His alternating-current ideas were not yet what carried him out the door, and they were not what he built next. What came next was failure, poverty, and a period of literal manual labor — which suggests that in 1885 even Tesla did not yet fully grasp, or could not yet finance, the true value of what was in his head. The clean legend has Tesla marching out of Edison's shop and immediately revolutionizing the world. The reality is that he marched out into some of the worst months of his life. The revolution was still three years away, and the road to it ran through a ditch.

You meet the "you don't understand our humor" moment, stripped of its specific dollars, in workplaces everywhere: the gap between what a manager casually says and what an employee earnestly hears, the promise that one party regarded as a bit of encouraging banter and the other treated as a debt of honor. Cultures clash, senses of humor misfire, and expectations set by a careless word curdle into lasting bitterness. The most consequential ruptures in history are not always the work of scoundrels; sometimes they are simply the wreckage left when two people of good faith, formed by different worlds, fail to understand each other's words — and then spend the rest of their lives each convinced that the other one lied. Tesla's tragedy was partly the tragedy of a literal, proud, foreign-born man in a rough joking culture he never quite decoded — a reminder that some of history's most consequential ruptures grow not from grand villainy but from ordinary misunderstanding magnified by wounded pride.

Whatever precisely passed between them in that Manhattan office, the result was clear and irreversible. Tesla was out, on the streets of New York, in his own words free but adrift, carrying the same head full of rotating fields he had carried off the boat a year before, and now without even the meager wages of the Machine Works to sustain him. The dreamer from Smiljan was about to discover how far a great idea will carry a man with no money and no backers — which is to say, into a hole in the ground, with a shovel in his hands.

Chapter 33: Digging Ditches

In the bitter winter of 1886 and into 1887, one of the most brilliant engineers alive could be found on the streets of New York with a shovel in his hands, digging trenches for a wage of about two dollars a day. Nikola Tesla, who carried inside his skull the complete design of the machine that would electrify the world, was laying cable and moving dirt for a labor gang, freezing through the workdays, half-starving, and — worst of all for a man of his temperament — surrounded by the very copper conduits of the electrical industry he knew he could transform, if only anyone would listen. He had come to America two years before to build the future. Now he was burying pipe in the frozen ground so that other men's currents could flow through it.

How does a genius end up in a ditch? The path there is instructive, because it corrects one of the most persistent myths about Tesla: that his troubles were simply the persecution of a pure soul by wicked men. The reality is more ordinary and more useful. Tesla fell into the ditch through a combination of bad luck, worse business judgment, and the ruthless normal workings of a capital-hungry industry — the same forces that swallow unwary inventors in every era.

When Tesla left Edison in 1885, he did not lack for admirers.

His reputation as an engineer was real, and a group of investors soon approached him to start a company under his own name. This should have been his launch. But there was a catch, and the catch reveals everything about why raw genius so often goes unrewarded. The backers were not interested in his alternating-current motor — the thing he most wanted to build, the idea that would eventually be worth a fortune. They wanted something with an immediate, obvious market: arc lighting. Arc lighting was the harsh, brilliant, sputtering illumination produced by drawing a spark across a gap between two carbon rods — the blinding light used for streets, factories, and public squares, too fierce for a home but perfect for lighting a city block. It was a proven, sellable technology, and the investors wanted a system of it they could take to market.

Tesla, needing money and a foothold, agreed. In 1885 he founded the Tesla Electric Light and Manufacturing Company and set to work, and he did the work superbly. He designed an original and effective arc-lighting system, developed the lamps and the dynamos, and even devised a novel commutator and some early forms of automatic control. His system was installed and it worked; he was granted several patents for it. By any measure of engineering, the venture was a success. By every measure that actually mattered to Tesla, it was a catastrophe — and the way it failed him is the second great lesson of these years.

Once Tesla had done the hard part, once the patents were filed and the system proven, his investors reorganized the business in a way that squeezed him out. They took the valuable patents, restructured the company to serve their own interests, and left the inventor holding almost nothing. Instead of cash or a meaningful stake, Tesla was reportedly paid off, in part, with a certificate of stock in the reorganized enterprise — a piece of paper that proved essentially worthless to a man who needed to eat. He had signed away the rights to his own work and been maneuvered out of the value it created. This was not melodramatic villainy; it was the standard, brutal arithmetic of a business world in which the man who owns the patents and the capital wins, and the man who merely has the ideas loses, unless he is as shrewd in the boardroom as he is inventive at the bench. Tesla, for all his genius, was not. He would repeat versions of this mistake, trusting the wrong partners and signing away the wrong rights, throughout his career.

It is worth dwelling for a moment on why a man of Tesla's brilliance kept walking into these traps, because the answer is neither stupidity nor mere bad luck. Tesla was, in a specific and costly way, naive about business, and the naivety flowed directly from his strengths. He believed, with an almost religious intensity, that the merit of an invention was self-evident and that the world would reward genius on its own recognizance. He assumed good faith in the men around him because he himself operated in good faith, and he found the grubby arithmetic of contracts, shares, and voting control beneath the dignity of a creator. He wanted to invent; he expected others to handle the money honestly and hand him his due. This is a beautiful way to be a scientist and a disastrous way to be an entrepreneur. The investors who took his arc-lighting patents were not necessarily monsters; they were ordinary businessmen doing what the rules of their game permitted, taking full advantage of a genius who could not be bothered to read the fine print. Tesla's contempt for that game would cost him, again and again, the fruits of his own mind — and it is one of the quiet tragedies of his life that the very purity of focus which made him a great inventor also made him an easy mark.

And so, in late 1886, Tesla found himself with no company, no money, no patents to his name that he still controlled, and a New York winter closing in. This is the period that gives the chapter its title. To survive, he took whatever manual labor he could get, and for a stretch of months he worked as a common ditch-digger, part of a gang laying electrical conduits and cables in the streets of Manhattan, earning around two dollars a day. It is one of the starkest images in the whole history of technology: the inventor of the polyphase alternating-current system, the man whose motors would one day spin in every factory on earth, standing in a trench in the cold, digging holes for other men's wires. Tesla later called this the lowest point of his life, a time of terrible hardships and, in his words, a period when his high education in various branches of science, mechanics, and literature seemed to him a mockery. He had the future in his head and a shovel in his hands, and the distance between the two very nearly broke him.

Here, though, the story turns, and it turns in a way that is almost too fitting to believe — which means, as always in Tesla's life, we should note both what is documented and what may have been dramatized. The traditional account holds that the foreman of Tesla's labor gang, struck by the strange, cultured, obviously overqualified man swinging a pick beside him, took an interest in his story. This foreman had a connection, and he introduced Tesla to someone who could help — in the usual telling, a manager at the Western Union Telegraph Company named Alfred S. Brown, who moved in circles that had money to invest. Brown, in turn, understood enough about electricity to grasp that Tesla was no ordinary laborer, and he brought in a lawyer and investor named Charles Peck. Together, Brown and Peck agreed to back Tesla properly.

The precise details of that lucky introduction — the sympathetic foreman, the chance conversation in the ditch — are the kind of pleasing narrative flourish that may have been smoothed and rounded in the retelling. What is solidly documented is the outcome: in April 1887, Tesla, Brown, and Peck founded the Tesla Electric Company, and this time the arrangement finally gave Tesla what he had never had before — the freedom and the funding to build the machine he had carried since the Budapest sunset. Brown and Peck set him up with a laboratory on Liberty Street in lower Manhattan, a real workshop of his own, and told him, in effect, to go and make his alternating-current motor real. For the first time since he had drawn the rotating field in the dirt of a Budapest park five years earlier, Tesla had the resources to translate the vision in his head into copper and iron.

The Liberty Street laboratory deserves a moment of its own, because of what came out of it in a stunningly short time. Freed at last to work on the problem he had carried for six years, Tesla did not merely build the single motor he had glimpsed in Budapest. In a burst of creativity that ranks among the most productive in the history of technology, he worked out an entire integrated system of alternating-current power — not just the induction motor but the generators to produce polyphase current, the transformers to change its voltage, and the transmission arrangements to carry it. He was, in effect, designing the whole architecture of modern electrical power in a small rented workshop in lower Manhattan, mostly in the space of a single year. The visions that had tormented him as a child now served him perfectly: he could hold the entire system in his head at once, every coil and connection, and refine it in imagination before committing it to metal. When the patents began to issue in the spring of 1888, they described not a lone gadget but a coherent new way to electrify the world — and that completeness is precisely what would make the work so valuable to the industrialist who was about to come calling.

There is a surprising fact in this reversal that is easy to miss. The men who finally backed Tesla were not great industrialists or electrical visionaries; they were a telegraph manager and a lawyer, ordinary businessmen who took a gamble on a talented immigrant they had reason to think might make them rich. The alternating-current revolution — the technology that would light cities and power the twentieth century — was funded, at its critical moment, not by a titan of industry but by a modest partnership that had the wit to recognize genius and the nerve to bet a little money on it. Great turning points in the history of technology often hinge on exactly such small, contingent human decisions: a foreman who paid attention, an investor who took a meeting, a lawyer who signed a check.

The contrast between where Tesla stood in the winter of 1886 and where he would stand just eighteen months later is almost vertiginous. In the ditch he was an anonymous laborer, one cold and hungry immigrant among tens of thousands, his great idea worth exactly nothing because no one would fund it. By the middle of 1888 he would be lecturing the nation's foremost engineers on a system destined to reshape civilization, and selling his patents to one of America's great industrialists. Nothing about his genius had changed in those eighteen months; what changed was that a few people with capital finally chose to believe in it. That is the sobering lesson buried in the ditch-digging story. The distance between obscurity and triumph, for an inventor, is often not a matter of merit at all but of whether the right person, at the right moment, decides to take a chance. Tesla's genius was a constant. His fortunes swung entirely on the accident of who happened to be listening.

You meet the shape of Tesla's ditch-digging years in the life of nearly every inventor and founder who has ever been ground down by the gap between having a great idea and having the means to build it. The modern startup world is full of brilliant technologists who lost control of their own creations to the people who held the money and the paperwork, who watched their patents assigned away and their equity diluted to nothing, who did the hard work and reaped little of the reward. Tesla's arc-lighting company, in which he built a successful system and was left with a worthless stock certificate, is the nineteenth-century ancestor of a story that plays out in Silicon Valley every year. The lesson is old and unwelcome: in the world as it is, ideas are cheap and ownership is everything, and the inventor who does not guard his stake may end up, metaphorically or literally, back in the ditch.

But now, at last, Tesla was out of the ditch and back at a bench that was his own. In the little laboratory on Liberty

Street, through the last months of 1887 and into the spring of 1888, he set about building the machine he had seen whole in his mind six years before — and this time, when it was finished, the world would finally have to reckon with the rotating magnetic field.

Chapter 34: The Rotating Magnetic Field

On the evening of May 16, 1888, before an audience of engineers gathered at Columbia College in New York for a meeting of the American Institute of Electrical Engineers, Nikola Tesla stood up to read a paper titled "A New System of Alternate Current Motors and Transformers." He was thirty-one years old, four years removed from the ditches, and about to describe the machine he had carried in his head since a sunset in Budapest. What he laid before that room was not a refinement of anything that existed. It was a genuinely new kind of motor — one with no commutator, no brushes, no sparking contacts of any kind, a machine of almost insolent simplicity that ran on the invisible whirling of magnetism itself. The patents had been granted only weeks before, in a cluster dated to that spring of 1888. In the history of electrical power there are a handful of true hinge-points, and this lecture was one of them.

To understand why, we have to understand the rotating magnetic field, which is the single most important idea in this entire book and also one of the most beautiful. It is worth slowing all the way down, because if you grasp this one thing, you understand how the modern world is powered.

Start with the problem, which we met back in the lecture hall at Graz. A motor turns because magnetism pushes and pulls on it. The trouble with alternating current is that it reverses direction many times a second — the push becomes a pull and the pull becomes a push, back and forth, dozens of times every second. Feed that jittering current straight into a simple motor and the rotor is yanked one way, then immediately the other, and it does nothing but buzz and shudder in place, like a person trying to walk who is shoved forward and backward in rapid alternation. This is why Tesla's professor had called an AC motor impossible: a force that reverses itself cannot, it seemed, produce steady rotation in one direction.

Tesla's insight was to stop fighting the reversal and start using it. Here is the trick, and it is genuinely clever. Instead of running a single alternating current into the motor, you run two or more separate currents, and you time them so that they peak at different moments — one current reaching its maximum while another is at zero, and so on, staggered around the cycle. This is what "polyphase" means: multiple currents, out of step with one another in a regular rhythm. Tesla arranged coils of wire in a ring around the inside of the motor and fed these staggered currents into them. And when he did, something remarkable happened. As each current rose and fell in its turn, the point of strongest magnetism moved around the ring — first strong at the top, then strong at the upper right, then at the right, then the lower right, sweeping smoothly around the circle in a continuous rotation. The magnetic field itself was turning. Nothing mechanical moved to make it turn; the mere timing of the currents made the magnetism sweep around like the beam of a lighthouse.

Now picture what that does to a piece of metal placed at the center of the ring. Reach for the first everyday analogy: a carousel of magnetism. Imagine a merry-go-round spinning steadily, and imagine you set a small object on the ground at its center that gets dragged along by the passing horses. The rotating magnetic field is exactly such a carousel — an invisible spinning of magnetic force — and the rotor, the turning part of the motor, is dragged around behind it, chasing the field endlessly around the circle. The rotor tries to catch up to the sweeping magnetism and never quite can, so it just keeps turning, smoothly and continuously, for as long as the currents flow.

Here is a second analogy, and it captures the deepest mechanism. Think of a greyhound at a racetrack chasing the mechanical lure — the fake rabbit that runs just ahead of the dogs around the rail. The dog runs and runs, always reaching for the lure, never catching it, and its motion is driven entirely by the chase. In Tesla's motor the rotating magnetic field is the lure and the rotor is the greyhound. The field sweeps around a fraction faster than the rotor turns, and the rotor forever chases it. That small permanent gap between the speed of the field and the speed of the rotor is not a flaw; it is the very thing that makes the motor pull. Engineers call it "slip," and it is why this design is called an induction motor: the moving field *induces* currents in the rotor — creating them out of nothing but the changing magnetism, exactly as Faraday's law of induction says it must — and those induced currents make the rotor magnetic in turn, so that it is grabbed and dragged along by the field it is chasing. No electricity is fed to the rotor at all. It has no wires running to it, no brushes touching it, no commutator to spark. The magnetism reaches across empty air and turns it.

Sit with how radical that is, because it is the whole point. Every direct-current motor Tesla had ever hated — every dynamo he had rebuilt in Edison's shop, every machine that had sparked and stunk in the Graz lecture hall — depended on the commutator, that clattering split ring of contacts and brushes that flipped the current and wore itself out and threw off dangerous sparks. Tesla's induction motor threw all of that away. There was nothing to spark because no current was switched. There was nothing to wear out because nothing rubbed against anything. The rotor was often just a simple cage or drum of metal, with no delicate windings to burn out. It was rugged, cheap, quiet, and nearly immortal — a motor you could seal inside a machine and forget about for twenty years. Where Edison's DC motors were high-maintenance thoroughbreds, Tesla's AC motor was a mule that would work forever and never complain. The elegance of it was almost provocative: the hardest problem in the field, solved by making the magnetism do the one thing everyone had said magnetism from an alternating current could not do — turn in a circle.

But the true importance of Tesla's motor reaches beyond the motor itself, and this is the connection that decided the war of the currents. Recall the central technical fact of this whole story: direct current could not travel far. Edison's DC could not be cheaply pushed more than about a mile from its power station without wasting most of its energy as heat in the wires, which is why his system needed a generating plant in every neighborhood. Alternating current did not suffer this limitation, because AC could be run through a transformer — a device that steps voltage up for the long journey across country and back down again for safe use, something no simple transformer can do with steady direct current. AC could therefore be generated at a great waterfall or a distant coal plant, sent hundreds of miles at high voltage over thin wires with little loss, and delivered cheaply to a whole region. AC had always had that killer advantage for transmission.

What AC had always lacked was a good motor. You could light lamps with alternating current, but you could not easily run machinery with it, and a system that could not turn the wheels of industry was only half a system. This was the missing piece, the hole in the AC argument that Edison's camp could exploit. Tesla's induction motor filled it completely. Now alternating current could do everything: it could be sent economically across vast distances *and*, at the far end, it could drive the motors that ran factories, streetcars, pumps, and elevators. With the rotating-field motor, the alternating-current system became whole — a complete, integrated means of generating power in one place and doing useful work with it in another, far away. That completeness is what made long-distance electrification practical, and it is why Tesla's motor, more than any lawsuit or execution or propaganda campaign, is what ultimately won the war of the currents.

There is a further elegance to the polyphase idea that is worth making explicit, because it explains why Tesla's system, and not some rival, became the template for the world. Once you are already generating several staggered currents to make the field rotate, those same currents can be carried across the country together, over a set of wires, and put to work at the far end — not only spinning motors but, with the addition of transformers, lighting lamps and running every other kind of electrical device. The very feature that solved the motor problem also happened to make transmission smoother and more efficient, because the staggered currents deliver power to the load in a steady, continuous stream rather than in the pulsing surges of a single alternating current. It was as if Tesla, in solving one problem, had accidentally solved several others at the same time — the mark of an idea that is not merely clever but fundamentally right, aligned with the grain of nature rather than forced against it. Engineers eventually settled on a three-phase version of his scheme, three currents staggered a third of a cycle apart, and three-phase power remains, to this day, the way essentially all bulk electricity is generated and transmitted across the planet.

The engineers at Columbia College that May evening understood, at least dimly, that they were seeing something important, though few could have foreseen how completely it would reshape the world. Among those who understood best were men connected to George Westinghouse, the Pittsburgh industrialist who had already bet his company on alternating current and who was desperately in need of exactly the thing Tesla had just described: a workable AC motor to complete his system. Within weeks, Westinghouse's people came calling, and in July 1888 Westinghouse bought Tesla's polyphase patents — the deal that armed the AC alliance for the fight against Edison and that forms the subject of chapters soon to come.

A word of honest correction, because the internet has turned this invention into a cruder legend than it deserves. Tesla did not, single-handedly and from nothing, invent alternating current; AC and the transformer existed before him, and other able minds — the Italian Galileo Ferraris among them — arrived at the idea of a rotating magnetic field at nearly the same moment, independently. What Tesla did, and did first in practical, patented, demonstrable form, was to build a complete and workable polyphase system with an induction motor at its heart. That is a colossal achievement, and it needs no exaggeration. Inflating it into the myth of a lone wizard who conjured the electrical age out of pure genius does Tesla a backhanded disservice: it replaces a real, comprehensible, hard-won triumph of engineering with a fairy tale, and the truth is more impressive than the fairy tale.

You meet the rotating magnetic field every hour of every day, more intimately than almost any other invention in this book. It is spinning inside the compressor of your refrigerator right now, and inside your washing machine, your dishwasher, your furnace blower, and your air conditioner. It drives the pumps that bring water to your tap and the fans that move air through every large building. It powers electric trains and subway cars and a great many of the machines in every factory on earth. Newer designs even hum inside many electric vehicles, dragging their rotors around with the same invisible carousel of magnetism Tesla saw in a Budapest sunset. Most of these motors will run for decades without a technician ever opening them, because there is nothing inside to wear out — no brush, no commutator, no spark. Add up all the electric motors turning quietly on the planet at this very moment, and the overwhelming majority trace their working principle back to the diagram a penniless immigrant sketched with a stick in a Budapest park and patented in a New York spring. The thing a professor once called impossible is now the most widely used electric motor in the world.

With his patents sold to Westinghouse and his motor validated before the leading engineers of the age, the penniless ditch-digger of two years before had handed the alternating-current cause the one weapon it needed to win — and set the stage for the industrialist from Pittsburgh who would carry that weapon into open war against the Wizard of Menlo Park.

Part V

The Gentle Genius and the AC Alliance

From the air-brake fortune of a Pittsburgh magnate to the purchase of Tesla's patents — how George Westinghouse and the transformer armed alternating current for war.

Chapter 35: The Air-Brake Fortune

On a raw afternoon in the spring of 1868, a young man named George Westinghouse stood on a railway platform in western New York and watched, as he had watched a hundred times before, the clumsy ritual by which a moving train was persuaded to stop. When the engineer up front decided to slow down, he pulled a whistle. Then, all along the length of the train, men called brakemen scrambled from car to car across the swaying, ice-slick rooftops, spinning iron wheels by hand to clamp the brakes on each car in turn. It was brutal, dangerous, and slow. A brakeman who lost his footing fell under the wheels; a train that needed to stop *now* — because a wagon had wandered onto the crossing, or another train loomed on the same track — simply could not. Westinghouse had recently read of exactly such a collision, two trains on the same line, the crews unable to halt in time. The problem gnawed at him. There had to be a better way to reach down the whole length of a train at once and squeeze every brake in the same instant.

The man doing the watching was twenty-one years old, and he already carried himself with the settled confidence of someone twice his age. To understand where that confidence came from, we have to back up to a small machine shop in Schenectady, New York, and to a boy who had grown up inside the smell of hot iron and cutting oil.

George Westinghouse was born on October 6, 1846, in the village of Central Bridge, New York, the eighth of ten children. His father ran a shop that built agricultural machinery — threshers, mowers, small engines — and George was one of those children who is simply *at home* among tools. He was a poor student in the ordinary sense, restless and bored by books, but put a lathe or a length of pipe in front of him and his whole mind woke up. As a boy he would slip away from chores to tinker in a corner of the shop, and by the age of fifteen he had, by family legend, built a small rotary steam engine of his own design. That is the first surprising fact worth pausing on: the man who would arm alternating current for war, who would hire Nikola Tesla and challenge Thomas Edison, filed his very first patent — for a rotary steam engine — in 1865, at the age of nineteen. The engineering gift was there from the start.

But between the boy in the shop and the young man on the railway platform came the war. When the Civil War broke out, Westinghouse, still in his teens, ran off to enlist; his parents hauled him back, judging him too young, and then relented. He served first in the Union cavalry and later, transferring branches, in the Navy, as an engineering officer aboard a warship — an experience that put him inside the belly of real machinery under real pressure. This is a detail the popular story of the "War of the Currents" almost always leaves out, and it matters. Westinghouse was a combat veteran. He had seen how systems fail when the stakes are life and death, and he had learned the habit of command. When he later faced Edison's propaganda and Wall Street's suspicion, he did not flinch the way a sheltered inventor might have. He had already been shot at.

After the war he briefly enrolled at Union College and lasted, by most accounts, about three months before withdrawing; the classroom could not hold him any better than the schoolroom had. He went back to the family shop, back to the machinery, and back to the railroads — for the railroad was the great arena of American ambition in the 1860s, the industry that was stitching a broken country into a single market. His earliest inventions were railway devices: a "car replacer" for lifting derailed cars back onto the track, and a "frog," a fitting that let one rail cross another at a junction. Both sold. Both taught him the same lesson the light bulb would later teach Edison — that an invention is worth nothing until it is manufactured, sold, and serviced as part of a working whole.

Then came the brake. The idea that seized Westinghouse was to run a single continuous pipe the whole length of the train, and to use that pipe to command every brake at once. His first thought was to fill the pipe with steam from the locomotive, but steam cooled and condensed to useless water before it could reach the rear cars. His breakthrough — and it is the kind of lateral leap that marks a first-rate engineer — was to switch the working substance from steam to *compressed air*. Air did not condense. It could be pumped to high pressure by the locomotive, piped the length of the train, and used to drive a piston at every car that pressed the brake shoes against the wheels. In 1869, at the age of twenty-two, he received the patent for the railway air brake and founded the Westinghouse Air Brake Company to build it.

Here is where the story becomes genuinely clever, and where a common misconception is worth correcting. Most people who have heard of the air brake imagine it works the obvious way: push air into the pipe, and the pressure pushes the brakes on. Westinghouse's *first* design did work like that. But that design had a fatal flaw, and Westinghouse, being the sort of man he was, went back and fixed it with an inversion so elegant it is still in use on freight trains today. The problem: if the train broke apart — a coupling snapped, a car came loose — the air pipe would tear open, the pressure would vanish, and under the "push-on" design the runaway cars would have *no brakes at all*, precisely when they needed them most. So Westinghouse reversed the logic. In his improved "automatic" air brake, the compressed air's job is to hold the brakes *off*. Each car carries its own little reservoir of pressurized air, and the continuous pipe keeps them charged. Let the pressure drop — whether because the engineer commands a stop or because the train has ripped in two — and every car, drawing on its own reservoir, slams its own brakes on automatically. The failure of the system *is* the safety of the system. A train that comes apart stops itself.

Think of it as the difference between a man who must actively hold a spring-loaded door shut and a man who must actively hold it open. If the first man faints, the door stays shut; if the second faints, the door flies open. Westinghouse designed the brake so that fainting — total failure — meant *stopped*, not *runaway*. It is a philosophy engineers now call "fail-safe," and Westinghouse is one of its founding fathers. Another homely analogy: it works like the dead man's grip on a modern lawnmower or a subway train, the handle you must squeeze to keep the machine running, so that if you let go — or collapse — it halts on its own. Westinghouse had that idea in 1872.

The air brake did what railroad committees and safety reformers had failed to do for a generation: it made high-speed rail travel survivable. Before the air brake, a train's safe stopping speed was limited by how fast a scattering of men could crank hand-wheels on lurching rooftops. After it, one engineer's hand on one valve stopped the entire train, evenly and instantly. Trains could run faster, longer, and heavier with less risk. So powerful was the improvement that in 1893 the U.S. Congress passed the Railroad Safety Appliance Act, effectively making air brakes mandatory on American trains — one of the first federal laws to require a specific safety technology. A private invention had become a public standard.

The fortune followed. And from that fortune Westinghouse built not one company but an empire of them — by the end of his life he had organized more than sixty companies and held some 361 patents in his own name. He did not merely brake trains; he reinvented how they were controlled. He turned his mind to railway signaling and pioneered "interlocking" systems, in which signals and track switches are mechanically linked so that a signalman physically *cannot* set two trains onto a collision course — again the fail-safe instinct, designing the possibility of the accident out of the machine. He built a business piping natural gas, and in the process invented meters and regulators and safe-delivery systems for a dangerous, invisible fluid. Each of these ventures taught him the same discipline: how to move a hazardous, invisible something — high-pressure air, natural gas — safely through a network to a customer. It was the ideal apprenticeship for a man about to move a far stranger and more dangerous fluid: high-voltage electricity.

But the fortune is only half of what made Westinghouse remarkable. The other half was his character, and here the historical record is unusually kind — and unusually surprising, given the ferocity of the war he would soon be dragged into. Westinghouse was, by the testimony of nearly everyone who dealt with him, a fundamentally decent man in an age not famous for corporate decency. He was courtly, formal, physically imposing, with a walrus mustache and a habit of listening carefully before he spoke. He treated his engineers as collaborators rather than hired hands, and he treated his workers better than almost any industrialist of the Gilded Age. As early as 1871 he gave his employees a half-day holiday on Saturday — an astonishing benefit for the era. He later introduced pension plans and paid vacations, and he was known to walk his factory floors and know his men by name. When Tesla, years later, would recall his backers and rivals, the bitterness he reserved for others never touched Westinghouse; of the industrialist he spoke with genuine warmth, calling him a man of noble character.

This decency was not sentimentality. Westinghouse simply understood, in his practical way, that a loyal and well-treated workforce built better machines — and that a business worth having was one you could be proud of. It is the third surprising fact of his life, and the one most at odds with the villains-and-heroes cartoon the current war is often reduced to: the man who would be branded, in Edison's pamphlets, as a merchant of death, as the purveyor of an "executioner's current," was in private life about the gentlest and fairest of the great industrialists of his time. When we reach the electrocuted dogs and the electric chair, keep this man in mind. The propaganda aimed at him was not merely false; it was almost perfectly inverted.

One more dimension of Westinghouse's character deserves attention, because it would shape the coming war as surely as any patent: his attitude toward invention and inventors. Unlike some of the great industrialists of his age, who regarded inventors as hired hands to be squeezed and discarded, Westinghouse genuinely respected the creative mind and dealt with it generously. He paid well for good ideas, credited the people who had them, and built his companies as places where talented engineers could do their best work with resources and autonomy. This was partly temperament and partly strategy — a man who wants the best inventions must make himself the sort of person the best inventors want to work with — but it was sincere, and it would prove decisive when the time came to acquire a certain Serbian immigrant's patents. Where a harsher magnate might have tried to strong-arm or cheat a struggling inventor, Westinghouse would deal fairly, and that fairness would win him the single most valuable set of patents of the age. His decency, in other words, was not merely a personal virtue but a competitive advantage, drawing talent and ideas toward him that colder men could not attract. The gentleness and the shrewdness were not separate qualities; in Westinghouse they were the same quality, seen from two sides.

There is a final surprising fact about Westinghouse that reframes his whole career and sets up his entry into the electrical war. For all his wealth and his sixty-odd companies, he was never primarily a financier or a speculator, never a man who made money from money. He was, to the end of his life, an *engineer* — a man happiest with a technical problem in front of him, who understood his businesses from the machinery up, who could walk into any of his shops and discuss the details of the work with the men doing it. This is a crucial distinction, because it meant that when Westinghouse looked at the choice between alternating and direct current, he did not see it as an investor sees a stock, as a bet on which company's shares would rise. He saw it as an engineer sees a design problem, and he judged it on the merits of the physics. That engineer's eye, trained on air brakes and gas mains and signaling systems, was about to fix itself on the greatest technical and commercial problem of the age — and to reach a conclusion that would put him on a collision course with the most famous inventor in the world.

You meet George Westinghouse's mind every time you ride a train or a subway, or hear the long pneumatic hiss of a bus kneeling at a curb — that hiss is compressed air, and the brake it releases is a direct descendant of his 1869 patent. Modern freight trains across the world still stop on the automatic air brake he perfected, and the fail-safe philosophy he pioneered now governs everything from elevators to nuclear reactors. He is the forgotten third man of the electrical age precisely because his genius was for systems that work so well we stop noticing them.

By the mid-1880s, Westinghouse was rich, respected, and restless, a man who had already conquered the railroad and the gas main and was casting about for the next great network to master. The newest, most electric of frontiers was quite literally electricity itself — and the question that would draw him in was not how to *make* the current, which Edison had already shown, but how to *move* it, cheaply and far, across the miles that Edison's direct current could not cross.

Chapter 36: A Better Way to Move Power

In a laboratory in London in the early 1880s, a French engineer named Lucien Gaulard and his English business partner John Dixon Gibbs were passing an alternating current through a curious device — two coils of insulated copper wire wound around a common core of iron, neither coil touching the other. Current flowed into the first coil. No wire ran from the first coil to the second. And yet, when they connected a lamp to that second, separate coil, the lamp lit. Electricity had leapt across the gap between the coils without any wire to carry it, and — stranger still — it came out at a different voltage than it went in. Feed the first coil a modest current and the second could deliver a mighty one, or the reverse. The two men called their device a "secondary generator." We call it a transformer, and it is, without exaggeration, the single most important reason the world runs on alternating current rather than direct.

To see why this humble box of wire and iron changed history, we first have to understand the problem it solved — the problem that had trapped Edison's direct-current empire within a one-mile circle, the problem introduced a few chapters ago and now ready to be cracked open. The problem has a name, and the name is *loss.*

When electric current flows through a wire, the wire fights back. No conductor is perfect; even good copper resists the passage of current a little, the way a garden hose resists the water pushing through it. That resistance turns some of the electrical energy into heat, and that heat, radiating uselessly off the wire into the open air, is simply gone — power you paid to generate and will never deliver to the customer. Every foot of transmission wire is a small tax collector, skimming energy off the current as it passes and throwing it away as warmth.

Now here is the cruel part, the fact that doomed long-distance direct current, and it is worth stating precisely because it is the hinge of the whole story. The heat lost in a wire does not rise in step with the current — it rises with the *square* of the current. Double the current in the wire, and you do not double the wasted heat; you *quadruple* it. Triple the current, and the loss grows ninefold. Engineers write this relationship as "I-squared-R" loss, where I stands for the current and R for the resistance of the wire, and that little exponent, that innocent-looking "squared," is one of the most consequential facts in the history of technology. It means that pushing a large current down a long wire is ruinously wasteful. The current heats the wire so fiercely that most of your power cooks the countryside instead of reaching the town.

So how do you send a great deal of *power* down a wire without sending a great deal of *current*? To answer that, we need to be clear about what electrical power actually is, and here is where a plain definition and a good analogy earn their keep. The power delivered by electricity is, roughly, its voltage multiplied by its current. **Voltage** is the electrical "pressure," the force pushing the charges along; **current** is the *quantity* of charge flowing per second, the sheer volume of the flow. Picture water again. Power is how much work the water can do — and you can get the same amount of work from a thin jet at enormous pressure as from a wide, lazy river at gentle pressure. A narrow fire-hose stream at high pressure and a broad slow canal can carry the very same energy. Power is pressure times flow; hold the power fixed, and you can trade one for the other.

That trade is the whole secret. Since the wasteful heating depends only on the *current* — on the flow — and not on the voltage, the trick is to send your power as *high voltage and low current*: the thin, screaming, high-pressure jet rather than the fat, slow river. Crank the voltage way up and the current needed to carry the same power drops proportionately, and because loss goes as the current *squared*, dropping the current tenfold cuts the waste not tenfold but a hundredfold. Send your power at a hundred times the voltage and one-hundredth the current, and you lose ten thousand times less energy to heat. A line that would have wasted nearly all its power at low voltage delivers almost all of it at high voltage. This is not a marginal improvement. It is the difference between a power station that can serve a single neighborhood and one that can serve a state.

But there is a catch, and the catch is why Edison was stuck. High voltage is wonderful for *sending* power across country, but it is lethal and useless *inside a home*. You cannot run tens of thousands of volts into a table lamp; it would arc, burn, and kill. What you need is a way to push power out onto the long-distance lines at ferociously high voltage — the thin high-pressure jet — and then, at the far end, just before it enters the customer's house, gently step that voltage back down to something a lamp can safely use. You need, in other words, a voltage-changing machine at each end of the line: one to step *up* for the journey, one to step *down* for delivery.

And this is exactly, miraculously, what the transformer does — and exactly what direct current cannot be made to do easily. Here we arrive at the deepest reason the war of the currents came out the way it did, and it rests on a discovery we met many chapters ago: Michael Faraday's law of induction. Faraday found that a *changing* magnetic field induces a current in a nearby wire. Not a steady field — a *changing* one. Motion, or fluctuation, is everything; a magnetic field that simply sits there, unchanging, induces nothing at all.

Now recall the difference between the two currents. **Direct current**, Edison's current, flows steadily in one direction, like a river running always downhill — smooth, constant, unvarying. **Alternating current**, Westinghouse's current, does something that sounds absurd when you first hear it: it reverses direction many times a second, surging one way, stopping, surging back the other way, sixty times per second in the American system, over and over. It is less a river than a tide, or the back-and-forth of a saw, or a piston forever pumping in and out. To the newcomer this sounds like a defect — why would you want electricity that can't make up its mind which way to go? But that ceaseless reversal is precisely the gift, because a current that is always changing direction produces a magnetic field that is always changing — and a changing magnetic field is exactly what Faraday showed can leap across empty space and conjure a current in a second, unconnected coil.

That is how the transformer works, and once you see it, the whole architecture of the modern grid falls into place. Wind two coils of wire around the same iron core, one with many turns and one with few, and connect an alternating current to the first. The current, forever reversing, sets up a magnetic field in the iron that is forever reversing with it. That churning field threads through the second coil and induces an alternating current in *it*, with no wire connecting the two — the lamp lights across the gap, just as it did for Gaulard and Gibbs. And here is the elegant bookkeeping: the voltage you get out depends simply on the *ratio* of the turns. If the second coil has a hundred times as many turns as the first, the voltage comes out a hundred times higher (and the current a hundred times lower). Reverse the ratio, and you step the voltage down. A transformer is nothing but two coils sharing an iron core, and by the ratio of their windings it multiplies or divides voltage at will, with efficiencies that even in the crude early machines reached ninety-odd percent, and today exceed ninety-nine.

Two everyday pictures make it stick. First, imagine a pair of gears. A small gear driving a large gear turns slowly but with great force; a large gear driving a small one spins fast but weakly. The transformer is an electrical gearbox, trading voltage for current the way meshed gears trade speed for torque — and, like gears, it can only pass along motion, never conjure it from stillness. Second, think of two children's swings side by side, joined by a loose rope. Push the first swing rhythmically and the second begins to sway in sympathy, energy crossing the rope without the children ever touching; but only *rhythmic* pushing works — hold the first swing motionless off to one side and the second hangs dead. Alternating current is the rhythmic push. Direct current is the swing held still.

Now the fatal asymmetry between the two currents is plain. Direct current is steady, and a steady current makes a steady, unchanging magnetic field, and an unchanging field induces *nothing* in the second coil. A transformer fed direct current does simply nothing — the lamp stays dark. To change the voltage of direct current in the 1880s meant clumsy, lossy, expensive machinery — motor-generator sets, banks of batteries — nothing remotely like the silent, cheap, nearly lossless efficiency of a coil of wire around an iron core. Alternating current could be transformed. Direct current, for all practical purposes, could not. That single fact is the technical heart of everything that follows: it is why AC could be flung across a hundred miles and DC could not crawl past one.

This is the piece Westinghouse saw, and it is why he was drawn to the strange new field. He did not invent the transformer; Gaulard and Gibbs and others had built early versions, imperfect and unreliable. But Westinghouse, the master of moving invisible dangerous fluids through networks, recognized the transformer for what it was — the missing valve that would let electricity travel the way his compressed air and natural gas already did. He acquired the rights to the Gaulard-Gibbs design, brought it to America, and handed it to a gifted young engineer to make it *work*.

It is worth confronting directly the objection that any thoughtful reader will now be forming, because facing it squarely reveals just how decisive the transformer really was. If direct current is so much more intuitive — a steady flow in one direction, easy to understand, easy to use in motors and batteries — and if alternating current's constant reversal seems at first like a needless perversity, then why did the world not simply find a way to change the voltage of direct current and keep everything simple? The answer is that people tried, hard, for years, and the physics defeated them. To step direct current up or down in voltage, engineers of the 1880s had essentially two options, and both were miserable. They could use a "motor-generator set" — a direct-current motor at one voltage mechanically coupled to a direct-current generator at another, a whole spinning machine with bearings and brushes to wear out, consuming a chunk of the power just to turn itself, expensive to build and maintain and never very efficient. Or they could rig banks of batteries in clever switching arrangements, which was clumsy, costly, and hopeless at any real scale. There was simply no cheap, silent, reliable, nearly lossless way to change the voltage of direct current, and there still is not one as simple as a coil of wire around an iron core. The transformer's staggering advantage was that it did the voltage-changing job with no moving parts at all, wasting almost nothing, for a fraction of the cost. That is why the reversal that seems like AC's defect is in fact its superpower: it is the price of admission to the transformer, and the transformer is the thing that makes wide, cheap distribution possible.

There is a deeper irony here worth savoring, because it overturns a lazy assumption that runs through much of the popular story. The very feature of alternating current that Edison's propagandists would seize upon as evidence of its unnatural danger — its restless, ceaseless reversal, so different from the honest steady flow of direct current — was precisely the feature that made it useful. AC's "instability," its refusal to hold still, was not a bug to be tolerated but the essential mechanism of its triumph, because only a changing current can work a transformer, and only a transformer can send power economically across distance. Nature, in one of her characteristic jokes, had arranged things so that the current which seemed cruder and more chaotic was in fact the one better suited to civilization's needs, while the current that seemed steadier and more docile was trapped forever within a mile of its source. The lesson recurs throughout the history of technology: the property that looks like a flaw from one angle is often, from another, the very thing that makes a thing work. Alternating current's genius was hidden inside its apparent madness.

You meet the transformer thousands of times a day without ever seeing it. It hums inside the gray cylinders bolted to utility poles up and down every street, stepping the high-voltage current of the neighborhood line down to the safe voltage of your outlets. It sits in the big humming substations that ring every city, and in the block of plastic on your laptop cord, and inside every phone charger — each one a tiny descendant of the coils Gaulard and Gibbs wound in London, quietly trading voltage for current so the current in the wall can safely feed the machine in your hand.

The physics was now clear and the machine was in Westinghouse's hands. What remained was to prove, out in the real world of a real American town, that a transformer could take current from a distant generator, step it up, send it down a line, step it down, and light the lamps in people's homes — not in theory, but on a cold night, in a place with a name.

Chapter 37: Great Barrington

On the evening of March 20, 1886, in the small Berkshire town of Great Barrington, Massachusetts, a slight, dark-eyed engineer named William Stanley Jr. threw a switch and watched a row of shop windows along Main Street bloom into light. There were only a couple dozen lamps, glowing in a handful of stores, a post office, two doctors' offices, and a telephone exchange, strung along a few hundred yards of a sleepy New England street where the snow still lay in the gutters. To the townspeople strolling past, it must have looked like a modest novelty — pretty, but hardly a revolution. They could not have known that they were the first Americans in history to be lit by a complete alternating-current system built the modern way: a generator making current, a transformer stepping the voltage up for the journey, a line carrying it across town, and transformers at the far end stepping it back down to light the lamps. Everything the world now runs on was there in miniature on that cold Massachusetts night. Great Barrington was the acorn from which the entire oak of the electrical grid would grow.

The man who made it happen was, at that moment, a sick and somewhat exasperated young engineer of twenty-eight. William Stanley Jr. had been born in Brooklyn in 1858 to a comfortable family; his father was a lawyer who fully expected his son to become one too. Stanley made it as far as Yale before the pull of the workbench overpowered the pull of the law, and he dropped out to apprentice himself to the electrical trade — an act of rebellion in his household roughly on the order of running away to join the circus. He was talented, meticulous, and physically frail, plagued for much of his life by ill health, and he had a temperament that did not always mix easily with the booming, impatient George Westinghouse. But he had a gift for the fussy, exacting work of making a laboratory principle survive contact with the real world, and in 1885 Westinghouse had hired him precisely for that gift.

The assignment was daunting. Westinghouse had acquired the American rights to the Gaulard-Gibbs transformer, but the imported design was, frankly, not good enough to build a business on. Its transformers were wired in a way that made the whole system fragile, and its efficiency and reliability fell short of what a paying utility would need. Westinghouse handed Stanley the pieces — the transformer, some imported Siemens alternators, a pile of problems — and told him, in effect, to make it into something that worked. It is worth correcting a common impression here: the story of AC in America is often told as though Westinghouse or Tesla simply *had* the transformer and switched it on. In truth the early transformers were balky, and the crucial engineering — redesigning the core, rethinking how the coils were wound, arranging the units so that a fault in one shop did not darken the whole street — was done by Stanley and a small circle of Westinghouse engineers over months of frustrating trial and error. The revolution was assembled by hand, one improved coil at a time.

Stanley did much of that work not in Pittsburgh but in Great Barrington, and the reason is a small human detail that the grand histories usually omit. Stanley's health had collapsed; he was seriously ill and had retreated to the Berkshires, where the mountain air was thought to be restorative and where his family had roots. Rather than lose his best transformer man, Westinghouse let Stanley set up a laboratory near his convalescence, in a disused rubber mill and a barn-like building in the town. So the first great demonstration of alternating current in America happened where it did partly by medical accident — because a brilliant, sickly young engineer needed clean air, and his boss was wise enough to bring the work to the man rather than drag the man to the work. Genius, once again, kept strange company with circumstance.

Through the winter of 1885 and into 1886, Stanley labored over his apparatus. He redesigned the transformer core into a form that ran cooler and wasted less energy, and — this was the decisive conceptual step — he wired his transformers *in parallel* rather than in series. The distinction sounds like plumbing arcana, but it made all the difference between a laboratory stunt and a usable utility. In the older series arrangement, all the lamps sat on a single loop like beads on one string, so that trouble anywhere on the string disturbed everything on it, and each customer's lamps affected the next. In Stanley's parallel arrangement, each customer's transformer tapped the main line independently, like houses each with their own faucet off a common water main; one shop could switch its lights on or off, or suffer a fault, without dimming or dousing its neighbors. This is exactly how every building on your street is wired today. Stanley did not merely light a few lamps; he worked out the *architecture* by which a whole town of independent customers could each draw the electricity they needed, when they needed it, from a shared line — the fundamental floor plan of the modern grid.

By March the system was ready for a real trial. Stanley's generator sat in the old rubber mill. From it, current at a moderate voltage was fed into a transformer that stepped it up to five hundred volts for transmission — far too high to run a lamp safely, but far more efficient for sending power down the line, exactly the high-voltage, low-current trick that the previous chapter laid out. That five-hundred-volt current traveled along wires strung on poles up Main Street, perhaps four thousand feet in all. At each shop, a step-down transformer took the five hundred volts and reduced it to the hundred or so volts that the incandescent lamps could burn without exploding. Step up for the journey, step down for delivery: the entire logic of alternating-current distribution, demonstrated in a town small enough to walk across in ten minutes.

On the night of March 20, it worked. The lamps in the stores along Main Street came on and stayed on, fed from a generator hundreds of yards away through wires carrying a voltage that would have been suicidal to run straight into a lamp. Over the following weeks Stanley expanded the little system, adding more lamps until some two dozen establishments were lit. There is a lovely, humanizing detail preserved in the accounts: Stanley, still weak, is said to have watched the first lighting with the quiet satisfaction of a man who has been proven right after a long, sick, exhausting winter, while the merchants of Great Barrington enjoyed the practical novelty of light that did not flicker, smoke, or need a match. Here is a surprising fact to sit with: this pioneering system, the technical ancestor of a planet's worth of power lines, served a town of only a few thousand people and lit fewer lamps than a modern living room, and it was almost entirely ignored by the national press, which was still transfixed by Edison's glamorous, well-publicized direct-current stations in the great cities. The future arrived quietly, in the provinces, unremarked.

It is worth being honest about the scale and the limits of what Great Barrington proved, because the mythology tends to inflate it. This was not yet long-distance transmission in the sense that would later stun the world at Niagara; four thousand feet is not a hundred miles. Nor did it yet include an AC *motor* — the lamps glowed, but nothing turned, and the want of a workable alternating-current motor would remain AC's great weakness for two more years, until a Serbian immigrant supplied the missing piece. What Great

Barrington *did* prove, unambiguously and in public, was the thing Edison's engineers kept insisting was impractical: that alternating current could be generated, transformed up, sent over a distance, transformed back down, and used to serve a scatter of independent customers safely and reliably. It was proof of principle for the whole distribution scheme, the first working American answer to the question of how to break out of Edison's one-mile prison.

For Westinghouse, watching from Pittsburgh, the demonstration was the green light. He was not a man who bet on hunches; he was a man who bet on tested machinery, and Great Barrington was the test passing. Within months he had committed his fortune and his formidable organization to alternating current, ordering equipment, filing for improvements, and moving to build commercial systems in real cities. The gentle giant of the air brake had found his next network. And notice the quiet symmetry with his earlier life: just as he had once run compressed air down a pipe the length of a train and used pressure differences to command every brake at once, he was now preparing to run high-voltage current down a line the length of a region and use transformers to deliver usable power to every home at once. It was the same instinct — move an invisible force efficiently through a network to serve many endpoints — applied to a new and far more powerful fluid.

A word about a lingering misconception. Stanley is sometimes cast, in the more romantic tellings, as the true inventor of the transformer, cheated of his due; and elsewhere the transformer is credited wholly to Gaulard and Gibbs, or to the Hungarian engineers of the Ganz Works who were doing parallel work in Europe at the very same time. The honest answer is that the transformer, like most great inventions, had many parents on two continents working almost simultaneously, and that Stanley's real and considerable achievement was not the raw idea but the *practical American system* — the improved, reliable, parallel-wired installation that actually lit a town and pointed the way to a business. He later fell out with Westinghouse and went on to found his own successful electrical-manufacturing company, and his transformer patents made him wealthy. History has been stingy with his name, which is a pity, because you live inside his handiwork.

It is worth lingering on just how modest and improvised the whole demonstration was, because the contrast between its humble reality and its world-historical consequence is one of the quiet marvels of the story. This was not a corporate showcase backed by a great budget and a publicity machine. It was a sick young engineer working in a converted rubber mill in a small town, with imported and jury-rigged equipment, running wires up a village street on poles, powering a couple dozen lamps in shops whose owners were mostly curious rather than awestruck. There were no reporters from the great metropolitan dailies, no dignitaries, no ceremony to match the fanfare that had greeted Edison's Pearl Street station four years earlier. The most consequential electrical demonstration of its decade happened almost in secret, witnessed by a handful of townspeople and a convalescent inventor. There is a lesson in this about how the future actually arrives: not always with trumpets in the capital, but sometimes quietly, in the provinces, built by tired people with imperfect tools, unnoticed by the very newspapers that were busy celebrating the technology it would eventually replace. The grandeur of an achievement and the grandeur of its unveiling are two different things, and history is full of revolutions that began in rubber mills.

Consider also a subtle point about what Stanley's parallel wiring really accomplished, because it is easy to underrate as mere technical housekeeping when it was in fact the key to making electricity a *utility* rather than a stunt. A utility, by its nature, must serve many independent customers whose needs vary from moment to moment — this shop wants light now, that one is closed, a third turns half its lamps off at midnight. A system in which every customer's usage disturbed every other customer's supply could never function as a real public service; it would be like a water system in which your neighbor drawing a bath emptied your tap. Stanley's parallel arrangement, in which each customer tapped the common line independently through their own transformer, was what made it possible for a single distant station to serve a whole community of customers who came and went as they pleased, each drawing exactly what they needed without regard to the others. This is the fundamental architecture of every public utility that followed, and it is why Great Barrington deserves to be remembered not merely as the first American AC lighting demonstration but as an early working model of the electric utility itself — the business form that would, more than any single invention, actually electrify the world.

You meet Great Barrington's legacy every single day, though the town's name means nothing to most people. Every time the lights in your house stay lit while your neighbor flips theirs off, you are enjoying the parallel wiring Stanley championed. Every gray transformer on a pole outside, stepping the street's high voltage down to your outlets, is a direct descendant of the step-down units he set on Main Street in 1886. The little Berkshire demonstration was, in the most literal sense, the pattern from which the electrified world was cut.

With the transformer proven and the distribution scheme working, Westinghouse now had the means to move power across the distances that had defeated Edison. What he needed next was a company built expressly to sell it — and a decision, made with his whole fortune on the table, to stake everything on the current his great rival had branded as dangerous.

Chapter 38: Westinghouse Electric Bets on AC

In the summer of 1886, in the smoke-hazed industrial city of Pittsburgh, George Westinghouse signed the papers founding a new enterprise and gave it his own name: the Westinghouse Electric Company. He was thirty-nine years old, already rich from the air brake, already the head of a sprawling industrial house, and he had just decided to gamble a substantial part of his reputation and capital on a form of electricity that the most famous inventor in America was busy denouncing as a public menace. It was the kind of decision that looks either visionary or reckless depending on which way the future breaks, and in the summer of 1886 no one — not Westinghouse, not Edison, not the bankers watching from New York — could yet say which it would be.

To appreciate the size of the bet, you have to appreciate the size of the incumbent. Thomas Edison was not merely a competitor; he was, by the mid-1880s, an institution. His name was synonymous with the electric light itself. His Pearl Street station in Manhattan had been running since 1882, and Edison direct-current stations were spreading through the wealthy cores of American cities, financed by the most respectable money on Wall Street, blessed by the aura of the Wizard of Menlo Park. To go into the electric-light business at all in 1886 was to challenge Edison. To go into it with a *rival kind of current* — to tell customers that the great Edison had backed the wrong horse — was something close to effrontery. And yet that is precisely what Westinghouse resolved to do, and he did it not out of contrarian spite but out of a cold engineer's reading of the physics laid out in the last two chapters. He had seen the arithmetic of I-squared-R loss. He had watched Stanley light Great Barrington. He knew that Edison's direct current, for all its prestige, was trapped inside a circle roughly a mile across, and that whoever solved long-distance distribution would ultimately serve the whole country while Edison served only the dense, expensive hearts of a few cities.

Here it helps to be precise about the strategic logic, because the popular story flattens it into "AC was simply better," a phrase this book has already promised to resist. Alternating current was not *simply* better; each current had its domain, and direct current had genuine advantages Edison was right to prize — it was well understood, it ran the early electric motors of the day, it was arguably safer at the low voltages used inside homes, and it charged batteries, which alternating current could not. What alternating current was better at was exactly *one* thing, but that one thing turned out to be decisive: moving power economically across distance. Direct current at a usable voltage bled away its energy within a mile; alternating current could be stepped up to high voltage, flung across many miles with little loss, and stepped back down at the far end. In a country as vast and thinly settled as the United States, where a single generating station in a river valley might economically serve towns scattered across a whole county, the ability to travel was not one advantage among many. It was the advantage that ate all the others. Westinghouse bet on distance.

He also bet on his own peculiar competence. Consider what Westinghouse actually knew how to do, better perhaps than any man alive: he knew how to take an invisible, dangerous, high-pressure fluid — compressed air, natural gas — and distribute it safely through a network of pipes and valves and regulators to thousands of ordinary customers, with meters to bill them and safety devices to keep them alive. High-voltage electricity was, from a systems point of view, one more invisible dangerous fluid to be moved through a network. The engineering culture he had already built — the instinct for fail-safe design, the discipline of metering and regulation, the willingness to solve the unglamorous problems of delivery rather than just the glamorous problem of generation — was almost tailor-made for the AC challenge. Edison, for all his genius, had built his organization around invention and around a particular technical choice already frozen into thousands of installed dynamos and lamps. Westinghouse was building his organization, from scratch, around *distribution*. It was a difference in DNA that would matter enormously.

Through 1886 and 1887 Westinghouse moved with characteristic speed and thoroughness. He gathered the pieces: the Gaulard-Gibbs transformer rights, Stanley's improved designs, imported and then domestically built alternators, and a growing cadre of first-rate engineers whom he treated, as always, as valued colleagues. He did not merely dabble; he committed the manufacturing muscle of his industrial empire to building alternating-current equipment at scale. And the market responded. The economic case was too strong to ignore, especially for the countless American towns that Edison's system could never profitably reach — places too small and too spread out to justify a DC station on every square mile. For such towns, a single AC station serving a wide radius was the only affordable path to electric light, and Westinghouse offered it. Here is a fact that startles people raised on the Edison legend: within just a couple of years of its founding, Westinghouse Electric had sold and installed AC systems in dozens upon dozens of towns and cities across the country, and by the end of the decade its installed lighting capacity was racing to rival, and in raw numbers of lamps served would soon threaten to overtake, Edison's own. The upstart was not nibbling at the edges of Edison's empire; it was building a parallel empire of its own, out in the vast provincial hinterland the Wizard had left uncovered.

But the bet had a flaw at its very center, a gap in the system that Westinghouse understood perfectly well and that kept him up at night. His AC system could light lamps beautifully. It could not, in 1886, turn a *motor.* And a lighting system that cannot also run motors is a system with a fatal ceiling on its future.

To see why this mattered so much, step back and ask what electricity is actually *for.* Light is the glamorous first use, the one that sells newspapers and dazzles investors. But the deeper, larger market — the one that would ultimately consume the overwhelming majority of all the electricity ever generated — is *motion.* Electric motors run streetcars and elevators and factory machinery, pumps and fans and mills, and eventually the refrigerators and washing machines and countless devices of ordinary life. A current that can only make light is a current confined to the evening hours and the parlor. A current that can also make things *turn* can run the whole economy around the clock. Edison's direct current, whatever its distance limits, could drive the motors of the day; that was one of its real strengths, and Edison's people never tired of pointing it out. Westinghouse's alternating current, at that moment, could not, because the ordinary motors of the 1880s depended on a device called a commutator — a set of sliding contacts that flips the current's direction inside the motor at just the right instants — and the commutator, sparking and wearing and fussy even on direct current, simply would not cooperate with a current that was already reversing sixty times a second of its own accord. Alternating current and the existing motor design were, in effect, at war with each other. Feed a standard motor alternating current and it hummed, heated, and refused to spin usefully.

This was the missing piece, and it was not a small one. It meant that as brilliantly as Westinghouse's lighting business was growing, it was growing on a foundation that everyone could see was incomplete. Edison's propagandists could and did point at the gap: *their* current ran motors, *ours* only lit lamps; theirs was a complete system, ours a half-built one. Westinghouse needed a motor that would run on alternating current directly, without a commutator, without sparking, a motor that would take his reversing current not as an obstacle to be fought but as the very thing that made it spin. Such a motor would complete his system, close the gap, and turn a lighting company into a power company. But no one had built one. The best electrical minds of two continents had wrestled with the problem of an AC motor and come up largely empty; the reversing current seemed to fight every scheme they tried.

There is a poignant irony worth naming here, because it sets up everything that follows. At the very moment Westinghouse was staking his fortune on alternating current and worrying over its missing motor, the man who had already solved that exact problem — who had, years earlier, in a park in Budapest, seen in a flash of vision how to make a motor spin using the reversing current itself, with no commutator at all — was living in New York, recently emerged from digging ditches for a living, filing patents on a system almost no one yet understood. Nikola Tesla had the missing piece. He had had it, in his mind, for years. What he did not have was the capital, the factory, and the industrial partner to build it into the world. And what Westinghouse had — the capital, the factory, the network, the desperate need for exactly that motor — was the perfect complement to what Tesla held. Two halves of a single machine were about to find each other.

Before we watch them meet, it is worth pausing on the character of the bet Westinghouse had made. It is easy, from the far side of history, to call it the obvious winning move — we know how the story ends, we know AC won, we live in an

AC world. But in 1886 and 1887 it was nothing of the kind. He was outspent and outnamed by Edison; he was betting on a current with a glaring hole in its capabilities; he was about to be smeared, in the ugliest publicity campaign of the age, as a merchant of death. Fortunes far larger than his had been lost backing sound technologies at the wrong moment. Westinghouse's willingness to commit anyway, on the strength of physics he trusted over prestige he did not, is the quiet heroism at the center of this half of the story — the gentle industrialist's nerve.

It is worth correcting a misconception about Westinghouse's decision that flatters hindsight at the expense of understanding. We are tempted, knowing how the story ends, to imagine that Westinghouse "saw the future" in some visionary, almost mystical sense — that he perceived, as through a window, the AC world that was coming and simply walked toward it. But this romanticizes what was in fact a hard-headed and genuinely risky business judgment made under real uncertainty. Westinghouse did not know AC would win; he *bet* that it would, which is a different thing entirely. He weighed the physics of transmission loss, which he trusted, against the enormous prestige and installed base of Edison, which was formidable and real, and against the glaring hole in his own system where a motor should be, and he decided the physics would ultimately prevail. It was the judgment of a superb engineer and a shrewd businessman, not the vision of a prophet, and it could have gone wrong in a dozen ways — a financial panic, a technical dead end, a failure to solve the motor problem, a successful destruction of AC's reputation by Edison's propaganda. That it proved right does not make it any less a gamble at the moment it was made. Honoring Westinghouse means seeing the courage of the bet, not pretending he possessed a certainty no one in 1886 could have had.

There is also a revealing contrast in *how* the two great rivals built their organizations, one that would matter enormously as the war unfolded. Edison had built his empire around a single towering figure — himself — and around a technology already frozen into thousands of installations; his organization was, in a sense, an extension of his own genius and his own past choices, magnificent but rigid. Westinghouse built differently. He assembled a company designed around *acquiring and integrating* the best ideas wherever they arose — the transformer from Europe, Stanley's improvements from Great Barrington, and soon the motor from Tesla — and around the unglamorous discipline of distribution and service that he had mastered in the air-brake and gas businesses. Where Edison's strength was invention concentrated in one extraordinary mind, Westinghouse's strength was integration, the capacity to gather other people's breakthroughs into a working whole and deliver it reliably at scale. This difference in organizational DNA is one of the quiet reasons the war went as it did: Edison had bet everything on a technology he had personally perfected and could not bear to abandon, while Westinghouse had built a machine for adopting whatever technology proved best. In a fast-moving field, the flexible integrator beats the brilliant loyalist, and Westinghouse had made himself the integrator.

You meet the fruits of that bet every time you plug anything into a wall, because the wall socket delivers Westinghouse's alternating current, transformed down from a high-voltage line, on the grid architecture his company pioneered. The very fact that a single distant power plant can serve your whole region — rather than a coal-fired dynamo needing to sit within a mile of your house — is the payoff of the wager he placed in that Pittsburgh summer.

The bet was made; the company was building; the one missing gear was a working AC motor. And that gear existed, patented and waiting, in the mind and the portfolio of an eccentric Serbian immigrant whose polyphase system Westinghouse was about to buy — in the deal that would finally arm alternating current for war.

Chapter 39: Buying Tesla's Patents

On the sixteenth of May, 1888, in the meeting rooms of the American Institute of Electrical Engineers in New York, a tall, gaunt, immaculately dressed man of thirty-one rose to deliver a lecture that would change the shape of the electrical world. His name was Nikola Tesla, and the paper he read aloud bore the dry title "A New System of Alternating Current Motors and Transformers." To the assembled engineers it was anything but dry. Tesla described, and to some degree demonstrated, a motor that did the thing everyone had said could not be neatly done: it ran on alternating current directly, with no commutator, no sparking brushes, nothing but a set of stationary coils and a solid rotating drum of metal, spun by a magnetic field that seemed to chase itself around in a circle. It was rugged, simple, and almost magical to watch. In the audience, or soon informed of what had been shown there, were men who understood exactly what it meant. One of them, watching from Pittsburgh through his agents, was George Westinghouse — and Westinghouse understood that the missing piece of his entire enterprise had just been laid on the table.

Within weeks, Westinghouse acted. In July 1888 his company concluded an agreement with Tesla to acquire the rights to his polyphase alternating-current patents — the family of inventions covering the induction motor and the whole system of generating, transmitting, and using AC in the multi-phase form that made the self-starting motor possible. Tesla was also brought to Pittsburgh as a consultant to help Westinghouse's engineers turn his laboratory principles into manufacturable machines. It was the alliance that armed alternating current for the war that was already beginning: Westinghouse now had not just a way to *move* power over distance, but a way to put that power to *work*, turning motors as well as lighting lamps. The half-built system was, at last, whole.

Now we must handle the money carefully, because the terms of this deal are the subject of one of the most durable myths in the history of technology, and this book has promised to correct such myths rather than repeat them. The popular legend runs like this: Westinghouse agreed to pay Tesla a royalty of two dollars and fifty cents for every horsepower of electricity sold under his patents; the royalties swelled toward a fortune of millions; then, years later, when Westinghouse's company was in financial trouble and its bankers demanded the royalty clause be torn up, Tesla — out of pure noble gratitude to the one industrialist who had believed in him — dramatically ripped up the contract with his own hands, sacrificing a fortune to save his friend. It is a beautiful story. It is also, in its most dramatic details, unverified and almost certainly embellished, and the sober truth is both less theatrical and more interesting.

What the records actually support is roughly this. Tesla received, for his polyphase patents, a substantial payment in a mixture of cash and Westinghouse stock — the figure most often cited is on the order of sixty thousand dollars in combined cash, notes, and shares, a considerable sum in 1888, though not the instant vast fortune of legend — together with a royalty arrangement tied to the power sold under the patents. The royalty, had it run for the full life of the patents on the scale Westinghouse's business eventually reached, might indeed have grown into millions. And it is true that the royalty was later renegotiated downward when Westinghouse Electric hit financial turbulence around 1890 and 1891, under pressure from the financiers who were rescuing the company. Whether Tesla literally tore up a physical contract in a grand gesture, or whether — far more plausibly — the arrangement was renegotiated through lawyers and buy-outs in the ordinary way that distressed companies restructure their obligations, is precisely the sort of detail the legend inflates and the documents do not confirm. What is not in doubt is that Tesla surrendered a claim that could have made him one of the richest men of the age, and that he did so in the context of Westinghouse's genuine financial peril and his own genuine loyalty to the man. The nobility was real; the theatrical contract-ripping is probably folklore.

Why does the myth matter enough to correct? Because it feeds the larger legend of Tesla as the pure, unworldly genius forever robbed by grubbier men — a legend that flatters our love of martyrs but obscures the truer, sadder pattern of his life. Tesla was not robbed by Westinghouse; Westinghouse was the one major businessman who dealt with him generously and honorably, paid him real money, and stood by him. If anything undid Tesla financially, it was his own indifference to money as anything but a means to his next experiment, and the merciless logic of capital markets that treated his royalties as an obstacle to be cleared. The relationship between Tesla and Westinghouse was, by the standards of the Gilded Age, a warm and mutually respectful one, and Tesla spoke of Westinghouse with unfeigned admiration to the end of his days. To turn it into a tale of exploitation is to miss both the decency of the one man and the tragic self-neglect of the other.

Now, what exactly had Westinghouse bought? To understand the value of Tesla's patents you have to understand the trick at their heart, the trick Tesla had first glimpsed years earlier in a park in Budapest, reciting poetry at sunset when the whole scheme reportedly burst upon him at once — the trick of the *rotating magnetic field*. This was covered more fully in an earlier chapter, but it is worth recalling in brief here, because it is the reason the deal was worth making. An ordinary motor forces a current back and forth through a spinning armature and uses a commutator to keep flipping the current at the right instants — a clever but fussy and spark-prone arrangement. Tesla's insight was to throw the commutator away entirely and instead make the *magnetic field itself* rotate. He did this by feeding the motor not one alternating current but several, staggered slightly in their timing — two or three separate currents, each surging and ebbing a fraction of a beat behind the last. Arranged around a ring of coils, these staggered currents produce a magnetic field whose peak sweeps around the circle, like the illusion of motion in a ring of theater bulbs blinking on and off in sequence, or like the way a line of people doing "the wave" in a stadium sends a crest traveling around the arena though no single person moves from their seat. Drop a metal rotor into that sweeping field and it is dragged around after the field, chasing it forever without ever catching it, spinning smoothly with no brushes, no commutator, nothing to spark or wear.

That system of staggered currents is what "polyphase" means, and it is what Westinghouse had bought: not just a motor, but a complete and coherent scheme for generating, transmitting, and using alternating current in a form that could both light lamps and turn machinery. It closed the gap that had haunted the previous chapter. Westinghouse Electric was no longer a lighting company with a motor problem; it was, in principle, a full-service power company that could sell electricity for any purpose. Here is a surprising fact to weigh: the basic design of the AC induction motor Tesla patented in 1888 is so sound that motors built on the very same principle turn inside a staggering share of the world's machinery today — in your washing machine, your furnace fan, your refrigerator compressor, the pumps of your town's water supply, and the drive systems of countless factories. It is one of the longest-lived inventions in all of engineering, still doing exactly the job it was designed for more than a century and a third later.

And yet the Pittsburgh period, which ought to have been a triumph, was for Tesla mostly a frustration — and the honest telling of this alliance requires saying so. Tesla was a solitary visionary who worked best alone, following the images in his own extraordinary mind. Westinghouse's Pittsburgh works were a large, busy, practical shop full of capable engineers who had their own established ways of doing things, their own preferences, their own opinions about frequencies and machine designs. Tesla found himself in the maddening position of the theorist among the technicians, forever explaining why his system had to be built *his* way while the shop men, sure of their craft, pushed back. A particular sticking point was the frequency of the alternating current: Tesla's motors were designed to run best at a lower number of cycles per second than Westinghouse's lighting engineers, wedded to their existing higher-frequency lamp systems, wanted to standardize on. The two camps argued; compromises were struck; and Tesla, a man ill-suited to the grind of committee engineering, grew restless and unhappy. Within about a year he left Pittsburgh and returned to New York and his own laboratory, where he could once more be the sole master of his own visions. The frequency dispute, incidentally, echoes down to the present day: the standardization battles of that era are part of why North America eventually settled on sixty cycles per second while much of the world uses fifty — a fossil of these very arguments, embedded now in every wall socket on earth.

It would be easy, and wrong, to read the frustrations of the Pittsburgh period as a failure of the alliance. In fact the alliance succeeded completely at the thing that mattered. Whatever the friction between the lone genius and the industrial shop, Tesla's patents were now Westinghouse's to build, and Westinghouse's formidable manufacturing organization set about turning them into real generators and real motors at real scale. The intellectual property that would ultimately power the Chicago World's Fair and harness Niagara Falls was now in the hands of the one company with the nerve, the capital, and the systems expertise to deploy it against Edison. Tesla had supplied the vision; Westinghouse would supply the world. That division of labor was exactly what each man was built for, and their brief, imperfect partnership armed alternating current with everything it needed to win.

It is worth weighing, fairly, why Tesla ultimately gained so little in wealth from an invention that generated so much, because the honest answer indicts neither Westinghouse nor some shadowy conspiracy but something in Tesla himself and in the nature of invention. Tesla's polyphase system made fortunes — for Westinghouse, for the investors, for General Electric once it too adopted AC, and eventually for the whole world that runs on it. Tesla saw comparatively little of that river of money, and the popular explanation is theft: that grasping capitalists robbed the noble genius. But the truer explanation is sadder and more universal. Tesla surrendered his royalty claim voluntarily, in the context of Westinghouse's real financial distress, out of loyalty and out of an almost total indifference to money as anything but fuel for the next experiment. He was a man for whom the beauty of the idea was the reward, and the commercial exploitation of it a tedious afterthought he was happy to sign away. Combined with his later habit of pouring every dollar he did earn into grand, unremunerative dreams, this temperament, far more than any villainy, is why the inventor of the modern power system died in comparative poverty while the system he invented enriched everyone else. It is a pattern that recurs throughout the history of invention: the person who conceives a world-changing idea and the person who captures its wealth are very often not the same, and the gap between them is usually explained less by crime than by the different talents that conceiving and capturing require.

There is a further irony in the Pittsburgh period worth drawing out, because it illuminates the fundamental mismatch between the two kinds of genius the alliance had joined.

Tesla's mind worked by vision — he saw complete machines in his imagination, turning and glowing in full detail, and regarded the mere business of building them as a formality. Westinghouse's shop worked by iteration — by building, testing, failing, and refining, the patient empirical grind by which a laboratory principle is beaten into a manufacturable product. Neither approach was wrong; both were necessary; but they grated against each other constantly. Tesla found the shop men maddeningly slow to accept what was obvious to him; the shop men found Tesla maddeningly airy about the practical difficulties that were their daily bread. The frequency dispute was only the most visible symptom of this deeper temperamental divide between the visionary and the engineers. And yet — this is the crucial point — the alliance worked precisely *because* it joined these opposites. Tesla's vision without Westinghouse's shop would have remained a beautiful set of patents; Westinghouse's shop without Tesla's vision would have had no motor to build. The friction was the price of a partnership that neither man could have done without, and the electrified world is the product of two incompatible kinds of genius forced, briefly and unhappily, to share a workbench.

You meet the fruit of this 1888 bargain constantly, and almost always without knowing it. The near-silent motor humming in your appliances, the vast induction motors driving industry, the whole notion that a single kind of current can both light your room and run your machines — all of it traces to the patents Westinghouse acquired that summer and the rotating field Tesla had first seen in a Budapest park. Two men who could barely stand to work in the same building had, between them, built the electrical foundation of the modern age.

Now the sides were fully drawn. Edison had his direct-current empire, prestigious, profitable, and physically limited. Westinghouse had his alternating-current system, complete at last with Tesla's motor, and hungry for the whole continent. Two incompatible ways of electrifying a nation now faced each other across an unbridgeable technical divide — and only one of them could win.

Chapter 40: Two Systems, One Country

By the closing months of 1888, a traveler moving across the United States could have seen, if he knew what to look for, two entirely different electrical futures being built at once in the same country. In the dense, wealthy hearts of the great cities — lower Manhattan, parts of Boston, Chicago's Loop — copper wires carried Thomas Edison's direct current from squat coal-fired stations to the mansions, offices, and shops within a mile or so of each plant, feeding steady, reliable, one-directional power to incandescent lamps and the motors of the day. And in scores of smaller towns scattered across the vast American interior — the county seats, the mill towns, the places Edison's system could never profitably reach — a different set of wires, strung higher and humming with a current that reversed itself dozens of times a second, carried George Westinghouse's alternating current from a single distant station across miles of countryside to light streets and homes that would otherwise have stayed dark. Two systems. One country. And a growing, unavoidable recognition on all sides that the country could not, in the long run, run on both.

This is the moment to lay out clearly what was actually at stake, because the "War of the Currents" that was about to erupt is so often told as a simple morality play — good genius versus bad genius, safe versus deadly, future versus past — that its real character gets lost. What was at stake was nothing less than the technical standard on which an entire civilization's power supply would be built, and with it fortunes, reputations, and the shape of the twentieth century. To grasp the ferocity of what followed, you have to grasp three things: why the two systems could not simply coexist, why the stakes were so enormous, and why the men involved could not walk away.

Begin with incompatibility, because it is the physical fact underneath the whole conflict. Edison's direct current and Westinghouse's alternating current were not two flavors of the same product, easily mixed and matched like different brands of lamp oil. They were fundamentally different at the level of the electricity itself. Direct current flows steadily in one direction, like a river; alternating current surges back and forth, like a tide. A motor built for one would not run properly on the other. A meter designed for one would not correctly bill the other. The generators, the wiring standards, the safety practices, the very devices in people's homes were different and largely incompatible. You could not feed Edison's lamps from Westinghouse's lines without adaptation, nor Westinghouse's motors from Edison's stations. Picture two railroads laying track across the same nation with different gauges — different widths between the rails — so that no train from one line could ever run on the other's tracks. Every mile of the wrong-gauge track laid today is a mile that will have to be ripped up or stranded tomorrow, and every locomotive built is a bet that *this* gauge, not the other, will be the one the nation settles on. That was the electrical situation. Each system installed made the eventual reckoning larger, because the more capital sank into one standard, the more painful it became to switch — and the more its owners had to *win*.

Now the stakes, which were staggering and growing by the month. Electrification was not a niche business; it was, everyone could see, going to be one of the largest industries in the history of the world. Light was only the beginning. Beyond light lay electric streetcars, elevators that would make the skyscraper possible, factory motors, and eventually a flood of devices no one had yet imagined, all of them consuming current, all of them requiring a standard. Whoever's system became the national standard would collect, in effect, a toll on a substantial share of the entire future economy — selling the generators, the motors, the meters, the wire, the expertise, and the power itself, decade after decade, to a country and eventually a planet. This was not a fight over a market; it was a fight over *the* market, the foundational infrastructure on which countless other markets would be built. The money at issue was not thousands or millions but, across the arc of the coming century, effectively incalculable. A surprising way to feel the scale: the electric-power industry that grew out of this standards war would become so large and so essential that, more than a hundred years later, a national panel of engineers would name electrification not merely a great achievement but *the* single greatest engineering achievement of the entire twentieth century, ahead of the automobile, the airplane, and the computer. The two men in Pittsburgh and New Jersey were fighting over the keystone of the modern world.

Then there were the personalities, and here the human drama sharpens, because the combatants were almost perfectly matched in stature and almost perfectly opposite in temperament. On one side stood Thomas Edison: the most famous inventor alive, a self-made man of prodigious energy and showmanship, a genius of *invention* and of *integration* who had built not just a bulb but a whole working system and, crucially, had staked his identity, his fortune, and his name on direct current. Edison's direct current was not just a business decision to him; it was, by 1888, woven into his sense of who he was. He had built the first great electric utility in the world on it. His investors had poured their money into it. His engineers had spent a decade perfecting it. To concede that alternating current was the future was, for Edison, to concede that he had backed the wrong horse in the most public and consequential technological race of the age — an admission his pride, his commercial interest, and his genuine (if convenient) safety convictions all forbade. This is essential to understanding the ugliness that followed: Edison did not merely disagree with alternating current; he had every possible motive, financial and psychological, to *destroy* it, and destroying it meant destroying Westinghouse.

On the other side stood George Westinghouse: not an inventor of Edison's flamboyant celebrity but an industrialist of the first rank, the courtly, fair-dealing Pittsburgh magnate we have come to know, who had made his fortune on the air brake and staked his second act on alternating current out of a sober reading of the physics. Where Edison was combative and theatrical, Westinghouse was reserved and methodical. Where Edison invented, Westinghouse organized and distributed. And behind Westinghouse, or rather within his system, stood the third figure, the strange visionary whose motor had made alternating current whole: Nikola Tesla, prophet of the rotating field, who cared nothing for the commercial war and everything for the beauty of the system itself. It is worth pausing to correct the popular framing here, which pits "Edison versus Tesla" as though the two were the great antagonists. They were not. Tesla was never Edison's true opponent in the War of the Currents; he had left Edison's employ years earlier and was, during the fight, mostly absorbed in his own laboratory and his lecture tours. The real war was *Edison versus Westinghouse* — the two organizations, the two industrial systems, the two rival standards — with Tesla's patents as the decisive weapon in Westinghouse's arsenal but Tesla himself standing largely to the side. The "Edison vs. Tesla" duel beloved of the internet is a retrospective simplification; the men who actually went to war were the Wizard and the air-brake magnate.

Why could none of them simply walk away, or compromise, or agree to split the country between them? Partly because the incompatibility made a split unstable — a divided country would be a permanently inefficient one, and market pressure would relentlessly push toward a single standard. Partly because the money was too large and the investors too committed to accept a truce. But mostly because of the peculiar logic of a standards war, which is different from an ordinary business rivalry and worth understanding on its own terms. In most competitions, two firms can share a market indefinitely, each serving its customers. But in a *standards* war, where the whole value lies in everyone using the same system, there is a powerful tendency toward winner-take-all. The more customers adopt one standard, the more valuable it becomes to adopt that same standard — because your motors, your appliances, your trained electricians, your spare parts all work with everyone else's. This is the same logic that later made the world converge on a single kind of railway gauge, a single telephone network, a single internet. It pushes relentlessly toward *one* winner, and it makes the fight existential, because second place in a standards war is not a smaller share of the prize; it is often no prize at all, a stranded system that the market slowly abandons. Edison and Westinghouse both understood this in their bones. Only one current could power the nation's future. The loser would not merely earn less. The loser would be erased.

And so the stage was set for a conflict that would descend, over the next two years, into a spectacle uglier than anything the young electrical industry had yet produced — a campaign of fear and misdirection in which a great man would behave badly, in which animals would be killed in public to make a commercial point, and in which the awful new technology of judicial electrocution would be dragged into a battle over corporate branding. It is a story with real villainy in it, and this book will not soften the villainy or pretend Edison's hands stayed clean. But it is also a story worth telling honestly rather than as caricature, because the men in it were neither saints nor monsters but brilliant, proud, frightened human beings with enormous fortunes and reputations riding on the outcome, making choices under pressures most of us will never face.

It is worth correcting one more misconception before the fighting begins in earnest, because it colors how the whole war is remembered: the idea that the conflict was, at bottom, a clean contest between a "safe" current and a "dangerous" one, with the safety question being the true and honest heart of the matter. The safety question was real, as later chapters will show, but it was not the true axis of the war, and treating it as such accepts the framing that one side deliberately imposed. The true axis of the war was *distribution economics* — the plain question of which system could deliver power to more customers at lower cost — and on that axis alternating current held a decisive and widening advantage. The safety argument was, for the most part, a weapon deployed by the side that was losing the economic argument, a way of changing the subject from "which system is cheaper and reaches farther" (a question with an inconvenient answer) to "which system is more likely to kill your children" (a question that could be dramatized to terrifying effect regardless of the underlying facts). To understand the War of the Currents, one must keep the real axis and the propaganda axis distinct: the war was fought, in the newspapers, over safety, but it was decided, in the marketplace and the physics, over distance and cost. Confusing the battlefield where it was fought with the ground on which it was actually won is one of the most common errors in the popular telling.

Consider, too, the peculiar tragedy that the incompatibility of the two systems imposed on the whole young industry, because it explains the desperation that drove men to such extremes. In most industries, a firm that backs a losing technology can adapt gradually, retooling over time, cushioned by the fact that its old and new products can coexist in the market. But the incompatibility of AC and DC allowed no such gentle transition. Every dollar Edison had sunk into direct-current infrastructure was a dollar that would be worth little or nothing if AC won; every DC dynamo installed was a stranded asset waiting to happen. This meant that the losing side faced not a manageable decline but a potential catastrophe, a wholesale writing-off of enormous invested capital. When the penalty for losing is not merely reduced profits but the near-total destruction of what you have built, men will fight with a ferocity out of all proportion to ordinary commercial rivalry. The incompatibility of the two currents did not merely set the stage for war; it loaded the stakes so heavily toward annihilation that it made the war's eventual ugliness almost inevitable. There was no room for a graceful loser, and so there was no restraint on the fighting.

You meet the resolution of this conflict every time you plug something in, because the war was decided in favor of alternating current, and the wall socket in your home is Westinghouse's victory made permanent — though, in one of history's finer ironies that a later chapter will explore, Edison's direct current would stage a quiet, century-delayed comeback in the long-distance transmission lines and the electronics of the digital age, so that the "loser" now runs inside every device you own. The war chose a standard for its era; it did not, as the myth of "AC simply won" would have it, settle the question of the two currents for all time.

Two systems now faced each other across one country, incompatible, enormously valuable, and championed by men who could not afford to lose. The propaganda was about to begin, the battle lines to be drawn — and the first shots would be fired not with generators or patents but with pamphlets, fear, and the bodies of dead animals.

Part VI

The War of the Currents

From a fear campaign of electrocuted animals to the first man in the electric chair — the ugliest and most consequential business war of the age, fought over the current that would power the world.

Chapter 41: The Battle Lines

In the last weeks of 1887, Thomas Edison sat in his laboratory at West Orange, New Jersey, and did something he did rarely: he grew genuinely afraid. The reports crossing his desk were not the sort a man of his temperament could easily dismiss. Westinghouse's alternating-current systems were spreading through the towns of America faster than anyone at the Edison companies had predicted — dozens of installations in a matter of months, lighting places Edison's own system could never economically reach. The upstart from Pittsburgh had, in barely a year, built something that looked less like a competitor nibbling at the edges and more like a rival empire rising in the hinterland. And Edison, who had spent a decade and a fortune and the whole weight of his reputation building a direct-current world, could read the arithmetic as well as anyone. Alternating current was cheaper to distribute over distance. If nothing stopped it, it would keep spreading. And if it kept spreading, it would win.

It is worth sitting with Edison's position at this moment, because everything ugly that follows grows out of it, and because understanding it is the difference between caricaturing Edison as a cartoon villain and seeing him as what he actually was — a great and proud man cornered by a technology he had bet against. Edison was not a fool who failed to understand alternating current's advantages. He understood them perfectly. That was precisely the problem. He could see the physics of transmission loss as clearly as Westinghouse could; he simply had no good answer to it within the direct-current system he had built and staked everything on. His stations had to sit within roughly a mile of their customers. Westinghouse's could sit in a river valley and serve a county. There was no clever tweak, no incremental improvement to direct current, that could erase that gap. And so Edison faced the bitterest choice a builder can face: admit that the foundation he had laid was the wrong one and rebuild on his rival's principles, or refuse to admit it and fight to destroy the rival instead. He chose to fight.

Consider what was actually on the table for him. Edison's fortune, his fame, and his very identity were fused to direct current. He had not merely invented a bulb; he had, as an earlier chapter showed, invented an entire *system* — the sockets, the meters, the dynamos, the distribution network — and that system was direct current down to its bones. The Edison Electric Light Company and its affiliates had raised enormous capital on the premise of a direct-current future; thousands of installed dynamos and hundreds of thousands of lamps embodied that premise in copper and glass. Wall Street's most respectable money, including that of the financier J. P. Morgan, sat behind it. To concede alternating current was to write down all of that at once — to tell his investors they had backed a losing standard, to tell the public that the Wizard had erred, to tell himself that the defining achievement of his life rested on the wrong choice. For a man of Edison's pride and Edison's commercial exposure, the psychological and financial pressures pushed the same direction, and that direction was: alternating current is not a better technology to be adopted; it is an enemy to be annihilated.

There was also, and this must be said fairly, a kernel of genuine conviction mixed in with the self-interest. Alternating current, as Westinghouse deployed it, really did run at high voltages on its transmission lines — voltages that could kill, and did occasionally kill, when linemen or unlucky passersby made contact with faulty wires. Edison's direct-current distribution, running at a low and relatively survivable voltage inside its one-mile circle, was in a real sense safer for the people near the wires. Edison held, and had held for years before it became commercially convenient, a sincere belief that high-voltage current was dangerous and that low-voltage direct current was the humane choice. Early in the conflict he even wrote, with apparent sincerity, that he did not think anyone would need to run current at the lethal high voltages alternating current required, and that direct current's safety was one of its virtues. Here is a hard truth the honest telling of this story demands: Edison's safety argument was *not entirely wrong*. High-voltage lines were and are dangerous; that is why they are strung far overhead and buried and fenced and marked with warnings to this day. The tragedy — and the moral fault — was not that Edison raised the safety issue. It was the monstrous, dishonest, and cruel way he would go about *exploiting* it, magnifying a real but manageable hazard into a campaign of engineered terror, and doing so, unmistakably, because his business was losing on the merits.

The opening shots of the war were, at first, the ordinary weapons of commercial rivalry — pamphlets and words. In early 1888 the Edison camp began circulating warnings about the dangers of alternating current, culminating in a bound booklet issued under the Edison Electric Light Company's name with the ominous title "A Warning." Bound in red covers, it catalogued, with grim relish, every accident and death that could be attributed to alternating-current wires, and it named Westinghouse and the other AC companies directly, painting them as reckless purveyors of a deadly current who were stringing lethal high-voltage lines over the heads of an unsuspecting public in pursuit of profit. It was propaganda, but it was propaganda with a real grievance woven through it, which is what made it effective and what made it insidious. The next chapter will look hard at this "red pamphlet" and the fear campaign it launched; for now it is enough to mark it as the moment the conflict crossed a line — from competing in the market to campaigning to have the rival's product branded, in the public mind, as an instrument of death.

A misconception worth clearing away here, because it distorts the whole picture, is the idea that Edison personally masterminded and micromanaged every dark deed of the war from a swivel chair, a lone genius pulling the strings of a grand conspiracy. The reality was messier and, in a way, more damning. The Edison side of the war was waged by a cast of players with overlapping and not-quite-identical motives: Edison himself, whose name and prestige gave the campaign its authority and whose laboratory quietly lent it resources and legitimacy; his business managers and lawyers, defending the company's capital; and freelance actors like the electrical engineer Harold Brown, who would soon become the campaign's most notorious foot soldier and whose exact relationship to Edison remains one of the genuinely contested questions of the story. Edison did not need to script every horror personally. He needed only to set the direction, lend his name and his laboratory's tacit backing, and let ambitious men who understood which way the wind blew do the rest. When we come to weigh his moral responsibility, this distinction will matter — not because it exonerates him, but because it locates his guilt precisely: not in every grisly detail, but in choosing to unleash and bless a campaign he knew was aimed at fear rather than truth.

The stakes that made all this worth doing bear repeating in concrete terms, because they explain why sober businessmen were willing to descend into such tactics. This was not a squabble over a lucrative contract. It was a war over the standard on which an entire civilization's electricity would run, and therefore over a stream of profit that would flow, quite literally, for as long as the lights stayed on — which is to say, forever. Whoever won would sell the generators and motors and meters and wire and power to a nation and then a world, decade upon decade. Whoever lost would watch his system slowly stranded, his capital written off, his name attached to the losing side of the greatest technological contest of the age. When the prize is that large and the logic of a standards war is that unforgiving — winner takes all, loser takes nothing — men who would never dream of such conduct over an ordinary deal will reach for weapons they would otherwise be ashamed to touch. That is not an excuse. It is an explanation, and the honest reader should hold both the explanation and the judgment at once.

Here is a surprising fact that captures the strange doubleness of Edison in this period. The same man who was preparing to brand alternating current as a killer was, in his own laboratory, employing the best engineers of the age and running the most productive invention factory in the world; the same restless brilliance that had given humanity the practical light bulb and the phonograph was now bending itself toward a campaign of fear. Greatness and pettiness lived in the same skull, and the War of the Currents is the episode where the pettiness, driven by fear and money, briefly got the upper hand. It is a caution worth carrying out of this book: that genius confers no immunity from bad faith, and that the very traits which make a man a great builder — his conviction, his pride, his refusal to quit — can, under the right pressure, curdle into the traits that make him a dangerous adversary.

It is worth naming, before we leave the drawing of the battle lines, the strategic mistake at the heart of Edison's decision to fight rather than switch, because it is one of the great cautionary tales in the history of business. Edison had, in principle, every resource he needed to win the AC future as easily as the DC present. He commanded the finest industrial laboratory in the world, a peerless organization of engineers, the most trusted brand in electricity, and the backing of the deepest pockets on Wall Street. Had he chosen, in 1888, to say simply, "Alternating current is the better tool for distribution; we shall build the best AC systems in the world and dominate that market as we have dominated this one," he might well have crushed Westinghouse on the merits and made himself the titan of the AC age. Instead, his pride and his sunk investment in direct current led him to defend the indefensible, to pour his genius and his reputation into destroying the future rather than owning it. It is the classic error of the successful incumbent, repeated a thousand times since in a thousand industries: the leader so wedded to the technology that made him great that he fights the technology that will replace it, and is destroyed not by his rival's strength but by his own refusal to adapt. Edison, of all people, had the means to lead the transition. He chose to resist it, and the resistance is what ruined his position, not any deficiency in his resources.

A misconception worth dispelling here is the notion that Edison "didn't understand" alternating current, that his opposition sprang from technical ignorance or an inability to grasp the transformer and the induction motor. This is precisely backward, and it flatters Edison in the wrong way while insulting him in another. Edison understood alternating current perfectly well; he had brilliant engineers who could have built AC systems whenever he gave the word, and some of them privately urged him to do exactly that. His opposition was not ignorance but its opposite — a clear-eyed recognition that AC threatened the specific empire he had already built, combined with a refusal to sacrifice that empire for the larger future. The tragedy of Edison in the current war is not that he was too dim to see AC's merits. It is that he saw them clearly and fought them anyway, because seeing them meant admitting his own system was doomed. There is no comfort in imagining he simply failed to comprehend the physics; the harder truth is that he comprehended it and let interest and pride override comprehension. That is a more human failing than ignorance, and a more instructive one.

You meet the residue of these battle lines every time you notice a warning sign on a transformer or a buried-cable marker, or hear a utility fret publicly about safety — the whole modern vocabulary of electrical danger was shaped, in part, by a fear campaign that weaponized real hazards for commercial ends. And you meet it, too, in the enduring difficulty of telling this story straight: more than a century on, we still struggle to hold in one hand Edison the incandescent genius and in the other Edison the propagandist, because we prefer our heroes and villains unmixed, and he insists on being both.

The battle lines were drawn: Edison's threatened empire on one side, Westinghouse's rising system on the other, and between them a public that would have to be persuaded, by fair means or foul, which current to fear. Edison had chosen his weapon. It was fear itself — and the next phase of the war would show just how far he was willing to carry it.

Chapter 42: Edison's Fear Campaign

Sometime in early 1888, a slim booklet bound in red covers began to circulate among the newspaper editors, city officials, and businessmen of the United States. Its title was a single word designed to make a reader stop and shiver: "A Warning." Issued under the name of the Edison Electric Light Company, its pages did not argue the fine points of electrical engineering. They told stories — story after story of men and animals killed by contact with wires, of linemen burned black on the poles, of the sudden and grotesque deaths that could come from touching the high-voltage alternating current that Westinghouse and his imitators were stringing over the streets of America. The booklet named names. It pointed directly at the alternating-current companies and portrayed them as reckless merchants of death, hanging lethal wires above the heads of an unsuspecting public for the sake of profit. It was, in the plainest terms, a campaign to make Americans afraid of a technology — and it worked.

This chapter must do something the rest of this book has mostly been able to avoid: it must watch a great man behave badly, deliberately, and over a sustained period, and it must neither look away nor pretend the badness was something less than it was. The fear campaign that Thomas Edison waged, and blessed, and lent his laboratory and his priceless name to, against alternating current between 1888 and 1890, is the ugliest chapter in the life of one of history's most celebrated inventors. Telling it honestly is not Edison-bashing; it is the opposite of the hero-worship that has too often sanitized him. The point is not that Edison was secretly a monster. The point is that a genuinely great and creative human being, cornered by commercial fear, chose again and again to reach for terror and deception instead of the honest competition he would surely have won some arguments in and lost others. That is a more useful and more sobering truth than any cartoon of villainy.

Begin with what the campaign actually claimed, because part of what makes it insidious is that it was not simply a pack of lies. Its core factual premise was true: high-voltage alternating current, running at the thousands of volts used on transmission lines, could kill a person who contacted it, and in the early, loosely regulated years of the industry it did kill people, especially the linemen who worked among a chaotic tangle of poorly insulated wires overhead. Edison's genuine and long-standing conviction that high voltage was dangerous had a real basis. But a fear campaign does not work by lying outright; it works by taking a real and manageable risk and inflating it, stripping it of all context and proportion, until a hazard that could be handled with insulation and regulation and sensible practice is made to seem like an omnipresent, uncontrollable menace lurking in every wire. Edison's "Warning" did not say, truthfully, that high-voltage lines required careful engineering and regulation, as gas mains and steam boilers and railroads all did. It said, in effect, that alternating current was death itself, and that any city that permitted it was inviting the slaughter of its citizens. It converted a safety problem into a horror story.

The rhetorical masterstroke — and it was a stroke of dark genius — was to attach to alternating current a single, unforgettable, damning phrase. Edison and his allies worked to brand AC as the "executioner's current," the current fit only for killing. This was not idle name-calling; it was a calculated act of what a later age would call branding, an attempt to fuse, in the public mind, the very idea of alternating current with the idea of death, so that no one could think of Westinghouse's product without thinking of a corpse. The whole later campaign to have the new electric chair powered by alternating current, which the coming chapters will follow, was aimed at exactly this: to make the association permanent and official, to have the state itself certify that AC was the current used to kill condemned men. If Edison could make "alternating current" and "the current that executes criminals" mean the same thing in the public imagination, he would not need to win the engineering argument. He would win the argument in people's guts, where arguments are really won.

Now the ethics, squarely faced. There are defenses one could mount for Edison, and honesty requires stating them before rejecting them. He genuinely believed high voltage was dangerous; the campaign therefore had a foundation in sincere conviction, not pure cynicism. High-voltage lines genuinely did kill people in those years; the danger he pointed at was real. And every business in a mortal competition puts its best face forward and its rival's worst face forward; some of what Edison did was ordinary hardball. All of that is true, and none of it is enough. Because the honest measure of the campaign is not whether it contained true facts but whether it was aimed at *truth* or at *fear* — whether its purpose was to inform the public so they could make a sound judgment, or to frighten the public into a judgment against a rival Edison was losing to on the merits. And on that measure the verdict is not close. The campaign was designed to terrify. It magnified real risks past all proportion, it suppressed the context that would have made those risks manageable, and — as the animal electrocutions and the electric-chair maneuvering would soon show — it was willing to stage grotesque public spectacles of death purely to drive the fear home. A man who truly wanted only to protect the public from high voltage would have campaigned for regulation, insulation, and safety standards, all of which existed and all of which would in time make high-voltage AC as safe as any other well-managed hazard. Edison campaigned instead to make people afraid. The difference between those two things is the difference between public service and demagoguery, and Edison, brilliant enough to know exactly what he was doing, chose demagoguery because it served his balance sheet.

It is worth correcting, here, a misconception that runs in the *opposite* direction from Edison-worship and is equally unfair — the modern tendency, fed by internet memes, to render Edison as a pure villain and cartoon monster, a mustache-twirling fraud who stole everything and murdered animals for fun. That, too, distorts the record. Edison in this period was not a villain by nature; he was a builder who had spent his life making things people wanted, and who, under a specific and enormous commercial threat, made a series of genuinely wrong and cruel choices. The fear campaign was a moral failure, but it was the moral failure of a particular man at a particular pressure point, not the true face of a lifelong scoundrel. Holding both of these in mind — Edison the extraordinary inventor and public benefactor, and Edison the author of a shameful campaign of engineered terror — is harder than embracing either the hero or the villain alone, but it is the only honest posture. Great people do bad things. The badness does not cancel the greatness, and the greatness does not excuse the badness. Both are simply true.

A surprising and telling fact about the campaign is how *modern* it feels. Everything the reader recognizes from present-day disinformation is already present in embryo in Edison's fear campaign of the 1880s: the true-but-misleading fact, the emotionally overwhelming anecdote deployed to swamp statistics, the memorable damning label ("executioner's current"), the staged spectacle engineered for maximum horror and press coverage, the commercial interest hidden behind the mask of public concern. The War of the Currents was, among other things, one of the first great modern misinformation campaigns, run by one of the first modern celebrities, over one of the first mass technologies — and studying it is uncomfortably like reading tomorrow's news. Another surprising fact: the campaign genuinely rattled Westinghouse, who was by temperament a builder and not a brawler, and who found himself forced to fight not over the technical merits he was confident of but over a manufactured public terror he had never anticipated. He responded with facts, with statistics on gas explosions and other everyday dangers, with reasoned rebuttals — the tools of a rational engineer — and he found, as rational engineers so often do when facing a fear campaign, that facts are a slow and clumsy weapon against a vivid horror story. The truth, as the saying would later go, was still lacing its boots while the terror ran around the world.

It is worth dwelling for a moment on the particular tragedy of Edison being the author of this campaign, because it sharpens the moral point rather than blunting it. Edison was, more than almost any figure of his century, a *popularizer* of electricity — the man who had taken a dangerous, mysterious force and packaged it into the friendly, domestic form of the incandescent lamp, who had put electric light into homes and made it seem safe and homely and good. His entire career had been an argument that electricity was humanity's servant, not its enemy. And now, out of commercial fear, he turned that same authority inside out, using his unrivaled credibility as the man who had domesticated electricity to persuade the public that a particular form of it was an instrument of death. He was uniquely positioned to reassure, and he chose instead to terrify. It is as though the person who had taught a nation that fire could be tamed into a warm hearth now went about warning them that their neighbor's hearth would burn their children alive. The betrayal was not only of Westinghouse but of Edison's own life's work, of the very message of safety and progress he had spent decades building. That is what makes the fear campaign not merely a business tactic gone too far but a genuine corruption of a man's better self.

Consider, too, a surprising asymmetry that the campaign exploited and that remains one of its most instructive features. Edison's direct current was genuinely safer at the point where ordinary people touched it — inside the home, at low voltage — while Westinghouse's alternating current was genuinely more dangerous at the point where few people ever touched it, on the high-voltage transmission lines strung overhead. A fair-minded observer would have concluded that both currents were safe when properly handled and dangerous when abused, and that the sensible response was regulation and good engineering, not panic. But the fear campaign performed a subtle sleight of hand: it took the danger that lived out on the distant high-voltage lines and made the public feel it in their own parlors, as though the executioner's current were coiled inside their table lamps waiting to strike. It transferred a real but remote hazard into the intimate space of the home, where it did not belong, and in doing so it manufactured a dread wildly out of proportion to the actual risk any householder faced. This relocation of fear — taking a genuine danger from where it exists and planting it where it will frighten most — is among the oldest and most effective tricks in the propagandist's book, and Edison's campaign is one of its purest early specimens.

You meet the legacy of Edison's fear campaign in more places than you might think. You meet it in the enduring public dread of "high voltage," a dread that is partly rational and partly the inherited residue of a deliberate scare. You meet it every time a company today runs a fear-based campaign against a rival's product, dressing commercial interest in the costume of public safety — a genre of persuasion Edison did not invent but did much to perfect. And you meet it, most importantly, as a permanent caution about how easily fear can be manufactured and sold, even by brilliant and otherwise admirable people, when enough money rides on frightening you.

But the fear campaign was, so far, mostly words and pamphlets. To make the terror unforgettable, Edison's side would need something more visceral than a red booklet — something the public could see with their own eyes and never forget. They would need a demonstration. And so the campaign moved from the printed page to the laboratory bench, where a man named Harold Brown was about to begin killing animals in public to prove that alternating current was death.

Chapter 43: Harold Brown and the Electrocuted Dogs

On the afternoon of July 30, 1888, in a lecture hall at Columbia College in New York, a crowd of invited spectators, reporters, and members of a state commission watched a man named Harold P. Brown lead a large black retriever into a wooden cage. The dog weighed some seventy-six pounds. Brown, a self-styled electrical expert with an air of scientific authority, wired the animal up and, before the assembled witnesses, applied direct current to it — Edison's current — in rising doses. The dog howled and strained and suffered, but at the voltages Brown used, it lived. Then Brown switched to alternating current — Westinghouse's current — at a fraction of the voltage, and the animal died almost at once. The lesson Brown wanted his audience to draw was crude and unmistakable: direct current hurts but spares; alternating current kills. He would repeat this grim performance, with variations, in the months that followed, working his way from dogs up through calves and eventually a horse, staging in public a series of electrical killings whose entire purpose was to brand alternating current, in the most visceral way imaginable, as the current of death.

Harold Brown is one of the strange, dark figures of this story, and to understand the electrocuted dogs we have to understand what he was and what he was not. He was not, at least not openly, an employee of Thomas Edison. He presented himself as an independent electrical engineer acting out of public-spirited concern for safety, a private citizen alarmed by the deadly wires going up over America's streets. That was the public story. The reality, which emerged in fragments then and more fully later, was murkier and more damning. Brown had quiet, material support from Edison's organization. He was given access to Edison's West Orange laboratory and its resources; Edison's own chief experimenters assisted him; the instruments and animals and space that made his demonstrations possible flowed, at least in part, from the Edison establishment. The precise degree of Edison's personal involvement in Brown's work remains one of the genuinely contested questions of the whole affair, and honesty requires flagging it rather than papering over it. What is not contested is that Brown could not have done what he did without Edison-side backing, and that Edison, at minimum, knew of it, permitted it, and let his laboratory serve as its stage. The arrangement gave Edison a precious thing: deniability. The killings could be presented as the work of a concerned independent expert while the Edison camp reaped the propaganda benefit at arm's length.

This is the place to correct one of the most persistent and colorful misconceptions attached to the entire War of the Currents, because it has attached itself specifically to this theme of animal electrocution and it badly distorts the public understanding of what happened. Almost everyone who has heard anything about Edison and animal electrocutions has heard about Topsy the elephant — the circus elephant supposedly killed by Edison with alternating current to smear Westinghouse, often with the detail that Edison filmed it for propaganda. The story is a mess of error. Topsy the elephant was indeed electrocuted, and the event was indeed filmed by the Edison film company — but it happened in 1903, at a Coney Island amusement park, more than a decade *after* the War of the Currents had already been decided and alternating current had already won. By 1903 there was no current war left to fight; the elephant's death had nothing to do with the AC-versus-DC struggle. Topsy was a dangerous animal that had killed a man and was condemned to die; her electrocution was a grisly public spectacle staged by the amusement park, not a battle in a commercial war, and Edison's personal role in ordering or attending it is, at best, greatly overstated and probably nonexistent. The Edison company filmed many sensational events; filming Topsy's death no more proves Edison masterminded it than filming a fire proves he set it. The real, documented, morally serious story of animal electrocution in the War of the Currents is not the famous elephant at all. It is the dogs, calves, and horses that Harold Brown killed in 1888, years earlier, while the war was actually being fought. Remember the dogs; forget the elephant. The truth is grimmer and less cinematic than the myth, and it belongs to a different decade.

Return, then, to what Brown actually did, and to the sheer bad faith engineered into his demonstrations, because the details matter for judging them. Brown's performances were rigged to produce a predetermined conclusion. He applied direct current to his animals at levels chosen to make them suffer without killing, and then applied alternating current at levels chosen to kill, and presented the different outcomes as proof of the two currents' essential natures — as though the currents themselves, and not the voltages and durations Brown selected, were doing the talking. It was the electrical equivalent of dosing one glass of water with a spoonful of salt and another with a spoonful of cyanide and concluding that water is safe but seawater deadly. Any competent electrician knew that either current, at sufficient voltage and duration, would kill, and that either, properly handled, could be survived; the lethal variable was the electricity's magnitude and how it passed through the body, not the mere fact of its alternating or being direct. Brown's demonstrations proved nothing about the relative safety of the two systems as actually deployed. They were not experiments seeking a truth; they were theater staging a conclusion. But they were *effective* theater, because a dying dog in a cage speaks to the gut in a language that no statistical table about transmission voltages can answer, and Brown understood that as well as any propagandist ever has.

There was a personal edge to Brown's crusade, a genuine and venomous rivalry between him and George Westinghouse that gave the campaign its persistence. Brown and Westinghouse traded public accusations and challenges; at one point Brown, in a fit of bravado, publicly dared Westinghouse to a kind of electrical duel, proposing that the two of them submit their own bodies to their respective currents to prove which was safe — Westinghouse to take Brown's direct current, Brown to take alternating, in escalating doses until one man cried off. Westinghouse, sensibly and with the dignity of a man who refused to be dragged into a carnival, declined to participate in such a stunt. The proposed duel captures the whole grotesque flavor of the affair: a public fight over a foundational technology conducted not through data and demonstration of the ordinary kind, but through spectacle, dare, and the threat of electrocution. It was as if the future of the electrical world were to be settled by a sideshow.

A surprising fact to weigh: for all the horror of Brown's demonstrations, and for all Edison's backing of them, the campaign's most consequential achievement was not to kill animals but to shape a *policy* — for it was Brown, armed with his demonstrations and his claims of AC's peculiar lethality, who became a central figure in the state of New York's move to adopt electrocution as its method of capital punishment and, crucially, to power its new electric chair with alternating current. The dogs in the cage at Columbia were, in a sense, rehearsals for a larger and darker performance: the electrocution of a human being, staged so as to stamp Westinghouse's current forever with the mark of the executioner. Brown would work, quietly and through intermediaries, to make sure the state's killing machine ran on Westinghouse's dynamos. The next chapters will follow that maneuver to its horrifying conclusion.

It is worth pausing on the human cost of these demonstrations in a way the historical accounts usually rush past, because the dogs were not abstractions. Brown obtained many of his animals from the streets and from local boys who were paid to round up strays — a bounty on neighborhood dogs, in effect, to feed a propaganda machine. The animals were terrified, and they suffered; the early demonstrations in particular, before Brown had calibrated his lethal doses, involved dogs that were shocked repeatedly without dying, howling and struggling in the cage while an audience of respectable men took notes. Some observers, even in an era far more callous about animals than our own, were sickened by what they saw and said so, and a current of public revulsion began to build against the spectacles even among people who had no stake in the war between the currents. There is a bitter irony buried here: the campaign meant to prove that alternating current was uniquely cruel ended up demonstrating, more than anything, the cruelty of the men conducting it. The horror the audiences carried away was as much about Brown as about the current in his wires.

A second point deserves emphasis because it cuts against the grain of the propaganda and is often missed. Brown's demonstrations were designed to prove that alternating current was categorically more dangerous than direct current — a difference of *kind*, as though AC were a poison and DC merely a nuisance. But the actual physics tells a subtler and more symmetrical story. What determines whether an electric shock kills is chiefly the amount of current that passes through the body and the path it takes, especially whether it crosses the heart and disrupts its rhythm. Both alternating and direct current can kill; both, at low enough levels, can be survived. There are real differences in *how* they affect tissue — alternating current at the low frequencies used in power systems is, at certain levels, somewhat more likely to throw the heart into the fatal quiver called fibrillation, which is a genuine and not-trivial fact — but this is a difference of degree and detail, heavily dependent on circumstances, not the black-and-white difference of nature that Brown staged. He took a nuanced, conditional truth and inflated it into a categorical horror, which is the essential move of all effective propaganda: find a real difference, then magnify it past all proportion and strip away every condition that would let the audience judge it fairly.

The moral reckoning here should be plain but not overdrawn. Harold Brown was a willing instrument, a man who chose to build a career on public cruelty in the service of a commercial cause, and he bears his own guilt for it. But he was an instrument, and the hand that supplied him — the laboratory, the resources, the tacit blessing, the invaluable cover of a great man's name — was Edison's. Edison did not, so far as the record shows, personally wire the dogs. He did something in its way worse: he created and sustained the conditions in which a Harold Brown could flourish, and he took the benefit while keeping his own hands ostensibly clean. That is the characteristic moral shape of the whole fear campaign, and it is why the honest verdict on Edison in these years is neither exoneration nor caricature but something harder — that a great man knowingly lent his greatness to a squalid cause, and let others do the squalid work.

You meet the legacy of Brown's demonstrations, uncomfortably, in the ethics of every laboratory that now uses animals, in the regulations and review boards that exist precisely because a less scrupulous age once killed animals for publicity as casually as Brown killed his dogs. And you meet it in the enduring power of the visceral demonstration over the reasoned argument — the reason a single vivid image still outruns a page of sober data in any public fight, exactly as a dying dog once outran the truth about two currents.

Brown had proven he could kill an animal in public and make the newspapers watch. The next step was to make the killing official, permanent, and human — to arrange that when the state of New York first put a man to death by electricity, the current that did it would be Westinghouse's, so that "to be electrocuted" and "to be Westinghoused" might come to mean the same terrible thing.

Chapter 44: "To Westinghouse a Man"

In the summer of 1888, the state of New York found itself with a problem that no government had ever faced before: it had decided to kill people with electricity, and it did not know how. A commission had been convened to find a more humane alternative to hanging, that ancient and frequently botched method in which the condemned might strangle slowly or, if the drop was misjudged, be decapitated outright. Electricity, the marvel of the age, promised something cleaner — a swift, painless, modern death at the flip of a switch. The legislature had passed a law; from the first of January, 1889, New York would execute its condemned criminals by electricity. But the law did not specify *how much* electricity, or of *what kind*, or delivered by *what machine*. Those questions were left to the experts. And into that vacuum stepped Harold Brown and, behind him, the whole apparatus of Edison's fear campaign, with an agenda that had nothing to do with humane execution and everything to do with a word they hoped to plant in the American language: to "Westinghouse" a man.

The scheme was as cynical as anything in the whole sorry war, and it was breathtaking in its ambition. The Edison camp did not merely want to win a commercial contest. They wanted to fuse, permanently and officially, the name of their rival with the act of judicial killing. If the state's new execution device could be made to run on alternating current — on Westinghouse's dynamos specifically — then every execution would be, in effect, a state-sponsored advertisement for the lethality of Westinghouse's product. The propaganda would no longer be a red pamphlet or a dead dog in a lecture hall; it would be the government of New York itself, solemnly certifying, every time it put a criminal to death, that alternating current was the current that kills. And the campaigners went further still. They actively promoted the idea that the verb for this new form of execution should be "to Westinghouse" — that to electrocute a condemned man should be called "to Westinghouse" him, the way one might say "to guillotine." Imagine the branding catastrophe: a company whose very name becomes the common word for being executed. No fear campaign in history had ever aimed at so total a poisoning of a rival's identity. They were trying to turn Westinghouse's name into a synonym for death.

To make it happen, Brown needed Westinghouse's machines, and here the story takes a turn into something close to farce, if farce could be this dark. Westinghouse, understanding perfectly well what was being attempted, refused to sell his dynamos for use in executions. He would not knowingly furnish the rope for his own hanging. So Brown and his allies resorted to subterfuge. Working through intermediaries and cutouts, they quietly acquired several Westinghouse alternating-current generators on the secondhand market — buying them through third parties so that Westinghouse could not block the sale — and delivered them to the New York prison authorities for use in the electric chair. It is one of the peculiar, grimly comic facts of the affair that the Edison side had to go sneaking around buying their rival's equipment on the sly, precisely because that rival's equipment was the whole point: only a killing performed with a genuine Westinghouse dynamo would carry the intended message. The machine had to bear the right name. And so, through back channels and secondhand dealers, Westinghouse generators were procured for the express purpose of killing a man in a way that would stain their maker's brand.

The word Brown chose to use for the new machine's current was itself a weapon. Around this time the term "electrocution" was coined — a blend of "electricity" and "execution" — and Brown and his allies pushed hard to have the killing described specifically as being "Westinghoused." The campaign to attach the rival's name to the act was conducted openly in the press and in the technical debates of the day. It was a bid to write the outcome of a commercial war directly into the dictionary, to make the English language itself carry Edison's propaganda forward long after the particular contracts and installations of 1888 were forgotten. Had it fully succeeded, we might today speak of criminals being "Westinghoused" the way we speak of them being "guillotined," and the name of a decent and gifted industrialist would have become a byword for state killing. That it did not fully succeed — that "electrocution," not "Westinghousing," entered the language — is one of the small mercies of the story, and a reminder that even the most determined propaganda does not always take.

It is worth pausing to correct a misconception about the electric chair's origins, because the popular version oversimplifies. The electric chair is often described flatly as "Edison's invention" or "Edison's idea," conjured up wholesale as a weapon against Westinghouse. This is not quite right, and the truth is more tangled. The impulse to find a humane replacement for hanging came from reformers and from the New York legislature, not from Edison; the underlying desire was genuine, and predated the current war's involvement. What the Edison camp did was not invent the electric chair from nothing but *capture* an existing reform effort and bend it to their commercial purpose — steering the choice of current toward alternating current, supplying the technical testimony that AC was the swift and certain killer, and working to attach Westinghouse's name to the result. The distinction matters because it locates the guilt accurately. Edison's people did not dream up capital electrocution as a plot; they hijacked a real and independently motivated policy and weaponized its details. It is a subtler crime than the myth suggests, and in some ways a more insidious one, because it worked by corrupting a genuine public-safety reform rather than by fabricating something out of whole cloth.

Edison himself lent the effort his most valuable asset, which was his testimony and his name. When the legal questions arose — was electrocution constitutional, was it humane, which current was appropriate — Edison was consulted as the great authority on all things electrical, and he gave testimony that supported the use of alternating current for the purpose, framing it as the current best suited to producing death. Coming from the most famous inventor in America, this carried enormous weight. It is a chilling thing to contemplate: the man who had given the world the friendly glow of the incandescent lamp now testifying, in effect, that his rival's current was ideally suited to killing human beings, and doing so in the knowledge that this testimony served his own commercial interest. Whatever residue of sincere safety conviction remained in Edison at this point — and some surely did — it had become impossible to separate from the naked commercial motive. He was helping to build a machine to kill a man, and helping to ensure it bore his rival's name.

A surprising fact that captures the moral vertigo of the moment: Edison had, for most of his life, professed a personal opposition to capital punishment. He did not, by his own account, believe the state should be killing people at all. And yet, when the state resolved to do so anyway, he threw his expertise and his prestige into ensuring it did so with alternating current — participating in the machinery of an execution he claimed to find abhorrent, because participating served his war against Westinghouse. One can hold contradictory convictions; human beings do it constantly. But there is something especially bleak in a man overriding his stated moral opposition to the death penalty in order to score a commercial point off a competitor. It is perhaps the clearest single window into how far the fear campaign had carried him from his better self.

To feel the full strangeness of the maneuver, consider it from Westinghouse's side of the table. Here was a man who had spent his life building machines to *save* human life — the air brake that kept trains from crushing their crews, the fail-safe signaling that kept them from colliding, the metering and regulation that made dangerous fuels safe to pipe into homes. His entire engineering philosophy was organized around the principle that a well-designed system should protect the people who touched it. And now his greatest commercial creation, the alternating-current dynamo, was being covertly bought up and repurposed into a machine whose sole function was to kill a strapped and helpless man, expressly in order to smear the very name he had built that reputation on. It is hard to imagine a more precisely aimed insult. Westinghouse understood the maneuver perfectly, and his refusal to sell his machines for the purpose, followed by his funding of the condemned man's appeals, was not merely commercial self-defense; it was the response of a man whose sense of what his life's work was *for* had been directly assaulted. The gentle industrialist of the air brake found himself dragged into a fight over a death chamber, and he fought it with the same fail-safe instinct he had brought to everything else — trying, this time, to design the killing out of existence altogether.

There is a further layer of cynicism worth exposing, because it reveals how thoroughly the technical question had been subordinated to the commercial one. The choice of alternating current for the electric chair was justified, publicly, on the grounds that AC was the swifter and more certain killer, and therefore, paradoxically, the more *humane* choice for an execution meant to be instantaneous. But this framing was itself a piece of the propaganda war, not a neutral finding of science. If AC was uniquely deadly enough to be the ideal current for humane execution, then, the argument implicitly ran, it was surely too deadly to be strung over the streets where ordinary citizens lived and worked. The same claim was made to serve two opposite conclusions at once — AC is the best current to kill a condemned man mercifully, *and* AC is the worst current to permit near innocent families — and the contradiction went largely unremarked because both conclusions pointed the same commercial direction: away from Westinghouse. It was a bravura performance of having it both ways, and it worked precisely because the public had no way to see that the "humane execution" argument and the "public menace" argument could not both be fully true.

You meet the residue of this episode, oddly, in the very word we use. "Electrocution" — that blend of electricity and execution that Brown's circle helped popularize — is now a common English word, applied to any death by electric shock, accidental or deliberate. Every time you read that someone was "electrocuted" by a downed power line or a faulty appliance, you are using a piece of vocabulary forged in the propaganda furnace of the War of the Currents. The word outlived the war that made it. And the near-miss of "Westinghousing" survives as a historical curiosity, a reminder of how close a great man's name came to being permanently defiled by a rival's cunning.

It is worth stepping back to marvel, grimly, at the sheer audacity of the whole conception, because there is nothing quite like it in the history of commercial rivalry. Businesses have always attacked their competitors' products, disparaged their quality, warned of their dangers. But the scheme to power the electric chair with Westinghouse's dynamos and to make "Westinghousing" the word for execution went far beyond ordinary disparagement into something closer to an attempt at permanent character assassination through the machinery of the state itself. The goal was not merely to say that Westinghouse's product was dangerous but to enlist the government of New York as an unwitting partner in a branding campaign, to have every future execution serve as a state-certified advertisement for the lethality of a rival's goods, and to poison the language so thoroughly that the rival's name could never again be spoken without a shudder. One struggles to think of a modern parallel — it would be as if a company contrived to have a competitor's brand name become the official term for a method of capital punishment, printed in statutes and spoken in courtrooms forever. That serious, respected businessmen conceived and pursued such a scheme is a measure of how the enormous stakes of the current war had unmoored the ordinary restraints of commercial conduct. When the prize is the electrification of a nation, apparently, even the death chamber becomes a marketing opportunity.

The machines were procured, the current was chosen, the testimony was given, the word was coined. All that remained was a condemned man to test the apparatus on — and in a jail in Buffalo sat a convicted murderer named William Kemmler, whose wretched death would become the horrifying climax of the whole campaign, and the moment when the War of the Currents reached its moral nadir.

Chapter 45: Kemmler

At a little after six o'clock on the morning of August 6, 1890, in a small brick outbuilding at Auburn Prison in upstate New York, a thirty-year-old convicted murderer named William Kemmler was led into a room where a heavy oak chair stood waiting, trailing wires. He was calm — witnesses remarked on his composure — and as the guards began to strap him in, he is reported to have said, in effect, take your time and do it right, I am in no hurry. They fastened the leather straps across his chest and arms and legs, fixed an electrode to the crown of his shaved head and another to the base of his spine, and made ready to demonstrate to the world that electricity could deliver a swift and painless death. What followed instead was one of the most horrifying scenes in the history of American justice, a botched, prolonged, and sickening killing that turned the electric chair's triumphant debut into a public catastrophe — and that stands, fittingly, as the moral nadir of the entire War of the Currents.

William Kemmler was not a sympathetic man. He was a Buffalo peddler, an alcoholic and largely illiterate, who had murdered his common-law wife with a hatchet in a drunken rage — a brutal, squalid crime with no ambiguity about his guilt. He is worth remembering by name precisely because the men who fought over his method of death cared so little about him as a human being; to the combatants in the current war, Kemmler was not a person but a demonstration, a body on which to test a proposition about electricity. That is part of what makes his story a nadir. He became the first human being ever executed by electric chair, and the manner of his dying was shaped less by any concern for justice or mercy than by a commercial war being waged over his condemned head.

For into Kemmler's case the War of the Currents intruded directly and shamelessly. George Westinghouse, watching the Edison camp's maneuver to power the electric chair with his dynamos and to brand his current as the executioner's, did something that has often been misread and deserves to be understood clearly. He funded Kemmler's legal appeals. Westinghouse's money paid for a formidable legal effort to have electrocution declared unconstitutional as a cruel and unusual punishment — an effort that climbed through the courts and reached, in the case known as *In re Kemmler*, a landmark ruling. Now, the cynical reading is that Westinghouse simply wanted to keep his dynamos out of the death chamber and protect his brand, and there was surely an element of self-interest in it. But it is worth being fair to him: Westinghouse also had a genuine humanitarian streak, a real revulsion at seeing his creation turned into a killing machine, and a sincere belief that the whole enterprise of electrocution was barbarous. The self-interest and the principle pointed the same way, as they often do, and it is neither necessary nor honest to reduce his action to one or the other. What is certain is the strange spectacle it produced: the maker of the current fighting in the highest courts to prevent the state from using his current to kill a man, while the maker of the rival current worked to ensure it did.

The courts ruled against Kemmler. In *In re Kemmler*, decided in 1890, the U.S. Supreme Court declined to find that electrocution violated the constitutional prohibition on cruel and unusual punishment, and the execution was allowed to proceed. The ruling turned in part on a definition of "cruel" that meant something like torture or lingering death, and the court accepted the assurances of the experts — assurances shaped, let us not forget, by the very men with a commercial interest in the outcome — that electrocution would be instantaneous and painless. That confidence was about to be shattered in the most public way imaginable.

What happened in that room at Auburn should be told plainly, because sanitizing it would betray the whole point of telling this story honestly. When all was ready, the current was switched on. Kemmler's body went rigid against the straps, and for seventeen seconds the current passed through him; then it was shut off, and the attending physicians pronounced him dead. But he was not dead. As the witnesses watched in dawning horror, Kemmler's chest began to heave; he was breathing, or gasping; a sound came from him; foam or saliva appeared at his mouth, and by some accounts a small wound on his skin began to bleed. The man was alive. In something close to panic, the executioners scrambled to restart the generator — which took agonizing seconds to build up its charge again — and sent the current through him a second time, for far longer, a minute or more. This time the accounts grow truly ghastly: the smell of burning flesh filled the room, smoke or steam rose from the electrode on his spine, and some witnesses reported that his body began to char where the current entered. Reporters fainted or fled the room to be sick. What had been promised as a swift and humane and modern death had taken something on the order of eight minutes and had been, by the unanimous testimony of those who saw it, an abomination — a slow, smoking, twice-administered killing that made hanging look merciful by comparison.

George Westinghouse, when he learned of it, delivered the verdict that history has remembered, and it was as devastating in its brevity as anything said in the whole war. "They would have done better," he said, "using an axe." The line cut straight to the heart of the matter: the great modern marvel, the clean electrical death that was to prove alternating current's peculiar lethality and Edison's humane superiority, had turned out to be more barbarous than the crude and ancient tools it was meant to replace. Kemmler had killed his wife with a hatchet; the state, with all the resources of modern electrical science and two rival geniuses' expertise brought to bear, had killed Kemmler worse.

It is important to correct a misconception the propaganda of the day tried to create and that lingers still — the idea that the botch somehow proved something about alternating current specifically, that AC's failure to kill cleanly showed it was either especially cruel or, conversely, that it vindicated one side or the other. It proved nothing of the kind. The horror of Kemmler's death was a failure of the whole grotesque enterprise: of the primitive and untested apparatus, of the executioners' inexperience, of the whole hubristic assumption that killing a man with electricity was a solved and gentle problem. It was not the current's fault in any way that reflected on its merits for lighting cities or turning motors. Both currents can kill cleanly or foully depending on how they are applied; the botch at Auburn was a botch of application, of a machine and a method rushed into use by men who understood far less than they claimed. To read it as a verdict on AC versus DC is to accept the very framing the propagandists imposed on a wretched man's death. The real lesson of Kemmler is not about currents at all. It is about what happens when human beings become props in other men's wars, and about the arrogance of assuming that a new technology has mastered a grave problem before it actually has.

The public reaction was revulsion, and it fell on all the parties. Newspapers that had covered the coming execution with ghoulish anticipation now recoiled from what they had helped stage. The reputation of electrocution as a humane method was, for a time, in ruins. And the War of the Currents, which had descended step by step from pamphlets to dead dogs to a secretly procured death machine, had now reached a place from which there was nowhere lower to go.

A man had been tortured to death in a smoking chair as the culmination of a fight over which company would sell America its electricity. Whatever else can be said about the combatants, they had, collectively, brought the industry to its moral floor.

A surprising fact about the aftermath: despite the horror of Kemmler's death, the electric chair was not abandoned. It was refined, and it spread; other states adopted it, and it remained a method of execution in the United States for more than a century, into our own era, killing thousands over its long career. The botch at Auburn was treated as a technical problem to be fixed rather than a moral verdict to be heeded — a grim testament to how a society, having committed itself to a machine, will improve the machine rather than reconsider the commitment. And a final, bitter irony: the branding campaign largely failed where it counted. Alternating current went on to win the war and power the world, and today no one flinches from their wall socket for fear of the electric chair. The attempt to poison AC's reputation through Kemmler's death succeeded only in poisoning the reputations of the men who attempted it.

It is worth understanding, in plain terms, why the execution went so catastrophically wrong, because the failure was not mysterious and it indicts the whole rushed enterprise. The men who designed and operated the chair simply did not know what they were doing, because no one had ever done it before. They did not know how much current was needed, or for how long, or how to place the electrodes so that the current would pass cleanly through the vital centers of the body. The electrodes themselves were crude, and the connection to Kemmler's skin was poor, so that much of the current met resistance and turned to heat — which is why he burned rather than died cleanly. The generator took time to build its charge, so that when the first jolt failed, there was a ghastly delay before a second could be delivered. Every one of these failures was a failure of inexperience and haste, of a society that had convinced itself, on the strength of a few animal demonstrations and some confident expert testimony, that killing a man with electricity was a solved problem when it was nothing of the kind. Harold Brown's dogs had died quickly in their small cages; the confident extrapolation from a seventy-pound dog to a grown man in an untested chair proved horribly wrong. The hubris of the whole affair — the assumption that the marvel of the age could be casually turned to a grave new purpose without hard-won knowledge — was written in Kemmler's burning flesh.

There is a further irony in the aftermath that deserves to be recorded, because it shows how the propaganda war twisted even the reaction to the horror. Both sides tried to spin the botched execution to their advantage, and the spectacle of them doing so over a man's smoking corpse is perhaps the final proof of how completely the human being at the center had been forgotten. The Edison camp and its allies insisted the execution had ultimately been successful — the man was dead, after all — and that any horror was a matter of squeamish observers and correctable technique. The Westinghouse side pointed to the ghastliness as proof of the whole enterprise's barbarism. Newspapers took sides. Experts issued competing statements. The argument over *whether Kemmler had suffered,* and *what his suffering proved,* became one more front in the commercial war, as though the man's agony were simply a data point to be entered into the ledger of AC versus DC. It is difficult to read the aftermath without concluding that the war of the currents had, by this point, corroded the moral sense of nearly everyone it touched, reducing a tortured death to ammunition. That corrosion — the way a fight over money and machines could make serious men argue publicly about how much a burning man had felt — is the truest measure of the nadir the war had reached.

You meet the shadow of Kemmler wherever the modern debate over humane execution is waged — the search for a "painless" method that recurs, decade after decade, from the electric chair to the gas chamber to lethal injection, has never escaped the essential problem his death exposed: that the promise of a clean, technological killing tends to shatter on contact with the messy reality of ending a human life. The chair at Auburn was the first broken promise in that long, unfinished series.

The current war had reached its lowest point, but the propaganda machine that produced it was still running — and to understand how public opinion was truly fought over in these years, we must widen our view from the death chamber to the whole sprawling battlefield of pamphlets, staged demonstrations, and expert testimony where the war for the public mind was really won and lost.

Chapter 46: Propaganda and the Press

Pick up a New York newspaper from almost any month in 1888 or 1889 and, somewhere among the columns of dense gray type, you are likely to find the War of the Currents being fought in ink. A letter to the editor from an "electrical expert" warning of the deadly wires overhead. A rebuttal from another expert dismissing the warnings as commercial scaremongering. A breathless account of a lineman killed on a pole, the current identified pointedly as alternating. A dry technical paper reprinted from an engineering journal, marshaling statistics to prove one side or the other. An advertisement, a pamphlet notice, a report of a demonstration at which animals were killed or, on the other side, at which a man safely passed current through his own body. The War of the Currents was, as much as anything, a war of *words and images*, fought across the newspapers and pamphlets and lecture halls of America for possession of something no dynamo could generate: public belief. This chapter steps back from the individual atrocities to survey that wider battlefield, because the current war was one of the first great struggles in which a foundational technology was decided as much in the arena of public opinion as in the laboratory

— and understanding how that struggle was waged tells us a great deal about how such struggles are still waged today.

Start with the sheer variety of weapons deployed, because the current war drew on nearly the whole arsenal of persuasion that a later, media-saturated age would come to know so well. There were the pamphlets, chief among them Edison's red-bound "A Warning," dressed in the sober costume of public safety while doing the work of commercial attack. There were the staged demonstrations, from Harold Brown's electrocuted dogs to the counter-demonstrations in which Westinghouse's people and their allies showed the current being handled safely. There was expert testimony, offered before legislatures and commissions and courts, in which men of science lent their credentials to one side or the other — credentials that the public, unable to judge the underlying physics, had little choice but to trust. There were the newspapers themselves, some genuinely trying to inform, others happy to run whatever was sensational, and a few effectively enlisted by one side or the other. And there was the raw manipulation of language, the campaign to make "electrocution" and "Westinghousing" carry a permanent charge of dread. It was a thoroughly modern propaganda war, and its participants understood, with a sophistication that can still surprise, that the technical question of AC versus DC would be settled in large part by whichever story the public came to believe.

The genius and the poison of the whole campaign lay in a single feature that deserves to be understood clearly, because it is the key to why the propaganda worked and why it was so hard to answer: the fear campaign was *not simply false*. This is the crucial and uncomfortable point. If Edison's side had merely lied — claimed, say, that alternating current caused disease or spoiled milk — the lies could have been exposed and dismissed. Instead, they built their campaign on a foundation of genuine fact. High-voltage alternating current on the transmission lines *was* dangerous. Linemen *did* die, sometimes horribly, in the tangle of poorly regulated wires that filled the skies of American cities in those years; there was a genuine public-safety scandal in the overhead wires of the late 1880s, culminating in shocking public deaths that had nothing to do with propaganda and everything to do with real, under-regulated hazard. So the fear campaign always had a true story to point to, a real corpse to display, a legitimate danger to invoke. What made it propaganda rather than public service was not falsehood but *proportion and purpose* — the deliberate inflation of a real, manageable risk into an existential terror, in the service of a commercial goal, with all the context that would have made the risk manageable carefully stripped away.

This is worth dwelling on because it corrects a misconception that flatters both sides and clarifies neither: the notion that the War of the Currents was a simple contest between honest safety concern on one side and cynical manipulation on the other. It was not. The truth is that *both* genuine safety concern and cynical manipulation were present on *both* sides, tangled together in a way that no clean line can separate. Edison's camp cynically exploited real dangers, yes — but the dangers were real, and some of Edison's people, perhaps Edison himself in part, sincerely believed high voltage was a menace. Westinghouse's camp defended a technology that really was more dangerous at its transmission voltages, and did so partly out of naked commercial interest — but Westinghouse also sincerely believed his system was safe when properly engineered, and he was, by the evidence, largely right. Each side had a genuine case and each side pushed it dishonestly when it suited them. The honest historian's job is not to award a white hat and a black hat but to show the mixture, because the mixture is the truth, and because the mixture is exactly what makes such campaigns so effective and so durable. A pure lie can be refuted. A true fact deployed with cynical purpose is nearly unanswerable, because every attempt to add the missing context sounds like special pleading, like a guilty party explaining away the corpse on the table.

Westinghouse's experience in this war illustrates the peculiar helplessness of the truth-teller against the fear-monger, and it is one of the quietly instructive threads of the whole story. Westinghouse was, by temperament and training, an engineer and a rationalist. When attacked, his instinct was to respond with facts: statistics on the actual number of deaths from alternating current compared with the deaths from gas explosions, from streetcars, from all the ordinary hazards of modern urban life that no one proposed to ban. He commissioned studies, wrote reasoned letters, marshaled data showing that electricity, properly handled, was no more dangerous than a dozen technologies the public accepted without a thought. And he found, as rational men in such fights so often find, that his careful statistics were no match for a single dead dog in a cage or a single vivid account of a lineman burning on a pole. A number cannot compete with an image; a probability cannot compete with a horror. The human mind is built to fear the vivid particular far more than the abstract general, and the fear campaign was engineered, whether by instinct or design, to exploit exactly that asymmetry. Westinghouse was bringing a spreadsheet to a ghost story, and for a while the ghost story was winning.

A surprising fact that captures the sophistication of the era's propaganda: the campaign to coin and popularize damning vocabulary — "electrocution," the attempted "Westinghousing" — represents one of the earliest self-conscious attempts in the history of commerce to win a business war by capturing *language itself*, by planting a word in the public mind that would carry the desired association forward automatically, without further effort, every time it was spoken. This is a technique we now associate with modern advertising and political spin, with the careful focus-grouped coinage of loaded phrases. Its appearance, fully formed, in a business dispute of the 1880s is a reminder that the tools of mass persuasion are older than we tend to think, and that they emerged, naturally enough, precisely when a mass press and a mass technology first made a mass audience worth capturing. Another surprising fact: for all the ink spilled, the propaganda war did not, in the end, determine the technical outcome. Alternating current won not because the public came to love it — the public, if anything, remained wary — but because its economic and engineering advantages in long-distance distribution were simply too great to be overcome by any story. The propaganda shaped the *experience* of the war, its ugliness and its human cost, more than it shaped the *result*. The dynamos, in the end, spoke louder than the pamphlets.

It is worth examining one more feature of the era's propaganda machinery, because it is the one most easily overlooked and most quietly powerful: the role of the expert. In a dispute over a technology that ordinary citizens could not possibly understand — where the underlying questions of voltage and current and physiology were genuinely beyond the reach of a newspaper reader or even a legislator — the public had little choice but to defer to men who claimed authority. And this created an enormous opening for manipulation, because expertise could be purchased, shaded, or manufactured. Harold Brown presented himself as a disinterested electrical expert; he was in fact an interested party with covert backing. Edison lent his own colossal authority as the greatest electrician of the age to conclusions that happened to serve his balance sheet. Westinghouse, in turn, marshaled his own engineers and their credentials. Each side dressed its commercial interest in the white coat of science, and the public, unable to evaluate the physics directly, was left to decide which authority to trust — a decision that turned as much on fame and showmanship as on truth. This is one of the most durable and troubling lessons of the current war: that when a question is too technical for the public to judge, the battle shifts to the credibility of experts, and credibility, unlike truth, can be bought and staged. The reader who has watched modern controversies play out around dueling experts will find the pattern wearily familiar.

A further, subtler mechanism was at work as well, one that made the fear campaign self-reinforcing. Every genuine accident — every lineman killed, every stray death from a fallen wire — became, in the charged atmosphere the campaign had created, immediate front-page confirmation of the danger, while the countless days on which the wires hummed along harmlessly were, of course, never news at all. Newspapers do not report the absence of catastrophe. So the very structure of the press guaranteed that alternating current's rare failures would be vivid and prominent while its constant successes stayed invisible, producing in the public mind a wildly distorted sense of how dangerous the technology actually was. The fear campaign did not have to fabricate anything; it merely had to lean on the natural tendency of news to amplify the dramatic and ignore the routine. This is a bias baked into the machinery of mass information itself, and Edison's people, whether by cunning or by instinct, learned to ride it. It is the same bias that makes a single plane crash more frightening than a year of safe flights, or a single rare side effect more alarming than a million quiet recoveries — a bias as active in our own information age as it was in the newspaper world of 1889.

You meet the descendants of this propaganda war constantly, in nearly every public controversy over a new or contested technology. The pattern established in the War of the Currents — the true-but-inflated risk, the vivid anecdote deployed to swamp the reassuring statistic, the expert testimony bought or shaded by interest, the loaded vocabulary designed to prejudge the question, the helplessness of careful data against visceral fear — recurs in the modern arguments over nuclear power, over vaccines, over chemicals and foods and technologies of every kind. The current war was, in this sense, a template, one of the first full-dress rehearsals of a genre of public struggle that now fills our own newsfeeds daily. To study how Edison and Westinghouse fought for the public mind in 1889 is to study, in a simpler and more legible form, how such fights are conducted still.

There is a final irony worth naming before we move on. The propaganda war, for all its ferocity, was expensive — in money, in reputation, in energy — and it was ultimately a distraction from the real contest, which was being decided by the relentless economics of transmission and the steady spread of AC installations across the country. While Edison poured resources and prestige into pamphlets and demonstrations and death chambers, the underlying business logic ground on indifferent to all of it. And that logic was, increasingly, a matter of *money* — of capital, of financing, of the enormous sums required to build electrical systems at national scale, and of the powerful men who controlled that capital. To understand how the war was really decided, we must follow the money, to J. P. Morgan and the financiers whose decisions, made in quiet offices far from the death chamber and the dog cage, would ultimately determine which current powered the world.

Chapter 47: The Money Behind the War

In a mansion at 219 Madison Avenue in New York City, in the years the current war was raging, there lived a man who had never invented a dynamo, never wound a transformer, never stood in a lecture hall to electrocute a dog or debate a rival — and who nonetheless held more real power over the outcome of the War of the Currents than Edison, Westinghouse, and Tesla combined. His name was John Pierpont Morgan, and he was the most powerful financier in America. Morgan's house on Madison Avenue holds a small but telling place in this story: it was among the first private residences in New York to be wired for Edison's electric light, its own generating plant installed in a cellar nearby, so that the great banker could show off the miracle he was funding. The installation was temperamental — it smoked, it smelled, it once reportedly caught fire — but Morgan kept it, because it was a symbol of the thing he had grasped earlier and more clearly than almost anyone: that electricity was going to be one of the great businesses of the coming century, and that whoever controlled its financing would control its future. The War of the Currents was fought by inventors, but it was decided, in the end, by money — and the money answered to men like Morgan.

This chapter widens the lens one final time, from the laboratory and the newspaper to the counting house, because no account of the current war is complete that treats it purely as a contest of engineering or even of propaganda. Underneath the physics and the pamphlets ran a current of capital, and it was the flow of that capital — where it pooled, where it dried up, who controlled it and on what terms — that ultimately shaped which system won and which men kept their companies. The romantic version of the story, the one that pits lone geniuses against each other, tends to leave the financiers offstage. But the financiers were never offstage. They were in the wings the whole time, and at the decisive moments they walked to the center and rearranged the players.

To understand their role, you have to understand the peculiar economics of the electrical business, which made it uniquely dependent on finance. Building an electrical system was staggeringly capital-intensive. Before a single customer paid a single bill, a company had to build generating stations, string miles of wire, manufacture or buy thousands of dynamos and lamps and meters, and sink enormous sums into infrastructure that would only pay back slowly, over years. This is not like opening a shop that starts taking money on its first day; it is like digging a canal or laying a railroad, an undertaking that devours capital up front and returns it only over a long horizon. A company in such a business is perpetually hungry for money, perpetually dependent on investors willing to fund the build-out and patient enough to wait for the return. And that dependence gave the men who controlled large pools of capital — the great investment bankers, Morgan foremost among them — enormous leverage over the companies they financed. He who feeds the furnace can, when he chooses, decide who tends it.

On Edison's side, the money was formidable and it was increasingly Morgan's. Morgan and his associates had backed

Edison's electrical ventures from early on, and as those ventures grew and consolidated into Edison General Electric, Morgan's influence over them grew with them. This mattered enormously, and in a way that cuts against the popular image of Edison as the master of his own destiny. Edison was a genius and a titan, but he was not, in the end, in control of the company that bore his name. The financiers were. And the financiers, being financiers, cared about one thing above all: return on capital. They were not sentimentally attached to direct current the way Edison was; they had no identity fused to the copper. If alternating current was going to win the market — and the economics were making that outcome clearer every year — then the rational move for a Morgan was not to go down fighting for DC out of loyalty to Edison, but to position his capital on the winning side, whatever the inventor's feelings. This dispassion of capital, this indifference to everything but the balance sheet, would prove decisive, and it would prove, for Edison personally, bitter.

On Westinghouse's side, the money was also considerable but the situation was more precarious, and here lies one of the war's central financial dramas. Westinghouse had committed enormous sums to building his alternating-current empire — buying Tesla's patents, licensing the transformer, manufacturing equipment, and installing systems in town after town, often on aggressive terms to win the business. He was expanding fast, and fast expansion in a capital-hungry business means debt, and debt means dependence on lenders. By around 1890 and 1891, Westinghouse Electric, for all its technical success, was financially overextended, stretched thin by its own rapid growth and squeezed by the tight money of a jittery market. The company came under real pressure from its financial backers, who demanded retrenchment and restructuring as the price of continued support. It was in this squeeze, as an earlier chapter noted, that Tesla's royalty arrangement was renegotiated — not out of villainy but out of the ordinary, merciless logic of a company fighting to satisfy its creditors. The financiers who rescued Westinghouse did so on their own terms, and those terms clipped his wings and constrained his freedom of action at the very moment the war was reaching its climax.

Here is a surprising fact that reframes the whole conflict: the two great combatants, Edison and Westinghouse, were both, in a sense, prisoners of their financiers, and neither man ultimately controlled his own company's fate. Edison would be pushed aside by the very consolidation his backers engineered; Westinghouse would spend years struggling to keep control of his firm against the pressures of finance, and would eventually, in a later panic, lose that control too. The War of the Currents is usually told as a duel between two strong-willed inventors, but at the level of raw power it was increasingly a matter of what the bankers decided, and the bankers decided by the cold light of the ledger, not the warm loyalties of invention. The men whose names are on the companies were, at the decisive moments, less free than the men whose names are on the loans.

This financial reality set the stage for the consolidation that would effectively end the war, and it is worth previewing here because it is the logical culmination of everything this chapter describes. If capital cared only about returns, and if alternating current was winning the economic argument, and if maintaining two rival, incompatible, capital-devouring systems in cutthroat competition was ruinously expensive for everyone — then the rational move, from the financier's godlike vantage, was not to keep fighting but to *consolidate*. To merge the warring companies, rationalize the industry, end the wasteful competition, and adopt whichever technology the market favored, all under the calm management of finance. This is precisely what would happen. The looming merger that would create General Electric, and the broader consolidation of the electrical industry into a small number of giant firms, was not primarily an engineering event or a propaganda outcome. It was a *financial* event, driven by men like Morgan who looked at the chaos of the current war and saw, above all, an inefficient use of capital to be corrected. The war would end not with a decisive battle but with a merger, and the merger would be the financiers' verdict.

It is worth correcting a misconception that the drama of the inventors encourages: the idea that the best technology simply won because it was best, in a kind of pure meritocratic triumph. Technology mattered enormously — alternating current's advantages in distribution were real and ultimately decisive. But technology does not deploy itself; it is deployed by capital, and capital chose alternating current not because capital admired Tesla's rotating field but because capital calculated that AC would earn more money over the long run. Had the financing gone differently — had the great pools of capital, for whatever reason, thrown their weight behind direct current and starved AC of funds at a critical moment — the outcome could have been delayed, distorted, or reshaped, whatever the underlying physics. The physics set the possibilities; the money chose among them. To understand why the world runs on AC, you must understand not only Faraday's induction and Tesla's motor but also the ledgers of Madison Avenue.

To appreciate just how much the financiers' logic differed from the inventors', consider the peculiar figure of J. P. Morgan more closely, because he embodies the whole dispassionate machinery of capital. Morgan was not an electrical man; he neither understood nor much cared about the physics of the two currents. What he understood, with a clarity that few of his contemporaries could match, was *order* — the imposition of rational, profitable structure on chaotic, wasteful competition. Morgan's genius, across his whole career, lay in taking fractured, ruinously competing industries — railroads, steel, and now electricity — and consolidating them into large, stable, dominant enterprises that would stop bleeding capital in destructive rivalry and start returning steady profits. To such a man, the War of the Currents was not a heroic clash of visions; it was a mess, an inefficiency, a spectacle of two great companies burning money to destroy each other when they might, under proper management, be combined into a single dominant firm earning handsomely for its investors. This is the crucial point about the money behind the war: the most powerful financial actor in the story had no loyalty to either current and no romance about invention whatsoever. He wanted the fighting to stop and the profits to flow, and he had the power to make it so. The inventors thought they were fighting a war to the death; Morgan saw a bothersome inefficiency he would eventually correct with the stroke of a merger.

There is a surprising and rather poignant fact embedded in this financial reality, one that inverts the usual heroic telling. The men we remember — Edison, Tesla, Westinghouse — are remembered precisely because they cared, passionately and personally, about the technology and the fight. Edison loved his direct current; Tesla loved the beauty of his rotating field; Westinghouse believed in his system with an engineer's conviction. Their passion is what makes them dramatic and memorable. But passion, in the cold arithmetic of the current war, was largely a *liability*. It was Edison's passionate attachment to direct current that blinded him to the sensible course of adopting AC and kept him fighting a losing battle long past the point of reason. It was Westinghouse's passionate overextension in pursuit of victory that left him vulnerable to his creditors. The dispassionate financiers, who cared about none of it, ended up with the real power, precisely because their indifference let them see the situation clearly and act on it without sentiment. There is an uncomfortable lesson here about how the world actually works: that the people who care most about a thing are often not the people who end up controlling its fate, and that the cool logic of capital tends, in the end, to outlast the hot convictions of genius.

You meet the legacy of this financial dimension every time you deal with a modern electric utility, for the industry that emerged from the current war was shaped decisively by its origins in high finance: capital-intensive, consolidated into large firms, deeply entangled with the great banks, and eventually so central to public life that governments would step in to regulate it as a natural monopoly. The pattern of a few giant, finance-backed, heavily regulated power companies — a pattern that still largely defines how electricity reaches your home — was set in these years, in the counting houses as much as in the laboratories. And the deeper lesson endures wherever a transformative technology meets the need for enormous capital to scale: that the inventor may light the spark, but the financier decides whether it becomes a fire.

The money had spoken, or was about to. The financiers looked at the wasteful, ugly, capital-hungry war and prepared to end it the way they ended most things — with a merger. But even as the bankers moved toward consolidation, the technology itself was about to deliver a series of triumphant vindications that would settle the engineering argument beyond dispute, beginning with the dazzling public lectures of the prophet of alternating current himself, Nikola Tesla, at the very height of his fame.

Chapter 48: Tesla's Triumphant Lectures

On the evening of May 20, 1891, before a rapt audience of engineers and scientists at Columbia College in New York, Nikola Tesla did something that made the assembled experts gasp and then, when it was over, rise to their feet in astonishment. He took hold of wires carrying an alternating current at a fearsomely high voltage — hundreds of thousands of volts — but oscillating at an enormously high frequency, and he let that current pass through his own body. By every crude intuition of the age, and by everything Edison's fear campaign had drilled into the public mind, this should have killed him instantly, charring him like one of Harold Brown's dogs. Instead, Tesla stood serene and unharmed, and as the current flowed through him it made the hair on his body stand out and caused wireless lamps he held in his hands — glass tubes with no wires connected to them at all — to glow with an eerie light, so that the tall, gaunt inventor appeared before his audience wreathed in an aura of electric fire, a living demonstration that alternating current, understood and mastered, was not a monster but a marvel. It was one of the great theatrical moments in the history of science, and it marked Tesla's arrival at the absolute height of his fame, the prophet of the very current his rivals had tried to brand as death.

To understand why this was so devastating an answer to Edison's campaign, you have to understand a piece of physics that Tesla grasped more deeply than almost anyone alive: the strange behavior of alternating current at very high frequencies. Recall that alternating current reverses direction many times a second — sixty times, in the ordinary power system. Tesla had learned to generate currents that reversed direction not sixty but tens or hundreds of thousands of times a second, using the resonant high-voltage transformer he had devised, the device now known the world over as the Tesla coil. And such high-frequency currents do something counterintuitive and wonderful: they tend to travel along the *surface* of a conductor rather than through its depths, a phenomenon called the skin effect. When such a current passes through a human body, it flows largely over the skin rather than plunging through the heart and vital organs, and so — within limits, and with expertise Tesla possessed and most did not — it can be survived where an ordinary current would kill. Think of it like the difference between a river that soaks deep into the soil and floods a valley, and a fine spray that races across the surface of a pond without ever sinking in. The ordinary lethal current sinks in; Tesla's high-frequency current raced across the surface. He was not immune to electricity, and it is dangerous to imagine his stunts were safe for anyone to copy — but he understood the physics well enough to make himself, for a few dazzling minutes on a lecture stage, into a safe conductor of a current that terrified everyone else.

The wireless lamps were, if anything, even more astonishing, and they too rested on a principle Tesla understood and his audiences barely could. He would hold up a sealed glass tube containing gas, with no wire running to it, and it would light up simply from being near his high-frequency apparatus — the rapidly oscillating electric field reaching out through the empty air to excite the gas within the tube into a glow, much as the changing field in a transformer's core reaches across the gap to the second coil. It was electricity delivered without wires, light conjured from apparently nothing, and to an 1891 audience it looked very nearly like magic. Tesla, who had a genuine flair for showmanship beneath his fastidious, monkish exterior, understood exactly what he was doing: he was making the invisible visible, turning the abstract argument about alternating current into a spectacle of beauty and wonder that no red pamphlet could compete with. Where Edison's campaign had shown the public death, Tesla showed them light — light without wires, current without harm, the future glowing in a glass tube held in a living man's hand.

Over the two years that followed, Tesla carried this performance across the Atlantic and back, and became, briefly, one of the most celebrated scientific figures in the world. He lectured before the Institution of Electrical Engineers and the Royal Institution in London in 1892, dazzling the cream of British science; the venerable physicists who watched him — men steeped in the mathematics of electromagnetism — came away half-convinced they had seen a new kind of genius. He lectured in Paris to similar acclaim. He addressed the American Institute of Electrical Engineers, whose members already regarded him with something like awe after his 1888 motor lecture. Everywhere he went he demonstrated his wireless lamps, his currents passing harmlessly through his body, his sparks and glows and resonances, and everywhere he left audiences persuaded that alternating current, far from being the executioner's fluid, was the very medium of the electrical future — and that the strange, courtly, brilliant Serb before them was its prophet. It was, for a man who had a few years earlier been digging ditches in New York for a living, an extraordinary vindication.

It is worth pausing on the Tesla coil itself, the instrument at the heart of these performances, because it is one of those inventions whose fame has outrun public understanding of what it actually is. A Tesla coil is a resonant transformer — a special kind of transformer that produces very high voltages at very high frequencies by exploiting the phenomenon of resonance, the same phenomenon by which a child on a swing goes higher and higher with small, well-timed pushes, or by which a wine glass shatters at the exact pitch that matches its natural ring. Tesla tuned his coils so that small inputs built up, cycle by resonant cycle, into enormous voltages, producing the long, crackling artificial lightning that has made the device a fixture of science museums and science fiction ever since. Tesla built it not as a toy but as a serious instrument in his grand project of transmitting power and information without wires — a project that would, in the coming years, lead him to Colorado Springs and to the doomed tower at Wardenclyffe, and that a later chapter will follow to its melancholy end. But in 1891, at the height of his fame, the coil was pure triumph, the engine of his dazzling lectures and the foundation of much that would follow in radio and wireless technology.

Here it is important to correct a misconception that the modern Tesla cult has inflated to distorting proportions. The internet has remade Tesla into a kind of secular saint, a lone unappreciated genius who single-handedly invented the twentieth century and was robbed of his due by lesser, greedier men — and the triumphant lectures are often enlisted as evidence of this superhuman status. The truth is more human and more interesting. Tesla was a genuine genius, one of the most original electrical minds who ever lived, and his triumphs were real. But he was not a solitary miracle worker who owed nothing to anyone; he built on Faraday and Maxwell and the whole edifice of nineteenth-century physics, he depended utterly on Westinghouse's capital and factories to make his ideas real, and his showmanship, for all its brilliance, sometimes ran ahead of what he could actually deliver. Some of the wireless-power dreams he began to voice in these years would never be realized, then or since, and the gap between his soaring visions and their practical fulfillment would become the tragedy of his later life. To honor Tesla truly is not to worship an infallible saint but to see the actual man: brilliant, original, theatrical, visionary, and increasingly prone to promising more than physics would allow. The lectures were a real and dazzling triumph. They were not proof that he was a wizard.

A surprising fact worth carrying forward: those wireless lamps Tesla lit on stage in 1891, glowing without any wire connected to them, were not merely a stunt but a genuine early demonstration of principles that underlie technologies you use every day. The idea of transferring electrical energy through space without wires, which looked like magic to his audiences, is the same basic idea that now charges your phone when you set it on a wireless charging pad, and that runs the wireless tags and induction systems woven through modern life. Tesla was demonstrating, in crude and spectacular form, a future that would take more than a century to arrive in your pocket. And another surprising fact: the high-frequency currents he pioneered found early and lasting use not in power transmission, where they proved impractical, but in medicine and radio — the principles of his coils fed directly into the development of radio technology and into medical devices that used high-frequency current on the body. The prophet of power transmission turned out to have quietly seeded the age of wireless communication.

It is worth dwelling on the sheer strangeness of Tesla the performer, because the lectures reveal a side of him that sits oddly with the popular image of the reclusive, otherworldly genius. On stage, Tesla was a consummate showman, and he knew it. He dressed impeccably, in formal evening clothes, cutting a tall and dramatic figure; he timed his demonstrations for maximum astonishment, building from the merely impressive to the apparently impossible; he understood the theatrical value of danger, of appearing to court death and emerging serene. There was calculation in this, and courage too — the currents he passed through his body were genuinely capable of harm if his understanding of the skin effect had been even slightly wrong, and he was, in a real sense, betting his life on his physics every time he took the stage. But there was also a deep purpose beneath the showmanship. Tesla grasped, as few scientists of his age did, that a technology's fate is decided not only in the laboratory but in the imagination of the public and the profession, and that to win the future for alternating current he had to make people not merely accept it but *marvel* at it. Where Edison's camp had worked to make the public associate AC with a smoking corpse, Tesla worked to make them associate it with a haloed man wreathed in harmless fire. It was propaganda too, in its way — but propaganda in the service of wonder rather than fear, and grounded in real physics rather than rigged demonstrations.

A surprising and humanizing detail from this period is how the great established scientists of Europe received him, because it corrects the myth that Tesla was always an outsider spurned by the establishment. In London, the eminent physicists of the Royal Institution and the Institution of Electrical Engineers were so taken with his 1892 lecture that they reportedly begged him to repeat it, and the venerable Sir William Crookes, a giant of British science, became a genuine admirer and friend. Tesla was, for this brief shining interval, not a neglected crank but the toast of the international scientific elite, courted and celebrated on two continents, his lectures reprinted and pored over, his name spoken in the same breath as the great physicists of the age. This makes the later arc of his life all the more poignant, for the man who would die alone and half-forgotten in a New York hotel room had once stood at the very summit of scientific celebrity, embraced by the finest minds of his era. The tragedy of Tesla is not that the world never recognized his genius; it is that it recognized his genius fully, and then, as his visions outran his ability to deliver them, slowly turned away. The triumphant lectures were the height from which the long descent began.

You meet Tesla's triumphant lectures every time you use anything wireless — every phone charger, every radio, every contactless payment — and every time a Tesla coil throws its crackling purple lightning in a science museum to the delight of children who have no idea they are watching a weapon from the War of the Currents. His demonstrations turned the abstract case for alternating current into visible wonder, and in doing so they did more to answer Edison's fear campaign than any statistic Westinghouse ever published. Where the propagandists had made people afraid of the current, Tesla made them love it.

But dazzling lectures, however persuasive, were still a matter of spectacle and stagecraft. What alternating current needed to win the engineering argument beyond all dispute was a hard, practical, undeniable demonstration that it could do the one thing its whole promise rested on: carry real power across a real distance, efficiently, in the open world. That demonstration was about to come, not on a New York stage, but across the hills and valleys of Germany, where an alternating current would leap more than a hundred miles and change the mind of every engineer who had doubted it.

Chapter 49: Lauffen to Frankfurt

In the late summer of 1891, in the German city of Frankfurt am Main, visitors to the International Electrotechnical Exhibition gathered before a striking display: a wall of a thousand incandescent lamps, blazing in a great illuminated sign, and beside them an artificial waterfall, its pump churning water uphill for the sheer spectacle of it. There was nothing unusual in electric lamps or electric pumps by 1891. What made this display one of the pivotal demonstrations in the history of technology was where the power came from. The electricity lighting those thousand lamps and driving that waterfall pump had not been generated anywhere near Frankfurt. It had been generated at a cement factory in the town of Lauffen am Neckar, and it had traveled to Frankfurt across roughly one hundred and seventy-five kilometers of overhead wire — well over a hundred miles — arriving with an efficiency that stunned the assembled engineers of the world. Alternating current had just done, in the open air of the German countryside, the one thing that would settle the great argument beyond dispute: it had carried real, substantial power across a great distance and delivered it, still useful, at the far end. The engineering case for AC was, after Lauffen to Frankfurt, essentially closed.

To feel the weight of this achievement, recall the central

problem that had defined the entire War of the Currents, the problem laid out many chapters ago and never really refuted: direct current could not travel. Edison's stations had to sit within about a mile of their customers, because sending direct current at usable voltages any farther bled the power away as waste heat in the wires. This was not a flaw to be engineered around; it was a hard limit rooted in the physics of transmission loss, the I-squared-R tax that punishes any attempt to push large currents down long wires. Alternating current's answer, in theory, was to use transformers to step the voltage way up for the journey — high voltage, low current, low loss — and step it back down at the destination. This was the theory Westinghouse had bet his fortune on and Stanley had proven in miniature at Great Barrington over a few thousand feet. But a demonstration over a few thousand feet, however sound in principle, still left room for doubt. Could the trick really scale? Could AC carry not a few lamps' worth of power across a village but a whole industrial load across a hundred miles of hills and weather? Skeptics, and there were many serious ones, insisted that some practical obstacle — some accumulation of losses, some instability, some engineering wall — would stop AC well short of the grand distances its champions promised. Lauffen to Frankfurt was the experiment that answered them, at full scale, in public, with the eyes of the world's electrical profession upon it.

The technical accomplishment deserves to be understood in its particulars, because they mark a genuine advance beyond even what Tesla and Westinghouse had so far deployed. The system used a form of alternating current called three-phase — three separate alternating currents running in the same set of wires, each staggered a third of a cycle behind the others, the very polyphase principle that Tesla had pioneered for his motors. This three-phase system, championed and engineered by talented men working for the German and Swiss firms behind the project, most notably the engineer Mikhail Dolivo-Dobrovolsky, proved to be an extraordinarily efficient way to both transmit power and run motors at the receiving end. At Lauffen, water turbines drove generators that produced the three-phase current; transformers stepped it up to a high voltage — in the range of tens of thousands of volts — for the long journey; the current flowed across a hundred and seventy-five kilometers of transmission line to Frankfurt; and there, transformers stepped it back down to drive the lamps and the pump. The efficiency of the whole chain — the fraction of the power sent from Lauffen that actually arrived usefully in Frankfurt — came out remarkably high, well above what many experts had predicted, in the neighborhood of three-quarters of the power surviving the journey. For a first-of-its-kind demonstration over such a distance, this was a triumph that exceeded expectations and silenced doubters.

Here is a surprising fact that captures how decisive the moment was: the Lauffen-to-Frankfurt demonstration did more to convince the world's engineers of AC's long-distance viability than all of Tesla's dazzling lectures combined, precisely because it was not a lecture-hall marvel but a brute practical fact out in the ordinary world. Tesla's wireless lamps had made audiences gasp, but a skeptic could always mutter that stagecraft is one thing and industry another. No one could mutter that about Lauffen. The power crossed a hundred and seventy-five kilometers of real German landscape and lit real lamps in a real exhibition hall, measurably, repeatably, efficiently. It was the difference between a magician's trick and a working machine, and engineers, being practical people, are moved by working machines. After Frankfurt, the technical argument was over. Whatever doubts remained about alternating current's viability for the long-distance transmission that Edison's direct current could never manage were, for any honest expert, laid to rest. The question was no longer *whether* AC could power the world across distance, but *how soon* and *how large*.

It is worth correcting a misconception that the American-centered telling of this story tends to create — the impression that the War of the Currents was a purely American affair, fought entirely between Edison and Westinghouse on American soil, with the rest of the world watching passively. In truth, the development of alternating current was an international effort, and some of its most decisive advances happened in Europe. The three-phase system perfected at Lauffen owed much to German and Swiss engineering; the transformer had roots in British and Hungarian and French work; the theoretical foundations were laid by British and continental physicists. Tesla himself was a European by birth and training who brought his continental education to America. The current war had an American center of gravity because America was the largest and most dynamic market, but the technology that won it was the product of many hands across many countries. Lauffen to Frankfurt is the reminder embedded in the story itself: the decisive proof of long-distance AC transmission happened not in New York or Pittsburgh but in Germany, engineered by men whose names most Americans have never heard. The electrification of the world was a global collaboration wearing an American costume.

Another surprising fact, one that points straight toward the war's climax: the success at Lauffen was watched with intense interest by the men then planning the most ambitious electrical project the world had yet conceived — the harnessing of Niagara Falls. The engineers and financiers behind Niagara faced exactly the question Lauffen answered: could power generated at the falls be transmitted to the industrial city of Buffalo, some twenty-odd miles away, or would it, like Edison's direct current, be trapped near its source? A hundred and seventy-five kilometers across Germany made twenty-six miles to Buffalo look eminently achievable. Lauffen to Frankfurt did not merely settle an abstract argument; it directly emboldened the decision, taken shortly afterward, to build the Niagara project as an alternating-current system using the polyphase technology of Tesla and Westinghouse. The German demonstration was, in a real sense, the proof of concept that unlocked the American colossus. The two are links in a single chain: Lauffen proved the principle, and Niagara would prove the industrial scale, and between them they would end the war for good.

It is worth pausing on the three-phase principle itself, because it is the quiet hero of this chapter and of the modern grid, and because understanding it makes sense of those steel towers you have seen marching across the landscape. Recall that a single alternating current pulses to a peak, falls to nothing, reverses to a peak the other way, and returns to nothing, over and over — which means that, momentarily, at the instants it passes through zero, it is delivering no power at all. A single-phase supply is therefore a bit like a one-cylinder engine that fires, then coasts, then fires again, delivering power in pulses with dead spots between them. Three-phase current solves this beautifully by running three separate alternating currents together, each staggered a third of a cycle behind the last, so that whenever one is passing through its momentary zero, the other two are carrying the load. The power arrives smooth and continuous instead of in pulses — like a three-cylinder engine in which one cylinder is always firing, so the crankshaft never coasts. This is why three-phase is the near-universal choice for both transmitting bulk power and running large motors: it delivers steady, gapless power, and it does so using conductors more efficiently than three separate single-phase circuits would. Those high-voltage lines on their towers almost always come in sets of three wires precisely because they carry the three staggered phases that Lauffen proved could span a hundred and seventy-five kilometers.

A word, too, about the men behind the German triumph, because their relative obscurity illustrates a recurring injustice in this history. Mikhail Dolivo-Dobrovolsky, the Russian-born engineer working for the German firm AEG who designed much of the three-phase apparatus, is a name almost no general reader knows, and yet his three-phase generators and motors were arguably as important to the practical electrification of the world as anything Tesla built. Charles Brown, the engineer of the Swiss firm Oerlikon who managed the transmission line, is likewise all but forgotten outside specialist circles. The current war's American mythology, with its titanic personalities, has crowded these European engineers almost entirely out of popular memory, even though the single most convincing demonstration of long-distance AC transmission in the entire nineteenth century was their achievement, not the achievement of any of the famous American antagonists. It is a useful corrective to the great-man theory of technological history: the decisive proof that alternating current could power a civilization across distance was delivered not by a household name in a lecture hall but by a handful of methodical continental engineers stringing wire across German hills, whose names have faded even as their handiwork hums on above every highway in the world.

You meet the legacy of Lauffen to Frankfurt every single time you use electricity, more directly than almost any other episode in this book, because the three-phase alternating-current transmission it demonstrated is, to this day, the fundamental technology by which power is carried across the world. The high-voltage lines marching on their great steel towers across the countryside, the ones you have seen a thousand times from a car window, carry three-phase alternating current in a direct and unbroken line of descent from what crossed the German hills in 1891. Nearly every long-distance power line on the planet, nearly every regional grid, rests on the principle proven at Lauffen. When you flip a switch and draw power from a plant that may be a hundred miles away, you are the beneficiary of that demonstration. It was one of those rare moments when a single experiment settles a great question and sets the pattern for a century and more of what follows.

There is a final reflection worth offering on why a demonstration like Lauffen to Frankfurt mattered in a way that no argument, however sound, ever could — because it speaks to how technological disputes are actually settled. For years the champions of alternating current had possessed the *the-*

ory of long-distance transmission; the physics of high voltage and low current and transformer efficiency was well understood by those who cared to follow it. And for years the skeptics had been able to reply that theory is one thing and practice another, that some unforeseen obstacle would surely intervene at large scale. This is the perennial deadlock of any argument about an untried technology: the proponents point to principle, the doubters point to the absence of proof, and neither can fully convince the other because the decisive evidence does not yet exist. What Lauffen supplied was that decisive evidence — not a better argument but an accomplished *fact*, a thing that had actually happened in the world and could not be argued away. A hundred and seventy-five kilometers of transmitted power was not a claim to be debated but an event to be reckoned with, and events, unlike arguments, end debates. This is why the great demonstrations — Great Barrington, Lauffen, and soon Chicago and Niagara — mattered so much more than the pamphlets and lectures: they converted contested theory into settled fact, and in doing so they did what no volume of reasoning could, which is to change the minds of people who did not want their minds changed. The engineer's deepest wisdom is that in the end you do not win the argument; you build the thing, and let the thing win the argument for you.

After Lauffen, the momentum that had been building — the spread of AC installations, the persuasive brilliance of Tesla's lectures, the relentless economics of distribution, the cold calculations of the financiers — became overwhelming. The engineering argument was won; the technical doubts were dead; the physics had spoken as plainly as physics ever speaks. What remained was for the business world, and Edison's own company, to catch up with what the technology had already decided. The tide, so long contested, was about to turn decisively — and even the mighty house that Edison had built would not be able to stand against it.

Chapter 50: The Tide Turns

In April 1892, Thomas Edison received a piece of news that must have landed like a physical blow. The company that bore his name — Edison General Electric, the great enterprise built on his inventions and his fame and his direct-current empire — was to be merged with its chief rival in the electrical-manufacturing business, the Thomson-Houston Company. And in the new corporation that would emerge from the merger, Edison's name was to be dropped. The company would be called, simply, General Electric. Not Edison General Electric. Not Edison anything. The most famous inventor in America, the man whose name had been synonymous with electric light for more than a decade, would see that name struck from the enterprise he had founded, and would himself be pushed to the margins of its control. It was, for Edison, a humiliation and a wound that he never entirely forgave. And it was, for the War of the Currents, the beginning of the end — the moment when the tide, so long and bitterly contested, turned decisively and irreversibly toward alternating current.

To understand how it came to this, we have to see the merger for what it really was, because the popular story often misreads it. The formation of General Electric in 1892 is sometimes told as a simple corporate transaction, a routine consolidation of two big companies. It was that, but it was also something more revealing: it was the financiers' verdict on the War of the Currents, rendered in the only language they truly spoke, the language of capital. As an earlier chapter showed, Edison did not control his own company; the financiers did, J. P. Morgan foremost among them. And the financiers had been watching the same evidence everyone else had been watching — the spread of AC installations, the triumph at Lauffen, the relentless economics of long-distance distribution — and they had drawn the obvious conclusion. Direct current, for all Edison's loyalty to it, was losing. Continuing to fight for it was a poor use of capital. The rational move was to consolidate the industry, end the wasteful competition, and position the merged company to compete in the alternating-current future that was plainly arriving. So Morgan engineered the merger, and in doing so he cast aside not only Edison's name but Edison's cherished technology, because sentiment about either was, to a financier, an expense he could not justify.

Here is the fact that makes the merger the true turning point of the war, and it is a fact of exquisite irony: the new General Electric, the company forged in part from Edison's own direct-current enterprise, promptly moved to embrace alternating current. It could hardly do otherwise; Thomson-Houston, one of the two merging partners, had already been building AC equipment, and the whole logic of the merger was to compete in the AC market. So the company that carried forward Edison's legacy — minus his name — became, in short order, a major manufacturer and champion of the very current Edison had spent years and a fortune and much of his good name trying to destroy. The house that Edison built went over to the enemy. There could be no clearer signal that the war was over. When your own company, refounded on your own inventions, adopts your rival's technology, the argument you have been waging has been lost not at the level of propaganda or spectacle but at the deepest level, the level where money and engineering meet and decide what the future will actually be built from. Edison had lost, and the proof of it was that his own successors were now building alternating-current systems.

It is important to be precise about what this turning of the tide did and did not mean, because misconceptions cluster here. It did *not* mean that Edison was ruined or that he faded into obscurity — far from it. Edison walked away from the electrical-power business with substantial wealth from the merger, and he simply turned his restless genius elsewhere, to phonographs, to motion pictures, to mining, to the ceaseless invention that was his true vocation. He remained a celebrated figure to the end of his long life, and when he died in 1931 the nation dimmed its lights in tribute. Losing the current war was a bitter personal defeat, a blow to his pride and a stain, in retrospect, on his conduct—but it was not the end of Edison, who was always larger than any single technology. Nor did the tide's turning mean that direct current vanished from the earth; DC persisted in many niches, and would, as a later chapter will relish explaining, stage an astonishing long-distance comeback a century later in the form of high-voltage direct current transmission. The turning of the tide meant something more specific and more historically decisive: that the great question of which current would form the *foundation* of the world's power distribution had been settled, and settled in favor of alternating current, on the merits, by the combined weight of physics, economics, and finance.

And it is worth insisting, one final time, on that phrase "on the merits," because this book has spent many chapters watching the war be fought by other, uglier means — by fear campaigns and electrocuted dogs and a man tortured to death in a chair, by pamphlets and propaganda and the manipulation of a grieving public's fears. Given all that squalor, there is a temptation to imagine that the outcome, too, was decided by squalor, that alternating current won because Westinghouse out-propagandized Edison, or that some trick or maneuver settled it. The truth is almost the opposite, and it is one of the quietly reassuring facts of the whole story. The propaganda war, for all its ferocity and cruelty, did not determine the outcome. Alternating current won because it could do the essential job — carry power across distance — that direct current could not, and because that capability was worth so much, economically, that no amount of fear-mongering could ultimately suppress it. Edison's campaign of terror delayed nothing and changed nothing in the final reckoning; it succeeded only in staining its author's reputation. The dynamos, in the end, spoke louder than the pamphlets, and the physics spoke loudest of all. There is a lesson in that worth carrying out of this dark part of the book: that fear campaigns, however vivid and however cruel, tend in the long run to break against the hard rock of what actually works.

A surprising fact to weigh as the war closes: by the time the tide turned in 1892, the two men whose names are most associated with the current war in the popular imagination — Edison and Tesla — were both, in different ways, already moving offstage from the power business. Edison was being pushed out of General Electric and turning to other pursuits. Tesla, having sold his patents to Westinghouse and left Pittsburgh, was absorbed in his own laboratory researches and his triumphant lectures, less a combatant in the corporate war than a celebrated prophet standing to one side of it. The war's final phase was fought and won by the organizations — Westinghouse Electric and the new General Electric — and by the financiers who controlled them, more than by the famous inventors whose names adorn the story. This is the pattern the whole latter half of the tale has revealed: that the great personalities set the drama in motion, but the outcome was ground out by companies, capital, and the inexorable logic of a superior distribution technology. The heroes lit the fuse; the institutions detonated the charge.

Another surprising fact, and a poignant one: the man who had done the most to make alternating current possible, whose rotating magnetic field and polyphase system had armed Westinghouse for the whole war, received almost none of the credit or the wealth as the tide turned in AC's favor. Tesla had signed away his royalties; the fortune that his patents generated flowed to Westinghouse and its investors, not to him. As alternating current swept to victory and began its conquest of the world, its true inventor was already drifting toward the financial precariousness and the grand, unrealizable dreams that would define his later years. The prophet would be celebrated and then, for a long span, half-forgotten, while the world he had made hummed on without much thought for the man who had imagined it in a Budapest park a decade before. There is no neat justice in the story. The tide turned in favor of Tesla's current, but not much in favor of Tesla.

It is worth being honest about a complication that the triumphant narrative of AC's victory tends to smooth over, because the tide turned less cleanly than the story usually suggests. Direct current did not simply surrender and vanish in 1892; it persisted, stubbornly, in the very cities where it had first been installed, for decades longer than most people realize. The dense urban cores that Edison had wired with direct current could not be economically converted overnight, and so DC lingered on in pockets of Manhattan and other cities well into the twentieth century — indeed, remarkably, the last direct-current customers in New York City were not fully cut over to alternating current until the early years of the twenty-first century, more than a hundred years after the war was supposedly decided. The tide turned in the sense that all *new* development went to alternating current and the fundamental question of the future was settled; but the physical legacy of Edison's system, sunk into the streets in copper and conduit, took generations to fully retire. This is how great infrastructure transitions actually happen — not with a clean switch thrown on a single day, but with a long, messy overlap in which the losing standard lingers in the installed base while the winning one claims all the growth. The reader who imagines that AC's victory meant DC's instant disappearance has mistaken the decision for its execution.

There is also a deeper irony worth planting here, one that a later chapter will harvest in full, because it guards against the tidy conclusion that alternating current was simply and permanently superior. The victory of AC in the 1890s was a victory about *distribution* — about the economics of carrying power across the modest distances between a generating station and its customers, using the transformer to step voltage up and down. That was the problem of the age, and AC solved it decisively. But it was never true that alternating current was superior for *all* purposes and *all* time, and the man who lost the war would, from beyond the grave, be partly vindicated. For carrying enormous quantities of power across truly vast distances — hundreds or thousands of miles — it would eventually turn out that high-voltage *direct* current is actually more efficient than alternating current, and the twenty-first-century grid has quietly returned to DC for exactly such long hauls, a century after Edison lost. And every electronic device you own, from your phone to your computer, runs internally on direct current, converting the AC from the wall the moment it arrives. So the honest verdict on the turning of the tide is not that DC was wrong and AC was right, but that AC won the specific battle that mattered in the 1890s — the battle over distribution — while DC waited, patiently, for the battles where its own strengths would tell. The war chose a foundation for its own era, not a final truth for all eras.

You meet the outcome of this turning tide in the most literal and constant way possible: in the electricity that runs your entire life. The alternating current in your walls, the grid that carries it from distant plants, the three-phase lines on their steel towers, the transformers on the poles — all of it is the settled verdict of the War of the Currents, the world that alternating current built after the tide turned against Edison in 1892. General Electric and Westinghouse, the two giants that emerged from the war, would dominate the American electrical industry for the better part of a century, and their AC systems would electrify first the nation and then, in pattern if not in ownership, most of the planet. The war that had descended into such ugliness produced, in the end, the clean and constant miracle you take for granted every time you flip a switch.

The engineering argument was settled; the corporate consolidation was underway; the tide had turned beyond any hope of turning back. But a settled argument is not the same as a finished story, and alternating current still had to prove its victory not in a lecture hall or a German field but on the grandest stage the age could offer — before the eyes of millions, at a great world's fair in Chicago, and then in the thunder of the mightiest waterfall on the continent, where the current that had won the war would finally be put to the work of powering a civilization.

Part VII

The World Electrified

From the White City's hundred thousand lamps to the thunder of Niagara and the humming grid that followed — how alternating current, having won the war, went on to wire a civilization.

Chapter 51: The Birth of General Electric

On a winter morning early in 1892, Thomas Edison read in the newspaper that the company bearing his name had ceased to exist. He had not been consulted in any way that mattered, and the decision that erased him had been made in the paneled offices of financiers who regarded sentiment as a line item to be crossed out. The firm he had built out of the incandescent lamp — Edison General Electric, a colossus that made dynamos and bulbs and sockets and the wire to string between them — was being fused with its chief rival, the Thomson-Houston Electric Company of Lynn, Massachusetts. The merged enterprise would be called, simply, General Electric. Not Edison General Electric. Not Edison anything. The most famous name in American industry, the name that had become a synonym for invention itself, was to be quietly dropped from the masthead of the greatest electrical manufacturer on earth.

Edison, by several accounts, took the news with a bruised sort of calm and one sharp private wound. A colleague remembered him saying that he had grown tired of the electric-light business anyway, that there were newer things to chase. But the man who had signed his letters "T.A. Edison" with a proud looping flourish, who had watched reporters turn his every utterance into gospel, could not have missed the meaning of it. He had been, in the plain language of the boardroom, pushed out. To understand how that happened — how the Wizard of Menlo Park lost control of the company that carried his own name — is to understand a truth this book keeps circling back to: that inventing a technology and owning the business built on it are two entirely different games, and the second is often won by men who never touched a wire.

Let us define the players plainly, because the story turns on them. **Edison General Electric** had been assembled in 1889 by consolidating Edison's scattered lamp works, machine works, and lighting companies into one large corporation. It was capitalized heavily, and much of that capital came from the banking house of J.P. Morgan, the financier who had bankrolled Edison for years and who sat, quietly and decisively, near the center of the web. **Thomson-Houston**, meanwhile, was the creation of two former schoolteachers, Elihu Thomson and Edwin Houston, but its real engine was a Massachusetts shoe-company executive named **Charles Coffin** — a businessman of formidable gifts who had turned Thomson's inventions into a fast-growing, ruthlessly well-managed firm. Here is the first surprising fact of the chapter: on paper, in 1891, Edison General Electric was the bigger company, with roughly ten million dollars a year in sales against Thomson-Houston's ten as well, but Thomson-Houston was more profitable and, crucially, better run. When two companies merge, the popular assumption is that the larger swallows the smaller. In this case, something closer to the reverse occurred. The smaller, sharper firm effectively took command of the larger, more famous one.

Why merge at all? The answer is a tangle of patents and exhaustion. By the early 1890s the electrical industry was choking on litigation. Edison General Electric and Thomson-Houston each held pieces of the crucial patents — on lamps, on generators, on the arc lights and the transformers that the two firms had partly copied from each other

— and each was spending fortunes suing the other and defending against Westinghouse besides. It is easy to forget, amid the drama of the current war, that the more mundane war of the period was fought by lawyers over who owned which idea. Morgan, who loathed waste and worshipped order, saw the obvious solution the way he always did: stop the two companies from fighting by making them one. A combined General Electric would hold an almost unbreakable patent position, dominate the domestic market, and — this mattered enormously to Morgan — stop bleeding money in the courts. Consolidation, not competition, was the banker's religion, and he practiced it on railroads and steel and now on electricity.

And so the merger was arranged in the first months of 1892, with a new capitalization of some fifty million dollars, and the presidency handed not to any Edison man but to Charles Coffin. Elihu Thomson, gracious and self-effacing, was content to remain the inventor rather than the boss. Edison's own lieutenants — Samuel Insull among them, of whom much more later — found themselves outmaneuvered. Edison held a large block of stock, but stock is not the same as control, and the board that governed General Electric answered to Morgan and to Coffin, not to the inventor whose name they had just retired. Here the second misconception worth correcting rears up. People often imagine that Edison was cheated, robbed in the night by scheming bankers. The reality is less lurid and more instructive. Edison had, over the previous years, steadily sold down his own holdings to raise cash for new experiments; he had traded ownership for freedom, and he could not then be surprised when the owners exercised what he had sold them. He was not swindled so much as he had, bit by bit, signed away the throne himself.

It is worth lingering on the peculiar role of J.P. Morgan, because he embodies a force in this story as powerful as any inventor and far less celebrated: the force of concentrated capital. Morgan did not know how to build a dynamo or wind a transformer, and he did not care to learn. What he understood, with a cold and formidable clarity, was that the electrical industry in its infancy was a chaos of small warring firms, each duplicating the others' effort, each burning money in patent suits, each too fragile to command the vast sums that a national electrical system would require. His instinct was always the same, whether applied to railroads or steel or electricity: eliminate the competition by absorbing it, impose order from above, and let a single dominant enterprise, safely financed, reap the efficiencies. To Morgan, the merger that erased Edison's name was not a drama of personalities at all. It was simply good housekeeping. That a genius's pride was trampled in the process was, to him, a detail beneath notice. There is a hard lesson in this that runs through the whole modern age: the men who invent the future and the men who own it are rarely the same men, and when their interests collide, it is very often the owner, not the inventor, who prevails. Edison had spent his life mastering nature; he had never quite troubled to master the balance sheet, and the balance sheet, in the end, mastered him.

One small, telling anecdote survives from the aftermath. When the new company's securities were issued and the name "Edison" quietly vanished from the corporate letterhead, some of Edison's loyal associates were more wounded on his behalf than he seemed to be on his own. Edison, for his part, affected a studied indifference, remarking in effect that he had gotten out of a business he no longer found interesting and into new ones that fascinated him. Whether this was genuine equanimity or a proud man's armor, no one can say for certain, and it was very likely both. What is certain is that he never again built a company that dominated an industry the way Edison General Electric had, and that the firm which shed his name went on to become one of the mightiest corporations in the history of the world. The inventor kept his laboratory and his fame; the financiers kept the empire.

There is a deeper irony still, and it sits at the heart of this chapter like a coiled spring. Edison had spent the late 1880s waging a savage propaganda war against alternating current — the animal electrocutions, the electric chair, the whispered slander that AC was the executioner's current. He had staked his professional identity on direct current, the safe, low-voltage, short-range system he had pioneered at Pearl Street. And yet the General Electric that rose from the merger in 1892 promptly, unsentimentally, went into the alternating-current business. Thomson-Houston had never shared Edison's ideological horror of AC; it built AC equipment and sold it profitably. Once the two firms became one, the new company simply adopted whatever technology the market demanded, and the market was demanding alternating current for the same reason it always had: because AC could be stepped up to high voltage, sent long distances cheaply, and stepped down again for use. Within a few years General Electric would be manufacturing AC generators and, before long, licensing the very polyphase principles that Tesla and Westinghouse had championed. The company that bore Edison's cause abandoned Edison's cause almost the moment his name came off the door. The general who had fought the war lost it twice: first in the marketplace, and then inside the walls of his own former citadel.

It is worth dwelling on Charles Coffin, because he is one of the forgotten architects of the modern world and a useful corrective to the myth that great companies are built by inventors alone. Coffin was not a scientist and never pretended to be. What he possessed was an almost preternatural sense for organization, finance, and talent. He understood, as Edison never quite did, that the future of electricity lay not in one more clever gadget but in the disciplined, industrial-scale manufacture of standardized equipment, backed by research, sold through a national organization, and financed by patient capital. Under Coffin, General Electric would establish, in 1900, one of the first industrial research laboratories in America — a descendant, in spirit, of Edison's own Menlo Park, but run as a permanent institution rather than the extension of one man's genius. Think of it as the difference between a brilliant chef and the founder of a restaurant chain: the first produces a masterpiece, the second produces a system that produces masterpieces reliably for a century. Coffin built the system. That his name is unknown to almost everyone, while Edison's is on the tip of every tongue, is precisely the point.

What did Edison do with the wound? Characteristically, he turned his back on it and walked toward something new. Here is the third surprising fact: the man who lost the electric-light business poured a good deal of the fortune he had left into an enormous, and ultimately disastrous, iron-ore mining and separation venture in the hills of northern New Jersey — a project that consumed years of his life and a great deal of money and largely failed. He also threw himself into the phonograph, into motion pictures, into the storage battery, into cement. The lesson he drew from being pushed out of General Electric was not to grieve but to invent his way to a fresh empire, and if some of those later ventures foundered, others — the movies above all — helped remake the culture. Edison was sixty-five when his name came off the company. He would live and work for nearly four more decades. Whatever else the merger of 1892 took from him, it did not take his restlessness.

Where do you meet General Electric today? Almost everywhere, and for well over a century you could hardly have avoided it. The company that erased Edison's name went on to make the light bulbs in American homes, the turbines in power plants, the engines under the wings of jetliners, the appliances in the kitchen, the medical scanners in hospitals, and — through a broadcasting subsidiary — the radio and television that entertained a nation. For most of the twentieth century, GE was one of the largest and most admired corporations on earth, a fixture of the stock indices, a symbol of American industrial might. It has since been broken up and diminished, its recent decades far rockier than its long ascent, and by the time you read this its shape may have changed again. But the ticker symbol it long traded under contained no hint of the man whose incandescent lamp started it all. That absence is itself a monument — to the fact that in the industrial age, the inventor who lights the world does not always get to keep it.

The merger closed one story and opened another. For with Edison sidelined and the corporate map redrawn, the stage was cleared for a spectacle that would settle, in the eyes of the public if not yet the ledgers, which current would light the coming century — a shimmering, unreal city of white palaces and a hundred thousand lamps, rising on the shore of Lake Michigan.

Chapter 52: The White City

On the evening of May 1, 1893, a young engineer stood at a switch inside the machinery hall of the World's Columbian Exposition in Chicago and waited for a signal from the president of the United States. Grover Cleveland, standing on a flag-draped platform before a crowd estimated at half a million people, laid his hand on a gold-and-ivory telegraph key and pressed it. In that instant, current surged through miles of copper, great engines lurched into motion, fountains leapt into the sky, a vast gilded statue was unveiled, and — as dusk gathered over the fairgrounds — nearly a hundred thousand incandescent lamps kindled at once along the cornices and colonnades of a city that had not existed two years before. Spectators gasped. Some wept. A reporter wrote that it was as though a piece of heaven had been lowered onto the shore of Lake Michigan. The lamps were powered by alternating current, generated by machines built by the Westinghouse Electric Company, and there has probably never been, before or since, a more spectacular single advertisement for a technology.

To grasp why this mattered so enormously, you have to understand what the fair was and what electric light still was not. The World's Columbian Exposition, nominally marking the four-hundredth anniversary of Columbus's voyage, was the grandest of the great nineteenth-century world's fairs — a temporary city of monumental neoclassical buildings, sheathed in a white plaster-like material and arranged around lagoons and grand basins. Because the buildings gleamed white by day and blazed with light by night, people began calling it, almost at once, the White City. And electric light, in 1893, was still a marvel most Americans had never personally seen. Fewer than one household in a hundred had electric lighting. The vast majority of people lit their homes with kerosene lamps and candles and gas, as their grandparents had. To walk into the White City after dark, then, was not to see a nicer version of something familiar. It was to step, for one dizzying evening, into the future — into a world made of light, at a scale that no city on earth could yet match.

The contract to build that world had been fought over bitterly, and the fight is where our two great combatants reenter the story. Lighting the exposition was the single largest electrical contract yet let anywhere, and both surviving giants of the American electrical industry wanted it: the newly formed General Electric, heir to Edison and standard-bearer for direct current, and the Westinghouse Electric Company, apostle of alternating current. General Electric bid first, and high — a figure that came in at roughly a million dollars, later trimmed under pressure to about $554,000. Then George Westinghouse, calculating that the publicity was worth more than the profit, undercut it savagely, offering to light the entire fair for $399,000. Here is the first surprising fact worth pausing on: Westinghouse very likely lost money, or made almost none, on the most important contract of his career, and he did it on purpose. He understood, as a natural showman-engineer understands, that the fair would be seen by a substantial fraction of the entire nation, and that a triumph in Chicago would be worth more than any advertisement money could buy. He was, in the language of a later age, buying market share with a loss leader — and he was right.

But winning the bid created a problem that nearly sank him, and it is one of the most delicious anecdotes in the whole history of the current war. General Electric, furious at being underbid, controlled the essential Edison patents on the one-piece, all-glass incandescent lamp. It refused to sell Westinghouse the bulbs. A hundred thousand lamps, and the man who had promised to supply them could not legally buy a single standard bulb from the one company that made them. Westinghouse's response was pure engineering defiance. His workers designed an entirely new kind of lamp — a two-piece "stopper" bulb, so called because the base was ground and fitted into the glass globe like a stopper into a bottle, a design that arguably slipped around the Edison patent. The stopper lamps were harder to make and pumped a poorer vacuum, so they burned out faster; Westinghouse had to keep a small army of workers manufacturing and replacing them throughout the fair, running the lamp works day and night just to keep the White City lit. The public saw a hundred thousand serene points of light. Behind the scenes was a frantic, improvised, round-the-clock scramble to keep them burning. It is a useful reminder that the great public triumphs of technology are almost always propped up, out of sight, by exhausted people solving impossible problems in real time.

Presiding over the electrical exhibits, in the great Hall of Electricity, was Nikola Tesla, and this was his fair as much as anyone's. Westinghouse's whole system rested on Tesla's polyphase patents — the alternating-current motors and generators that could take power from a spinning machine, send it along wires, and turn it back into motion at the far end without the fragile sparking commutators that plagued direct-current machinery. Let us define **polyphase alternating current** plainly, since it is the beating heart of everything the fair demonstrated. Ordinary single-phase AC is like a single person pushing a child on a swing: shove, wait, shove, wait — power arrives in pulses, and at the bottom of each pulse there is a dead spot. Polyphase AC is like two or three people standing around the swing, each pushing at a slightly different moment, so that the pushes overlap and the swing is driven smoothly and continuously with never a dead spot. That overlapping, rotating push is exactly what spins a motor without any need for the brushes and commutators of a DC machine — and it is what made large-scale AC practical. At the fair, Tesla put this principle on flamboyant display.

His exhibits were part physics lecture and part magic show. He built a demonstration of the rotating magnetic field using an egg-shaped copper ovoid that, placed atop a coil, would rise, spin, and balance on its end — "the Egg of Columbus," a sly nod to the fair's theme and a vivid proof that his invisible field could make metal dance. He lit long glass tubes into glowing letters spelling the names of distinguished electricians and the word "Westinghouse." Most theatrically of all, Tesla passed hundreds of thousands of volts of high-frequency current through his own body, so that it enveloped him in a halo of crackling sparks and lit lamps he held in his hands while, he assured the awed spectators, doing him no harm. Here we must pause to correct a misconception, because Tesla's stunt has fed a century of dangerous folklore. It is emphatically not true that high voltage is harmless. What made Tesla's demonstration survivable was that the current was of very high frequency, which tends to travel along the surface of the body rather than through the vital organs, and of relatively low actual power. Ordinary household current at ordinary frequency, at a tiny fraction of those voltages, kills people every year. Tesla knew exactly what he was doing; the lesson for everyone else is precisely the opposite of what the spectacle seemed to teach.

It is worth understanding just how fully the fair had been conceived as a demonstration of the electric future, because lighting was only the most visible part of it. Electricity did not merely illuminate the White City; it ran it. An electric intramural railway carried visitors around the enormous grounds. A moving sidewalk — one of the first the public had ever seen — glided people along a pier, driven by electric motors. Electric launches, silent and smokeless, ferried sightseers across the lagoons while their coal-fired rivals belched overhead. There were electric kitchens demonstrating that meals could be cooked without fire, electric fountains that threw colored water into the night in choreographed displays, and a whole building, the Electricity Building, devoted entirely to the young science and its wonders. For the ordinary visitor, the cumulative effect was a kind of controlled overwhelming: a glimpse, assembled in one place, of a whole way of life that did not yet exist anywhere on earth but that would, within their children's lifetimes, become simply ordinary. The fair was, in the most literal sense, a working model of the twentieth century, built two years early on the mudflats of Chicago.

There is a human detail here that deserves remembering, because it captures the strange mixture of triumph and precariousness that ran through the whole enterprise. George Westinghouse, having staked his reputation on the fair, is said to have worried himself sick over whether the improvised stopper lamps and the untested machinery would hold up under six months of continuous operation before millions of eyes. Any serious failure — a citywide blackout, a fire, a fatal accident — would have handed his enemies exactly the ammunition they had been manufacturing for years, and would have branded alternating current, before the largest audience imaginable, as unreliable and dangerous. That the system ran, night after night, month after month, without catastrophe, was itself the argument. The public did not need to understand transformers to absorb the lesson their eyes delivered: this new kind of electricity worked, beautifully and safely, at a scale nothing else could touch. In the war of perception, which is the war that ultimately decides such things, that quiet reliability was worth more than any single dazzling stunt.

The numbers of the fair remain staggering. Over the six months it ran, from May to October 1893, the exposition drew roughly twenty-seven million visits — this at a time when the entire population of the United States was about sixty-five million. A meaningful share of the whole nation walked through the White City, and a great many of them saw electric light, and electric motors, and electric novelties, for the first time in their lives. The Westinghouse system in the machinery hall generated several times the electrical output of any central station then operating in America; by some reckonings the fair briefly consumed more electric power than the entire city of Chicago surrounding it. And nearly all of it — the lamps, the motors, the moving sidewalk, the intramural railway, the fountains — ran on alternating current. Millions of ordinary people, who knew nothing of transformers or commutators or the sneering slander that AC was the executioner's current, went home having seen with their own eyes that alternating current was safe, beautiful, abundant, and modern. No pamphlet Edison's camp had ever printed could compete with that.

A word is owed here to the shadow the White City cast, because the fair's dazzle was not the whole truth of it, and honesty requires the darker half. The gleaming neoclassical palaces were made of a temporary plaster-like staff, not marble, and they were torn down or burned within a couple of years; the White City was, quite literally, a stage set, a vision that could not last. Beyond its luminous avenues, the real Chicago of 1893 was a city of brutal labor conditions, of a savage economic depression that struck that very year, and of stark inequalities that the fair's serene facades were designed to make invisible. Some contemporaries noted the irony bitterly: a shimmering electric utopia erected for the amusement of visitors, a few streetcar-miles from tenements and sweatshops. The point is not to sneer at the fair, which was a genuine wonder, but to hold both truths at once. The electric future it advertised was real and was coming, but it would arrive into a world of ordinary human injustice, and electricity, for all its miraculous power, would prove to be a tool that served that world's existing divisions as readily as it eased them. The light was real. So were the shadows it did not reach.

Where do you meet the White City today? You meet it every time a stadium blazes to life under the lights, every time a skyline throws its reflection across the water, every time a theme park or a boulevard or a bridge is strung with illumination for no reason but delight. The idea that light could be not merely useful but a public spectacle, a form of civic wonder and mass entertainment, was crystallized on those Chicago lagoons in 1893. The fair also seeded a hundred smaller futures — the Ferris wheel made its debut there, spun by a steam engine; a young man named Frank Baum reportedly drew on the shimmering white towers when he later imagined the Emerald City of Oz; the neoclassical style of the fair shaped American civic architecture for a generation. But its deepest legacy was psychological. It settled, in the public imagination, a question the engineers had already answered on the merits. After the White City, alternating current was no longer the frightening upstart. It was the light of the future, and everyone who had stood on those luminous avenues knew it.

The fair proved that AC could light a city of dreams. What remained was to prove it could carry the raw power of the physical world itself — and for that, the engineers turned their eyes to a wall of falling water on the border of New York State, thundering exactly as it had for ten thousand years, waiting to be put to work.

Chapter 53: The Thunder of Niagara

For as long as human beings had lived within earshot of it, Niagara Falls had been an argument that men could not win. Every second, something like six thousand tons of water hurled itself over the crest and fell nearly two hundred feet into the gorge below, throwing up a mist that could be seen for miles and a roar that could be felt in the chest. The Iroquois had regarded it with awe; the first European visitors described it in the language of the sublime, that eighteenth-century word for beauty so overwhelming it edged into terror. Poets came to be humbled by it. And engineers came, increasingly, to covet it — because every ton of that falling water represented energy, energy pouring uselessly into the river at a rate that dwarfed all the coal being burned in all the cities of the continent, and the nineteenth century was a century that could not abide waste on such a scale. The question that consumed a generation of ambitious men was brutally simple and maddeningly hard: how do you capture the power of a waterfall and send it where people actually live?

The dream was old. As early as the 1750s, a small canal had drawn water off above the falls to turn a mill wheel, and over the following century a modest industrial district grew up along a hydraulic canal, running sawmills and flour mills directly off the water's mechanical force. But this was capturing the falls only in the crudest way — a wheel turned by water turns a shaft, and the shaft can drive machinery in the same building, or perhaps the building next door. Beyond a few hundred feet, the power was simply gone. You cannot run a leather belt or an iron rod for twenty miles. The overwhelming majority of Niagara's energy thundered past, untouched, because there was no way to carry it to the factories and homes of the cities that needed it. The falls were a fortune locked in a vault to which no one had the key. What was needed was a way to convert mechanical power into some other form — a form that could travel — and then convert it back again at the far end. That form, though not everyone yet believed it, was electricity.

In 1886 a group of investors incorporated a company to do exactly this, and in 1889 the enterprise passed into the hands of a serious and methodical financier named **Edward Dean Adams**, who became the guiding force of the Niagara Falls power project. Adams was not an inventor; he was an organizer of capital and talent, and the scale of what he proposed to organize was without precedent. He and his backers meant to bore great tunnels and shafts, install turbines driven by the falling water, and build generators large enough to electrify not just the local mills but the industrial future of an entire region. The trouble was that in the late 1880s nobody actually knew which kind of electricity to generate, or whether long-distance electrical transmission on such a scale was even possible. Adams did something that reveals the peculiar honesty of the project: he admitted he did not know, and he went looking for the world's best minds to tell him.

To that end, in 1890, the backers established an international commission, chaired by the eminent British physicist Sir William Thomson — soon to be Lord Kelvin, one of the towering scientific figures of the age — and offered prizes for the best proposals on how to develop and transmit Niagara's power. Here is the first surprising fact of this chapter, and it is a humbling one. Kelvin, whose name is on the absolute temperature scale and who was as authoritative as any scientist alive, initially favored direct current, and was deeply skeptical of alternating current for the job. Some of the most distinguished experts consulted leaned toward DC, or toward compressed air, or toward transmitting the power mechanically by ropes and cables, or hydraulically through pipes. The decision that seems obvious in hindsight — generate polyphase alternating current, step it up to high voltage, send it down the wires — was, at the time, genuinely contested by the leading authorities on earth. The eventual triumph of AC at Niagara was not a foregone conclusion. It was a judgment, argued over for years, that could plausibly have gone another way.

What tipped it was accumulating evidence and a pair of decisive demonstrations. In 1891, in Germany, engineers had transmitted three-phase alternating current about 175 kilometers, from a power station at Lauffen to an electrical exhibition at Frankfurt — a feat that proved, on a scale nobody could dismiss, that high-voltage AC could carry substantial power across long distances with tolerable losses. And in 1893 the White City in Chicago showed that Westinghouse's Tesla-based polyphase system could light and drive an entire temporary metropolis reliably. Slowly, the commission and Adams came around. The choice, when it was finally made in the mid-1890s, was for a two-phase alternating-current system built on Nikola Tesla's polyphase patents, with the great electrical equipment to be manufactured by the Westinghouse Electric Company. It was the ultimate vindication. The system a Serbian immigrant had conceived while walking in a Budapest park, sketched in the dust with a stick because his companion did not believe him, was chosen to harness the mightiest waterfall in the Western world.

It helps to understand, plainly, why alternating current was the right answer, because the reasoning is the load-bearing wall of the entire modern grid. Electric power is, very roughly, voltage multiplied by current. To send a given amount of power down a wire, you can use high voltage and low current, or low voltage and high current — the product is the same. But the energy wasted as heat in the wire depends on the current, and rises with the square of it, so doubling the current quadruples the loss. This means that to send power a long way efficiently, you desperately want high voltage and low current. The genius of alternating current is that a **transformer** — a pair of coils wound on a shared iron core, with no moving parts — can effortlessly step AC voltage up for the long journey and step it back down for safe use at the other end. Think of it this way: sending power at low voltage is like trying to move a crowd through a doorway all at once, everyone jammed and shoving and generating heat; stepping the voltage up is like sending them single file down a long corridor, quick and cool, then opening the doors wide again at the destination. Direct current, in that era, had no comparable trick; there was no simple, efficient way to change its voltage. Niagara's power had to travel, and only AC could travel. That, in the end, is why Kelvin was wrong and Tesla was right.

The construction was audacity made of stone and iron. Rather than dam the river, the engineers dug a great wheelpit — a vertical shaft nearly two hundred feet deep — and a tunnel more than a mile long, lined with brick, to carry the spent water back to the river below the falls. Water was diverted from above the cataract down through enormous turbines seated at the bottom of the wheelpit; the turbines turned long vertical shafts that rose up to the powerhouse at the surface, where they drove the generators. The first powerhouse, known as **the Adams Power Plant** in honor of Edward Dean Adams, was designed with a dignity befitting the project's grandeur; the eminent architects Stanford White and his firm gave it the sober monumentality of a temple, and in a sense it was one — a temple to the proposition that human ingenuity could put a waterfall on a leash. The generators Westinghouse built for it were giants for their day, each rated at five thousand horsepower, far larger than any dynamo yet made, machines so novel that building them required inventing the methods to build them. Here is the second surprising fact: many observers, right up to the moment the switches were thrown, genuinely feared the whole thing would fail — that generators of such unprecedented size would overheat, or vibrate themselves apart, or that the long transmission line would simply not work. Fortunes and reputations were staked on machinery that had never been tested at anything near full scale.

Consider, too, the sheer audacity of committing to a design at that scale before the crucial questions had answers. When Adams and his engineers settled on the size of the generators, no generator of that capacity had ever been built or run; when they specified the transmission voltage, long-distance transmission at that scale was still largely theoretical; when they bored the tunnel, they were digging on the assumption that machinery not yet proven would justify the excavation. It is easy, from the comfort of a world where all of this simply works, to underestimate the nerve it took. The men who built Niagara were placing a colossal bet, with other people's fortunes, on the proposition that a cluster of new and partly untested technologies would all come together and function on the first try. Had they been wrong — had the generators failed, or the line proved unable to carry the load — the setback to alternating current, and to the whole idea of remote generation, might have delayed the electrified world by years. That they were right does not make the gamble less real. Great infrastructure is almost always built this way: not on certainty, but on a disciplined, well-informed leap of faith by people willing to stake everything on machinery that does not yet exist.

The choice of Tesla's system also carried a quiet vindication that the newspapers of the day largely missed. Only a few years earlier, Tesla had been an object of ridicule in some quarters and, in the propaganda of the current war, the theorist behind a "dangerous" and "deadly" technology. Now the most prestigious engineering project on the continent, vetted by an international commission of the world's leading physicists, had examined every alternative — direct current, compressed air, mechanical transmission by rope and cable, hydraulic pressure sent through pipes — and had chosen, deliberately and after years of study, the polyphase alternating-current system Tesla had patented. There is no more crushing rebuttal to a smear than a panel of Nobel-caliber minds staking the future of the largest power project on earth on the very thing that was smeared. The commission's verdict was not swayed by showmanship or slander. It was swayed by the physics, and the physics pointed unmistakably at alternating current.

And there is a quieter irony threaded through the triumph that deserves an honest word, because the popular story tends to varnish it. The harnessing of Niagara was also, unavoidably, a diminishment of it. Every gallon routed through the turbines was a gallon that did not go over the falls, and even in these first years, conservationists warned that the industrial development of Niagara threatened to reduce one of the natural wonders of the world to a spigot. That tension — between the falls as sublime spectacle and the falls as raw horsepower — has never fully gone away, and it is why the flow over Niagara is, to this day, regulated by international treaty, dialed up for the tourists during daylight hours and drawn down for the generators at night. The men who tamed the falls gave humanity a gift of almost incalculable value. They also, in the same act, tamed something that had been wild since the last ice age, and not everyone thought that was only a gain.

You meet the descendants of the Adams plant every time you flip a switch, but the more direct place to meet it is in the very idea of a power station distant from the people it serves — the notion that electricity is generated somewhere else, out of sight, perhaps far away, and delivered to you over wires you never think about. Niagara established that principle at grand scale for the first time. The original Adams powerhouse no longer generates; it was eventually demolished, though a portion and its legacy are preserved, and enormous hydroelectric stations built later still draw power from the same river. But the concept it proved — big remote generation, high-voltage transmission, distribution to a whole region — is the architecture of the electrical world you live in now.

Yet proving that the generators would spin, and that the current would flow, was only half the wager. The other half was whether that current, born in the thunder of the falls, could be flung across the countryside to a real, waiting city — and put to work in its factories and homes. That test came in the winter of 1896, and it fell on the streetcars and workshops of a place called Buffalo.

Chapter 54: Power to Buffalo

Just after midnight on November 16, 1896, in a switching station on the edge of Buffalo, New York, a group of engineers stood watching a bank of instruments and waiting for a telephone to ring. Twenty-six miles to the west, at the powerhouse beside Niagara Falls, another team stood by the great generators, one hand near the switches, the other on the line to Buffalo. At the agreed moment the word came down the wire, a switch was closed at the falls, and the current — alternating eddies of electricity born from the falling water, stepped up to some eleven thousand volts for the journey — rushed east through the wires strung on poles across the snow-dusted countryside. In Buffalo it was received, stepped back down, and fed into the lines of the local street-railway company. Somewhere in the city, a streetcar that had been dead on its rails hummed and lurched into motion, driven by the power of Niagara Falls flung across twenty-six miles of open country. It was, by the reckoning of most historians who have told this story, the moment the War of the Currents ended.

To feel the weight of that midnight, you have to remember how modest and local electricity had been only a decade and a half before. When Edison opened the Pearl Street station in lower Manhattan in 1882, his direct-current system could push power only about a mile before the voltage sagged and the light dimmed to a useless orange glow. His empire of light was, of physical necessity, an empire of small islands — a station here lighting a few blocks, a station there lighting a few more, each hemmed in by the stubborn refusal of low-voltage direct current to travel. To electrify a whole country on Edison's original model would have required a power station on nearly every block, an absurdity of coal and copper and expense. The dream of carrying power over real distances — of generating it where nature was generous and consuming it where people were many — had haunted the industry from the beginning. On that November night, the dream stopped being a dream. Twenty-six miles was not a laboratory stunt or a temporary fairground marvel. It was a permanent commercial transmission line, delivering real power to a real city to do real work, and it changed the scale of the possible forever.

Let us be precise about the sequence, because the anniversaries get muddled in the retelling. The Adams Power Plant at Niagara first generated alternating current in 1895, and at first that power was used locally, driving the works of a nearby electrochemical industry that had deliberately clustered around the plant to take advantage of cheap electricity. The Pittsburgh Reduction Company — the enterprise that would grow into the aluminum giant Alcoa — was among the first great customers, because refining aluminum by electrolysis devours enormous quantities of electricity, and Niagara offered it cheaper than anywhere on earth. That is a surprising fact worth holding onto: the first industry the harnessed falls truly transformed was not lighting or streetcars but the mass production of aluminum, a metal that had been, only a few years earlier, so hard to refine that it was counted a precious rarity, more valuable than silver. Cheap Niagara power helped turn aluminum from a jeweler's curiosity into the stuff of pots, foil, and eventually airplanes. But the symbolic and strategic milestone was the transmission to Buffalo in November 1896, because carrying the power the twenty-six miles to Buffalo was the thing skeptics had said could not be done economically, and doing it settled the argument for good.

The victory was total in a way that is easy to state and hard to overstate. Within a few years, the modest first plant at Niagara had been joined by additional powerhouses; the number of generators multiplied; and lines reached out not only to Buffalo but, in time, toward the distant glow of a market that included, eventually, the fringes of New York City's own hinterland. The technology that had been slandered as the executioner's current, that Lord Kelvin himself had doubted, that Edison had spent a fortune trying to strangle in its cradle, now powered the greatest engineering achievement of the age. Every serious new central station built anywhere in the world from that point forward was, overwhelmingly, an alternating-current station. General Electric, Edison's own orphaned company, built AC equipment as fast as it could. The debate was over. Alternating current had won not by decree but by the plain arithmetic of physics and money: it could go where direct current could not, and going where the power is needed is the entire business of an electrical system.

It helps to sit for a moment with the sheer physical romance of what that twenty-six-mile line represented, because familiarity has worn the wonder off it. Consider what had to happen, invisibly, in the space of a heartbeat, for that Buffalo streetcar to move. Water fell over a cliff at Niagara. Its weight and motion spun a turbine, which spun a generator, which converted that raw mechanical violence into alternating current. A transformer stepped the voltage up to some eleven thousand volts so it could travel. The current then raced east through the wires at nearly the speed of light, crossing twenty-six miles of frozen countryside in a fraction of an instant. At Buffalo another transformer stepped it back down to a usable level, and it flowed into the motor of a streetcar and became, once again, motion — the motion of a vehicle full of people rolling down a city street. The power of a waterfall had been turned into electricity, flung across the width of a county, and turned back into mechanical power to carry a commuter to work, all in less time than it takes to blink. Nothing remotely like this had ever been possible in the history of the world. For all of prior human existence, to use the power of a river you had to go and stand beside it. Now the river came to you.

There is also a lovely fact about what this made feasible for the first time: the deliberate placing of power generation where nature is generous and industry where people are convenient, with the two joined only by a line of wire. Before long, the Niagara district became a magnet for exactly the industries that craved cheap, abundant electricity — not only aluminum but abrasives like carborundum, and various electrochemical and electrometallurgical processes that had scarcely existed before because no one had ever had enough cheap power to run them. Whole new industries were, in effect, willed into being by the availability of Niagara's current, clustering around the plant like a town growing up around a well. This is a pattern that would repeat across the coming century wherever great blocks of cheap power appeared: abundant electricity does not merely serve existing demand, it creates entirely new possibilities, new products, and new ways of making things that were previously unthinkable. Cheap power is not just a convenience. It is a solvent that dissolves old limits.

It is worth pausing here to correct two misconceptions that cling to this triumphant moment. The first is the tidy notion that the War of the Currents "ended" on a single date, like a treaty signed at a ceremony. It did not. Direct-current systems already installed in the hearts of cities kept running for decades; parts of Manhattan were still served by DC into the second half of the twentieth century, the last customers not converted until surprisingly late. The war ended the way most wars actually end — not with a thunderclap but with a long, grinding decline of the loser, punctuated by symbolic defeats. Niagara-to-Buffalo was the decisive symbolic defeat, the moment after which no informed person could still argue that the future belonged to Edison's current. But "decisive" is not the same as "instantaneous," and the honest picture is one of a verdict rendered in 1896 and enforced over the following forty years.

The second misconception is more seductive and needs firm handling. It is the romantic idea, popular on the internet, that Tesla single-handedly harnessed Niagara and that he alone deserves the laurel. The truth is both less tidy and more interesting. Tesla's polyphase patents were indispensable; without his rotating magnetic field and induction motor, practical large-scale AC would not have existed, and the system chosen for Niagara was built squarely on his inventions. But turning those patents into five-thousand-horsepower generators, mile-long tunnels, and a working transmission line required the industrial muscle of George Westinghouse's company, the organizational drive of Edward Dean Adams, the capital of a syndicate of financiers, and the labor of thousands of workers and engineers whose names no one remembers. Niagara was not the work of one genius. It was the work of a genius's ideas married to a great company's competence and a nation's money — which is, in truth, how nearly all large technologies actually get built. To hand the whole garland to Tesla is to repeat, in reverse, the very error that once handed the light bulb entirely to Edison.

And yet Tesla did receive his moment of honor, and it moved him deeply. On a visit to the completed works, the man whose vision had made it possible stood inside the powerhouse among the humming machines built to his design, watching his abstraction made physical on a scale he could scarcely have imagined when he sketched it in a park. Later, at a banquet held in his honor at Niagara, Tesla spoke with an emotion he rarely showed in public. He had been ridiculed and doubted; he had, in the lean years after leaving Edison, dug ditches for a living. Now the greatest waterfall in the Western Hemisphere turned generators built to his patents, and the power flowed to a city. Here is a small, poignant fact: Tesla had reportedly told an uncle, back in the old country when he was a boy dreaming over a picture of Niagara Falls, that one day he would go to America and capture the power of that very cataract. Whether the memory was polished by hindsight or literally true, the shape of it was real. The boy who had promised to harness Niagara had lived to see it done, to his own design, and to stand inside the temple of it while the machines sang.

It is worth naming, too, the men and the reckoning that this victory did not reward, because triumph in the marketplace is not the same as triumph in the bank. Neither Tesla nor Westinghouse grew rich in proportion to what they had achieved at Niagara. Tesla, as we have seen, had years earlier surrendered the enormous royalties his polyphase patents might have earned him, so the very system he watched turning the falls to his design put comparatively little into his pocket. Westinghouse, for his part, had spent so heavily and priced so aggressively to win the great contracts — the fair, Niagara, and the market battles against General Electric — that his company was often financially stretched, and within little more than a decade a financial panic would cost him control of the firm entirely. Here is a sober fact that runs through this whole history like a dark thread: the people who make a technological revolution are frequently not the people who profit most from it. The profits tend to flow, over time, to the organizations and the financiers who consolidate and operate the mature industry, not to the inventors and pioneers who took the early risks. Niagara was a total victory for alternating current, and a total vindication for Tesla's vision. It was a far more equivocal victory for Tesla and Westinghouse the men.

Where do you meet this triumph today? You meet it in the plainest and most invisible fact of your daily life: that the electricity in your walls was generated somewhere you have never been, perhaps hundreds of miles away, and delivered to you over a web of high-voltage lines you drive past without a glance. That entire arrangement — remote generation, long-distance high-voltage transmission, local distribution, all on alternating current — is the direct descendant of what was proven on the Niagara-to-Buffalo line in 1896. Every high-tension tower marching across a landscape, every substation humming behind its chain-link fence, every transformer on a pole outside a house, traces its ancestry to that midnight. The falls still thunder, and enormous hydroelectric stations on both the American and Canadian sides still draw their power from the river, feeding the grid of a continent. The dream that had haunted the century — capture the power where nature gives it, and send it to where the people are — had been realized, and it has never been surrendered since.

But the truth Niagara concealed beneath its grandeur was that most of the electrified world would never be lucky enough to sit beside a waterfall. The power that would actually light the cities and turn the factories of the coming century would come not from falling water but from fire and steam — and it would take a quiet English aristocrat, and a strange little boat, to make that power cheap enough to change everything.

Chapter 55: Parsons and the Steam Turbine

On the afternoon of June 26, 1897, the Royal Navy assembled the most formidable collection of warships on earth in the waters off Spithead, near Portsmouth, to celebrate the Diamond Jubilee of Queen Victoria. Row upon row of gray steel battleships and cruisers rode at anchor, dressed with flags, crewed by sailors at attention, a floating monument to the supremacy of the British Empire. And then, unbidden and uninvited, a small, sleek, absurdly fast little vessel came tearing through the fleet at a speed no ship in the world could match — better than thirty knots, roughly thirty-four miles an hour on the water, a velocity that seemed physically impossible for a boat. Naval picket craft gave chase and were left wallowing in her wake. The interloper was named Turbinia, she was the private demonstration vessel of an engineer named Charles Parsons, and her outrageous, unauthorized dash through the assembled might of the Royal Navy was the most brazen product launch in the history of engineering. Parsons had invented a new kind of engine, the world had declined to take it seriously, and so he had decided to embarrass the world into paying attention. It worked.

To understand why Turbinia mattered — and why she belongs in a history of electricity at all — you have to understand the problem that had quietly become the central bottleneck of the electrified world. By the late 1890s the great questions of how to distribute electricity had largely been answered: alternating current, transformers, high-voltage transmission, all proven at the White City and Niagara. But most of the world did not sit beside a waterfall. For the overwhelming majority of cities and factories, electricity had to be generated by burning fuel — coal, mostly — to boil water into steam, and using that steam to drive a machine that spun a generator. And the machine that did the spinning, in 1890, was the reciprocating steam engine: the same basic device, refined and enlarged, that James Watt had perfected more than a century before. It worked by admitting steam into a cylinder to shove a piston back and forth, the piston turning a crank the way your leg turns a bicycle pedal. It had powered the entire Industrial Revolution. And for the specific job of generating electricity at scale, it was becoming a catastrophe.

The trouble with the reciprocating engine was written into its very motion. Back and forth, back and forth — every stroke meant stopping a heavy mass of metal dead and flinging it in the opposite direction, thousands of times a minute. That endless reversal wasted energy, shook the building, wore out the parts, and, most damning of all, imposed a hard ceiling on speed and size. A generator likes to spin fast and smooth and steady; a piston engine is, by nature, jerky and self-limiting. The largest reciprocating engines built to drive dynamos were monstrous three-story cathedrals of whirling iron, beautiful and ruinously expensive, and even they could not spin fast enough or smoothly enough to make electricity truly cheap. As long as this was the only way to turn coal into current, electricity would remain, for most purposes, a relative luxury. The waterfall at Niagara was magnificent, but you cannot put a Niagara in Manchester or Berlin or Chicago. What the world needed was a fire-powered engine as smooth and tireless and scalable as the falls themselves.

That is precisely what Charles Parsons built, and the idea behind it is beautifully simple once you see it. Parsons — the youngest son of an Irish earl, a genuine aristocrat who chose the workshop over the drawing room — reasoned that instead of using steam to shove a piston back and forth, you could use it to spin a wheel directly, continuously, in one smooth rotation. Let us define the **steam turbine** plainly, with the analogies that make it click. Imagine a child's pinwheel: blow on it and it spins. A turbine is a pinwheel for steam — a shaft studded with row after row of curved blades, enclosed in a casing, so that high-pressure steam blasting through the blades pushes them around and spins the shaft, fast and continuous, with no stopping and reversing at all. Or picture a waterwheel turned on its side and driven not by a stream of water but by a jet of steam. Parsons's crucial refinement, patented in 1884, was to arrange many rows of blades in series, so the steam gave up its energy gradually, stage by stage, as it passed down the length of the machine — like a river dropping through a long staircase of small waterfalls rather than one violent plunge, extracting the energy smoothly and efficiently at every step. The result was a machine with no reciprocating parts, no shaking, that could spin at thousands of revolutions a minute — exactly the fast, steady rotation a generator craves.

Here is the first surprising fact: Parsons did not invent the turbine to power ships or even primarily to make a splash. He invented it, from the very beginning, to generate electricity. His 1884 turbine was coupled directly to a small dynamo, and the whole point was efficient electrical generation. The reciprocating engine turned too slowly and roughly for the fast generators he envisioned, so he built an engine to suit the generator rather than forcing the generator to suit the engine. It was a rare case of an inventor starting from the electrical need and working backward to the mechanical solution. The very first Parsons turbo-generator produced only about seven and a half kilowatts — enough to light a few hundred lamps, trivial by later standards. But the principle scaled with astonishing generosity. Within a couple of decades, turbines were being built ten thousand times more powerful than that first little machine, and they kept growing.

The world, however, did not immediately grasp what Parsons had done, which is why he had to resort to piracy at the Queen's own naval review. He had adapted his turbine to drive ships, convinced it would revolutionize marine propulsion, and the Admiralty had shrugged. So he built Turbinia, tuned her turbines to a howl, and crashed the Jubilee fleet review in front of the assembled admirals, foreign dignitaries, and press of the world. The stunt was reckless and possibly illegal and completely irresistible, and it made the turbine famous overnight. Here is the second surprising fact: within just a few years, the Royal Navy — the very institution Parsons had humiliated — had adopted turbine propulsion for its newest and most powerful warships, and the great ocean liners soon followed. The technology that would go on to quietly power the electrical grid first conquered the public imagination as the engine of the fastest ships afloat.

It is worth understanding why the turbine's smoothness translated so directly into cheapness, because the connection is not obvious at first glance. A generator produces electricity most efficiently when it spins fast and at a perfectly steady speed; any jerk or vibration in the drive shows up as flicker in the lights and wear in the machinery. The reciprocating engine, forever slamming its pistons to a halt and hurling them back, could never provide that serene, high-speed rotation — it was like trying to draw a smooth straight line with a hand that trembles. The turbine, spinning in one continuous direction at thousands of revolutions a minute, was steadiness itself, a hand that never shook. This meant not only that the electricity was cleaner but that the whole machine could be made astonishingly compact for its power. A reciprocating engine capable of a given output was a vast, heavy, room-filling monster; a turbine of the same output could be a fraction of its size and weight. Less metal, less floor space, less foundation, less friction, less maintenance — every one of those savings flowed straight into the cost of the electricity, and every one of them compounded as the machines grew larger. The turbine did not merely do the old job better. It changed the economics of the entire enterprise.

Here a comparison makes the scale of the revolution vivid. The largest reciprocating steam engines ever coupled to generators — those cathedral-sized behemoths of whirling iron that briefly powered the first great central stations — represented the absolute limit of what that technology could achieve, and they topped out at a few thousand horsepower while shaking their buildings and demanding armies of oilers and machinists to keep running. The turbine blew straight through that ceiling as though it were not there. Within a few decades of Parsons's first little seven-and-a-half-kilowatt machine, engineers were building turbo-generators tens of thousands of times more powerful, and they kept scaling up through the twentieth century into the hundreds of thousands and eventually beyond a million kilowatts from a single unit. No reciprocating engine could have approached such figures; the very idea is absurd, like trying to reach orbit on a bicycle. The turbine's capacity to grow, seemingly without limit, was precisely what allowed power stations to grow, and the growth of the power station, as Samuel Insull would soon prove in Chicago, was the key that unlocked electricity cheap enough for everyone.

But it is the electrical story that matters most for us, and it is enormous. The steam turbine is, without exaggeration, the machine that made the electrified world affordable. Because a turbine could be built far larger, spun far faster, and run far more efficiently than any reciprocating engine, the cost of generating a unit of electricity from coal collapsed over the following decades. Bigger turbines meant bigger, more efficient power stations; more efficient stations meant cheaper electricity; cheaper electricity meant that light and power spread from the wealthy few to the ordinary many. And here we arrive at the central, counterintuitive truth this chapter exists to deliver, the third surprising fact and the correction of a stubborn misconception. When people picture the birth of large-scale electricity, they picture Niagara — the falling water, the romance of harnessed nature. But the falls were the exception, not the rule. The overwhelming majority of the electricity that lit the twentieth century was, and still is, generated not by water but by boiling water into steam and spinning turbines with it. The turbine, not the waterfall, powered the electrified world. Niagara got the poetry; Parsons got the planet.

Stand back and look at the shape of it. A power station burns a fuel — coal, or later oil, natural gas, or the heat of a nuclear reactor — to boil water. The steam rushes through a Parsons-style turbine, spinning it at thousands of revolutions a minute. The turbine's shaft turns an alternating-current generator, which converts the mechanical spinning into electricity. Transformers step the voltage up, the wires carry it away, transformers step it back down, and it arrives at your wall. That basic sequence — fuel, boiler, turbine, generator, grid — has been the beating heart of electricity generation for more than a century, and the turbine is the irreplaceable machine at its center. Change the fuel all you like; whether the heat comes from a mountain of coal or the fission of uranium, it is very often, in the end, boiling water to spin a turbine of a design Charles Parsons would instantly recognize.

You meet the turbine every time you use electricity generated by burning anything, which for most of the world remains the majority of the power you consume. It hides behind the reactor and the cooling tower and the smokestack; it is the reason a nuclear power plant, for all its exotic physics, is at bottom a very sophisticated machine for boiling water. And there is a poignant coda to Parsons's life that closes the circle of this whole book: in his last years he became obsessed, of all things, with trying to manufacture diamonds and to reproduce the work of earlier scientists, chasing that dream to his death aboard a ship in 1931 — the same year, as it happened, that Thomas Edison died. But his real monument was already everywhere, humming in every power station on earth, an aristocrat's pinwheel turning the fire of coal into the light of civilization.

The turbine solved how to make electricity cheaply and abundantly. But cheap power in a machine hall is not the same as electricity in every home and shop and factory in a nation. Turning abundant generation into universal service required a different kind of genius altogether — not a machine but a business, invented in Chicago by a small, precise Englishman who had once been Thomas Edison's private secretary.

Chapter 56: Samuel Insull Invents the Utility

In the summer of 1892, a slight, brisk, impeccably dressed Englishman of thirty-two stepped off a train in Chicago to take charge of a small and struggling electric company, and in doing so left behind the most glamorous position in American industry for one that most of his colleagues thought beneath him. His name was Samuel Insull. For more than a decade he had been Thomas Edison's private secretary and then his indispensable business manager — the man who knew where every dollar of Edison's empire was, who negotiated the contracts, who ran the machine works, who had been so close to the great inventor that he was practically a member of the family. When the merger of 1892 formed General Electric and pushed Edison aside, Insull was offered a senior post in the new colossus. He turned it down. Instead he took the presidency of the Chicago Edison Company, a minor utility with a single small generating station and a few thousand customers, in a city where dozens of tiny electric companies were clawing at one another for scraps. His friends were baffled. He had traded a throne room for a workshop. He had also, though no one yet knew it, gone to invent the business model that would actually electrify the world.

Because that is the thing the machines alone could never do. This book has spent many chapters on inventions — the filament, the transformer, the induction motor, the turbine — and every one of them was necessary. But an invention is not the same as an industry, and a working technology is not the same as universal service. In 1892 electricity was still, for almost everyone, an expensive novelty. The problem was no longer really scientific; it was economic. How do you take a thing that is costly to make and deliver it cheaply enough that ordinary families and shops and factories can afford it — cheaply enough, in fact, that nearly everyone will want it? The answer Insull worked out over the following decades was not a device but a set of ideas about cost, scale, and organization, and those ideas turned out to be as revolutionary, in their quiet way, as anything Tesla or Edison ever built.

The first idea was **economy of scale**, and Insull grasped it with a clarity that eluded almost everyone else. A power station has enormous fixed costs — the building, the turbines, the generators, the boilers — that must be paid whether it serves a hundred customers or a hundred thousand. But once the station is built, producing one more unit of electricity costs relatively little. This means that the bigger the station and the more customers it serves, the lower the cost of each unit of power. Insull's instinct, therefore, was radical: build far bigger than seemed prudent, drive the cost per unit down, cut prices, attract still more customers, and use the revenue to build bigger still. Where his competitors ran many small stations and charged high prices to a wealthy few, Insull bet everything on giant central stations serving the many at low prices. He bought the largest turbines Parsons and General Electric could build — at one point commissioning a machine so enormous the manufacturer feared it would fail — precisely because a bigger machine made cheaper power. Think of it like a printing press: printing a single book is absurdly expensive, but printing a hundred thousand copies makes each one cheap, and the press that can print a million makes them cheaper still. Insull applied that logic to electricity, and it worked.

The second idea was subtler and more beautiful: **load diversity**. Here we must define the problem plainly. Electricity, in that era, could not be stored in any large quantity — it had to be generated at the instant it was used. So a power company's nightmare was the "peak": the moment when everyone wanted power at once. If a company sold only to homes, everyone would switch on their lights at dusk and switch them off at bedtime, so the expensive generating equipment would sit idle most of the day and strain to the breaking point for a few hours each evening. That is a ruinous way to run a machine. Insull's insight was to mix customers whose demands peaked at different times — factories running in the daytime, streetcars drawing power through the day, offices lit in working hours, homes in the evening, icemakers and pumps at odd hours. Blend them together and the total demand smooths out across the day and night, so the same equipment serves far more customers and almost never sits idle. The analogy is a restaurant that would go broke serving only the dinner rush; add a breakfast crowd, a lunch crowd, and an afternoon coffee trade, and suddenly the same kitchen and staff earn money all day long. Insull chased customers of every kind not despite their differences but because of them, because a diverse load was a cheap load.

To measure and bill this sprawling, varied clientele, Insull championed the third pillar: **mass metering** and clever pricing. He promoted the wide use of the electric meter so that every customer paid for exactly what they used, and he pioneered rate structures — separating the cost of being connected from the cost of energy consumed — that made electricity attractive to small users while keeping the system solvent. Here is a genuinely surprising fact: Insull learned some of these pricing principles not in America but from a British engineer, and he imported and refined ideas about demand-based rates that most American utility men had never considered. He was, in the truest sense, an intellectual of the electricity business, treating billing and load management as seriously as others treated dynamos.

But the largest and most consequential of Insull's ideas was the boldest, and it reshaped the relationship between business and government for a century. Insull recognized that electricity supply was a **natural monopoly** — that it made no economic sense to have five companies stringing five sets of competing wires down the same street, digging up the same roads, duplicating the same expensive stations. The efficient arrangement was a single provider serving a whole territory. But an unregulated monopoly is a license to gouge, and the public would never, and should never, tolerate that. So Insull did something that stunned his fellow tycoons: he went to the government and asked to be regulated. He proposed a grand bargain — later called the regulatory compact — in which the utility would be granted an exclusive franchise to serve a territory, and in exchange would accept government oversight of its rates and service, earning a fair but limited, guaranteed return. Around 1898, in a now-famous speech to his industry association, Insull argued that state regulation of exclusive utility monopolies was not a threat to be fought but a framework to be embraced. His peers thought he had lost his mind, or turned socialist. In fact he had seen further than any of them: that a regulated monopoly could raise capital cheaply, plan for the long term, and expand service to everyone, precisely because the bargain made it stable and trusted. That grand bargain — exclusive territory in exchange for public oversight — is the legal skeleton of the electric utility to this day.

Put these ideas together and you have the modern utility, and its effect on ordinary life was staggering. Under Insull, the price of electricity in Chicago fell and fell again, even as the number of customers exploded from a few thousand to hundreds of thousands to millions. Electricity that had begun the era as a rich man's amenity became, within a generation, something a working family could afford. Insull built interconnected systems, extended lines into surrounding towns and countryside, and turned a patchwork of feeble little companies into vast, integrated power networks. He understood, too, that he had to create demand as well as satisfy it, and so his companies actively promoted electric appliances, easy payment plans, and every new use for power they could dream up, because every toaster and fan and motor sold was another customer helping to fill in the load curve. He was not merely selling electricity; he was manufacturing the electrified way of life and then supplying it.

It is worth dwelling on how radical Insull's embrace of regulation truly was, because it cuts against every stereotype of the Gilded Age tycoon. The robber barons of his era fought regulation with every weapon they had, regarding any government oversight as theft and tyranny. Insull, alone among the great magnates, walked into the arena and asked to be bound. His reasoning was not sentimental; it was ruthlessly practical, and it rested on a paradox he grasped and his rivals did not. An unregulated monopoly, he saw, would inevitably provoke a public backlash so fierce that it might end in the government simply seizing the utilities outright and running them as public enterprises, as many cities and countries were beginning to do. By accepting regulation voluntarily — a public commission setting fair rates and guaranteeing service — the private utility could defuse that threat, keep itself in private hands, and, crucially, gain something enormously valuable in return: predictability. A regulated monopoly with a guaranteed territory and a guaranteed fair return was the safest investment imaginable, and safe investments can borrow money cheaply. Cheap capital meant Insull could build his giant stations and string his endless lines at lower cost, which meant still cheaper electricity, which meant still more customers. The regulatory bargain was not a chain he reluctantly accepted. It was a tool he sharpened and wielded, and it became the foundation on which the entire American electric-utility industry was built.

There is a delicious irony worth savoring in Insull's whole career, and it circles back to the rivalry at the heart of this book. Samuel Insull had been Thomas Edison's man — his secretary, his manager, his fierce partisan through the darkest years of the current war, when the Edison camp was slandering alternating current as the executioner's current. And yet the utility empire Insull built in Chicago ran, of necessity, overwhelmingly on alternating current, because only AC could economically serve the sprawling territory his economies of scale demanded. The disciple of the great champion of direct current became one of the master builders of the alternating-current world. Insull, being a practical man rather than an ideologue, felt no contradiction in this at all; he had absorbed from Edison not a loyalty to any particular current but the far more valuable lesson that the technology was only ever a means, and that the real prize was the system that delivered it to paying customers. He simply took that lesson further than Edison ever had, all the way to its logical end: that the decisive invention in electricity was not a machine but a business.

And then, having built one of the great fortunes and empires of the age, Samuel Insull crashed more spectacularly than almost anyone in American history — and honesty requires that we tell it. Through the 1920s, Insull had assembled his sprawling utilities into a dizzying pyramid of holding companies, layer stacked upon layer, financed by mountains of debt and stock sold to a public that trusted his name almost religiously. It was a structure that worked wonderfully as long as times were good, and catastrophically when they were not. When the Great Depression struck after 1929, the pyramid buckled and then collapsed, wiping out the savings of hundreds of thousands of small investors who had bought Insull securities on his reputation. His empire, valued in the billions, disintegrated. Insull fled abroad, was pursued, extradited, and put on trial for fraud. Here is the final surprising fact: a jury acquitted him — the collapse, it turned out, was a matter of a flawed and overleveraged structure meeting a historic catastrophe, not simple criminality — but his name was ruined all the same. He died in 1938, reportedly with pocket change to his name, in a Paris subway station, a fallen titan whom the public had turned into a symbol of everything that had gone wrong with capitalism. His downfall directly inspired New Deal laws that broke up the great utility holding companies and tightened their regulation for decades to come.

You meet Samuel Insull's invention every single day, more intimately than you meet any single machine in this book. Every time your electric bill arrives, itemizing your usage; every time you take for granted that flipping a switch will simply work, at a price you can afford, supplied by one company that serves your whole neighborhood under the eye of a public regulator — you are living inside the model Insull built. The regulated utility, the natural-monopoly bargain, the giant interconnected system serving rich and poor alike at low unit cost, is his monument, and it is a far larger monument than most people realize, because it is the mechanism through which the marvels of Edison and Tesla actually reached the human race. The technology made electricity possible. Insull's business made it universal.

But even Insull's genius had a boundary, and it ran, quite literally, at the edge of town. Beyond the last streetlight, out where the farms began, the wires stopped and the darkness of the old world held on — and closing that final gap would take not a businessman but a nation.

Chapter 57: Wiring the Nation

In the spring of 1913, a crowd gathered on Fifth Avenue in New York to watch the sun set behind a building that had no business being as tall as it was. The Woolworth Building rose seven hundred and ninety-two feet into the Manhattan sky, the tallest structure in the world, a Gothic cathedral of commerce so lofty that people called it exactly that. And on the evening of its dedication that April, President Woodrow Wilson pressed a button in Washington and, by the closing of a circuit, lit the whole soaring tower at once, flooding it with electric light so that it stood against the darkening sky like a lantern the size of a mountain. What the crowd was really looking at, though most of them could not have said so, was a building made possible by electricity twice over — lit by it, and lifted by it, for no structure of that height could exist without the electric elevators that carried people to its upper floors and the electric pumps that pushed water up its spine. The skyscraper was not merely decorated with electricity. It was a creature of electricity, unthinkable without it. And in the three decades on either side of that evening, electricity was quietly performing the same transformation on nearly everything else in American life.

The years from roughly 1900 to the 1920s were the period in which electricity stopped being a marvel and became an infrastructure — the invisible substrate of modern existence, as fundamental and unremarked as the air. The pieces had all been invented: the generator, the transformer, the induction motor, the turbine, the utility business model. What happened now was that these pieces knit themselves together into a single vast system and spread across the map of the country, and as they spread they remade the physical texture of daily life so thoroughly that a person who fell asleep in 1900 and woke in 1930 would scarcely have recognized the world. To wire the nation was not simply to string up more lamps. It was to rebuild the factory, the home, the city, and the very rhythm of human labor from the ground up.

The first great theater of transformation was the factory, and the revolution there was the electric motor — a change so profound that historians rank it among the most important in the entire history of manufacturing. To feel its weight, you have to picture the old factory, the factory of steam. A single giant steam engine sat in the basement or the yard, and its power was distributed throughout the building by a nightmarish tangle of spinning shafts running along the ceilings, with leather belts dropping down from them to drive each machine. The whole enormous apparatus of shafts and belts, called line shafting, had to turn all day long even if only one machine was in use; it wasted power in friction, filled the air with flying belts that maimed and killed workers, made the factory dark and cramped because everything had to cluster beneath the shafts, and dictated the entire layout of the building. Now give every machine its own small electric motor, fed by wires that can run anywhere. Suddenly the belts and shafts vanish. Each machine runs only when needed, drawing power independently. The factory can be laid out for the logic of the work rather than the geometry of the driveshaft; it can spread out into a single well-lit floor; machines can be arranged in the sequence of production, which is the seed of the assembly line. Here is a genuinely surprising fact: the productivity gains from electrifying factories took a couple of decades to fully arrive, because at first manufacturers simply replaced the central steam engine with one big central electric motor and kept the old belt-and-shaft layout. Only when they grasped that they could give every machine its own motor and redesign the whole factory around that freedom did the enormous gains materialize. The lesson — that a new technology's real payoff comes only when we reorganize our lives around it, not when we bolt it onto the old ones — is one the world would relearn, much later, with the computer.

Meanwhile, the same current was invading the home, and transforming it in ways that reshaped how families lived and who did the work. The electric light came first and was itself revolutionary: clean, safe, and instant, it abolished the daily labor of trimming wicks and filling lamps with kerosene, banished the soot and the fire risk, and extended the useful day well past sunset for everyone, not just those who could afford abundant candles. But light was only the beginning. Into the electrified home marched a parade of machines that took over tasks that had consumed lives: the electric iron, the vacuum cleaner, the washing machine, the refrigerator, the fan, the toaster, the sewing machine with its little motor. Each was sold, tellingly, as a liberator, especially of women, on whom the crushing physical labor of the household had chiefly fallen. Here a misconception deserves gentle correction. It is often assumed that these appliances simply reduced housework, freeing hours of leisure. The reality, as historians have shown, was more complicated: standards of cleanliness rose to meet the new machines, and much of the time saved was reinvested in doing more, so the promised leisure partly evaporated. The appliances did not so much shrink domestic labor as transform its character — from hauling water and beating rugs to operating machines. Still, no one who had scrubbed laundry by hand on a washboard would have traded the electric washer back.

Binding all this together was the physical growth and, crucially, the standardization of the grid. In the early years, electricity had been a chaos of incompatible systems: different voltages, different frequencies, different everything, so that equipment made for one city's supply would not work in another's. Over these decades, order was imposed. In North America the alternating current settled on a standard frequency of sixty cycles per second — sixty hertz — while much of the rest of the world standardized on fifty; the United States settled on household voltages around one hundred ten to one hundred twenty, while much of the world adopted two hundred twenty to two hundred forty. These choices, some of them historical accidents frozen into permanence, are why to this day your appliances need adapters when you travel abroad, a small daily inheritance of decisions made a century ago. Let us define the key move plainly: **interconnection**. Instead of each power station serving its own isolated island of customers, utilities began linking their stations and territories together with transmission lines, so that power could flow between them. The payoff was enormous and worth an analogy. A single power station is like a lone farmer who eats only what his own field yields — a bad harvest means hunger, and any surplus rots. Interconnect many stations into a grid, and it becomes like a market with a thousand farms feeding a shared storehouse: if one station fails or one city's demand spikes, power flows in from elsewhere; the whole system is more reliable, more efficient, and able to draw on the cheapest available generation at any moment. The grid, in this sense, is less a machine than an ecosystem, and interconnection was the moment the separate islands became a continent.

The reliability that interconnection provided was not a minor convenience; it was the thing that let the whole civilization lean its weight on electricity in the first place. As long as power came from a single isolated station, every hospital, every factory, every streetcar line lived at the mercy of that one station's failures — a broken boiler or a burned-out generator meant darkness for everyone it served, with no recourse. A society cannot build itself around a utility it cannot trust. Interconnection changed the psychology as much as the engineering: once power could flow in from neighboring stations to cover any single failure, electricity became something a community could genuinely rely upon, could plan around, could build its schools and subways and surgical theaters upon without keeping a candle in the drawer. The grid grew reliable, and because it grew reliable, people entrusted more and more of life to it, until the entrusting became total and the grid became, without anyone quite deciding it, the indispensable nervous system of modern existence. That deep dependence, which we now take entirely for granted, was manufactured deliberately, decision by decision, interconnection by interconnection, across these two or three transformative decades.

A further surprising fact belongs here, because it corrects a common assumption that electrification was a smooth and universally welcomed march of progress. In truth it was contested, uneven, and often bitterly fought over questions of ownership and control. In many cities and towns, fierce battles erupted over whether the electricity supply should be owned privately, as Insull's model held, or owned by the public, as a municipal service like water or roads. Some communities built their own city-owned power plants; some countries chose national ownership outright; the American compromise of the regulated private monopoly was only one answer among several the world tried. And there were losers in the transition as surely as winners: the gas-lighting companies that electricity displaced fought a long rearguard action; the workers whose crafts the electric motor made obsolete found their skills devalued; whole trades vanished. Progress, seen up close, is never the tidy triumphal procession it becomes in retrospect. The wiring of the nation was a genuine revolution, with all the disruption, conflict, and creative destruction that the word revolution honestly implies.

The consequences rippled outward into the shape of cities themselves. We have met the skyscraper, made possible by the electric elevator — and here is a fact worth pausing on:

without the elevator, buildings are limited to the number of stairs people will willingly climb, roughly five or six stories, so the entire vertical city, the dense downtown of towers, is a gift of electric power. But the electric streetcar reshaped the horizontal city just as decisively, carrying workers cheaply from outlying neighborhoods into the center and thereby creating the very possibility of the suburb and the commute. The electric sign turned commercial districts into blazing nightscapes; by the 1920s, stretches of Broadway glittered so brilliantly with electric advertising that the district earned the nickname the Great White Way. Cities that had gone dark and quiet at sunset for all of human history now pulsed with light and motion around the clock. Electricity did not merely illuminate the modern city; it drew its floor plan, ran its transit, staffed its factories, and set its restless twenty-four-hour tempo.

One more transformation belongs in this account, quieter than the skyscraper or the factory but arguably as profound: the electrification of time itself, and of the human relationship to night. For the whole of prior history, the rhythm of life had been dictated by the sun. People rose with the light and, for want of any cheap alternative, largely surrendered the dark hours to sleep or to the dim, costly glow of flame. Abundant electric light broke that ancient bargain. The night shift became possible, and with it the factory that never stopped and the newspaper printed in the small hours; entertainment moved into the evening on a mass scale, filling theaters and, soon, cinemas after dark; the whole notion of a "nightlife" as an ordinary feature of urban existence, rather than a rare and often disreputable exception, took shape in these electrically lit decades. A subtler consequence followed: as artificial light unmoored human activity from the sun, it began to standardize and regiment time in new ways, synchronizing millions of lives to the clock and the shift-whistle rather than the dawn. The electric century did not merely light the dark. It colonized it, annexing the night for work and commerce and pleasure, and in doing so it reorganized the most basic rhythm of human existence, one that had held since before we were human at all.

You meet the wiring of the nation in the most literal way imaginable: in the outlet on your wall, the standardized shape of its holes, the specific voltage and frequency humming behind it, all of them descendants of the choices frozen into place in these decades. You meet it in the elevator you ride, the escalator you step onto, the motor in every device that hums or spins or cools, the very existence of buildings taller than a few floors and cities that never sleep. By the 1920s, in the industrialized world, electricity had crossed an invisible line: it had gone from being something remarkable that some people had to something ordinary that a modern person simply expected, like running water. The electrified century had truly begun, and its citizens had already started to forget that it had ever been otherwise.

And yet this transformation, for all its sweep, had drawn a hard and cruel boundary. The blazing cities and the humming factories and the appliance-filled homes belonged to the towns. Beyond the reach of the streetlights, across the vast rural expanse where a great many people still lived, the wires simply stopped — and there, in the dark, a different and slower reckoning was waiting.

Chapter 58: Light for the Farm

On a summer night in the mid-1930s, on a farm somewhere in the American countryside, a family sat waiting in a kitchen lit by the yellow flicker of a kerosene lamp, watching a single bare bulb that hung from the ceiling on a new-strung wire. They had been told the power would come on that evening, and they hardly believed it. Then, without ceremony, the bulb glowed — dim at first, then bright, a hard clean white light that filled the room to its corners and threw shadows the old lamp had never reached. Across the country, in scene after scene that the people who lived through them would describe for the rest of their lives, farm families wept at the sight of their own kitchens fully lit for the first time. Some ran outside to look at the light pouring from their windows. Some, it was said, left the lights burning all night, unwilling to switch off a miracle. One farm woman, asked years later what electricity had meant to her, is remembered to have answered simply that the greatest thing was to have light to read the Bible by in the evening. To understand why a light bulb could bring grown adults to tears, you have to understand the gulf it closed — the immense, humiliating, decades-long gap between the electrified city and the darkened countryside.

That gap is the central fact of this chapter, and it was vast. By the early 1930s, the great cities of America blazed with light and hummed with motors; roughly nine in ten urban homes had electricity. But out on the farms, where a substantial fraction of the nation still lived, the figure was reversed and then some: only about one farm in ten had electric power. Ninety percent of rural households, some six million farms, still lived essentially as their great-grandparents had — lighting their homes with kerosene and candles, drawing water by hand or by windmill, washing clothes on a board, keeping food in a cellar or an icebox, doing the endless heavy labor of the farm entirely by muscle, human and animal. The modern world glittered a few miles down the road, and it stopped, abruptly and completely, where the town ended. For the farm family, the twentieth century had, in the most tangible sense, not yet arrived.

Why had the wires stopped? The answer was brutally economic, and it circles back to everything Samuel Insull had understood about the business of power. Running electric lines is expensive — the poles, the wire, the transformers, the labor of stringing miles of it across the landscape. In a city, that cost is shared among hundreds of customers packed into every mile of line, so the price per household is low. In the country, a mile of line might reach only three or four farms, each a long way from the next, each using relatively little power. The math that made electricity cheap in the city made it ruinously unprofitable in the country. The private utilities, run as businesses answerable to their investors, looked at the sparse farms and the long distances and concluded, rationally by their own logic, that rural electrification could not pay. Where they would serve farms at all, they often demanded that the farmers themselves pay the entire staggering cost of construction up front — sums far beyond what any ordinary farm family could afford. Here a misconception deserves correcting: the countryside stayed dark not because the technology could not reach it, and not because farmers did not want it — they wanted it desperately — but because the private, profit-seeking model that had so brilliantly electrified the cities had no incentive to electrify the farms. The market that lit Chicago left the farmer in the dark, and left to itself, it always would.

Into that failure stepped the government, and the vehicle was one of the most successful programs of the New Deal. On May 11, 1935, President Franklin Roosevelt, confronting the Great Depression with a torrent of federal action, created by executive order the **Rural Electrification Administration**, universally known as the REA, and the following year Congress gave it firm statutory footing. The REA's method was clever and, in retrospect, transformative. It did not, for the most part, build power lines itself or try to force the reluctant private utilities to do so. Instead it offered low-interest, long-term federal loans, and it channeled them chiefly to a new kind of organization: the rural electric **cooperative**. Let us define that plainly, because the co-op was the human engine of the whole effort. A cooperative is a utility owned not by outside investors but by the very customers it serves — the farmers themselves banded together, each a member and part-owner, borrowing the federal money collectively to build the lines to their own homes, then paying the loan back out of their electric bills. It was self-help underwritten by the nation. The farmers signed up their neighbors, cleared the rights-of-way, sometimes dug the postholes and helped raise the poles with their own hands, and in exchange the light came to their kitchens.

The results came with a speed that astonished even the program's champions. When the REA began in 1935, roughly one in ten American farms had electricity. Within about a decade and a half, that figure had climbed to the great majority; by the early 1950s, something close to nine in ten farms were electrified, and within a few more years the darkness was, for practical purposes, banished from the American countryside. It was one of the fastest and most complete transformations of daily life ever achieved by public policy. And the transformation went far beyond the bulb on the ceiling, profound as that first light was. Electricity on the farm meant the electric pump, which meant running water indoors and the end of hauling it by the bucket; it meant refrigeration, which meant food that did not spoil and a dairy farmer who could keep milk cold enough to sell; it meant electric milking machines, feed grinders, and tools that multiplied the work a family could do; it meant the radio, which ended the deep isolation of rural life and brought news, music, markets, and the wider world into the farmhouse; it meant, eventually, the whole apparatus of modern domestic life that the cities had enjoyed for a generation. Here is a surprising fact worth holding onto: rural electrification did not merely make farm life more pleasant, it made it more productive and helped keep agriculture viable, even as it also, ironically, accelerated the mechanization that would eventually draw many people off the land altogether. The light that saved the farm also helped remake it forever.

It is worth pausing on the human labor and ingenuity that the cooperatives demanded, because the light did not simply appear; it was dragged across the countryside by the sweat of the very people who received it. Building a rural electric line was brutal work. Rights-of-way had to be cleared through woods and across fields; holes had to be dug, often by hand in stony or frozen ground; poles had to be hauled by wagon or truck over roads that were frequently little more than mud, then raised and set and strung with wire. In many communities the farmers themselves pitched in — signing up their reluctant neighbors, because a line needed enough customers per mile to justify itself, and lending their own backs and teams to the construction. The REA sent engineers and organizers to help, and developed clever, economical designs and standardized methods to bring the cost per mile down to something the loans could bear. There is a surprising and rather moving fact here: much of the early organizing was social as much as technical, a matter of convincing skeptical, cash-poor, often deeply independent farm families to take on debt and trust a distant government and a newfangled power they had never used. The county meetings where neighbors argued over whether to sign up, the door-to-door canvassing, the shared labor of raising the lines — this was democracy and engineering fused, an infrastructure built not merely for the people but, in large part, by them.

And the change, once the current arrived, reached into the deepest textures of daily life in ways that are easy to list but hard to feel from a fully electrified vantage point. Consider only the matter of water. On the unelectrified farm, water was carried, bucket by heavy bucket, from a well or a pump, often by the women and children of the household, hundreds of pounds of it every day, for cooking and washing and livestock. The electric pump ended that ancient drudgery at a stroke, sending water through pipes into the house and barn at the turn of a tap. Consider light itself: the kerosene lamp not only gave a feeble, flickering glow but had to be filled, its wick trimmed, its sooty chimney cleaned, and it carried a constant risk of fire that burned down countless farmhouses over the generations. Consider the radio, which shattered the profound isolation that had defined rural life for all of history, ending the long silent evenings and connecting the remotest farmhouse to the news, the weather, the markets, the music, and the voice of a wider world. Each of these was, individually, a transformation; together they amounted to nothing less than the arrival of the modern age in places the modern age had passed by.

There is a poignant political texture to this story that is worth an honest word. The private utilities had spent years insisting that rural electrification was impossible, uneconomic, a fantasy. When the REA and the cooperatives proved otherwise — when the lines went up and the farms lit up and the loans were paid back at remarkably low rates of default — the same utilities that had refused the farms suddenly grew interested, and in some places rushed to string "spite lines" through the most profitable rural areas to skim the best customers before the co-ops could reach them. The farmers, who remembered exactly who had left them in the dark, mostly stayed loyal to their cooperatives. Many of those co-ops still exist today, still owned by their members, still humming power out across the countryside, a living legacy of the New Deal in tens of millions of acres of rural America. It is a rare and instructive case of a problem the market genuinely could not solve being solved, decisively and durably, by collective action.

And America was not alone in this, though its co-op model was distinctive. The gap between the electrified city and the dark countryside was a universal feature of the electrifying world, and nation after nation confronted it in its own way — through state-owned utilities, national grids, and deliberate government campaigns to push the wires past the point where private profit gave out. The great truth that the American REA demonstrated so vividly held everywhere: full electrification, reaching the last farm and the remotest village, was almost never accomplished by the market alone. It required a public decision that light and power were not merely commodities to be sold where profitable, but something closer to a right, or at least a national purpose worth subsidizing. That principle — that the grid should reach everyone, not merely everyone who was cheap to reach — is one of the quiet moral inheritances of the electrical age, and it is being tested anew in our own time. As you read this, well over a tenth of the human race, hundreds of millions of people, mostly in the poorest regions of the world, still live without reliable electricity, waiting on the far side of the very same gap that once separated the American farmhouse from the town. The work the REA began is, globally, not yet finished.

You meet the legacy of light for the farm whenever you drive through open country and see the power lines marching alongside the road, out past where any city logic would justify them, threading toward a distant farmhouse or a lone barn. You meet it in the simple, now-invisible assumption that electricity is something everyone has, town and country alike — an assumption that was not true within living memory, and that had to be deliberately made true, farm by farm, pole by pole, against the plain verdict of the balance sheet. The light that a farm family once wept to see is now so ordinary that its arrival is forgotten. But it was one of the last and greatest acts in the electrification of a civilization: the closing of the gap, the wiring of the whole, the extension of the current to the very edges of the map.

With that, the story of building the electrified world reaches its completion — the cities lit, the factories turning, the farms glowing, the grid spanning the continent from the falls to the farthest field. But the three men whose rivalry set all of this in motion were, by now, moving toward their endings, and one of them was about to gamble everything he had left on a dream so vast that it would break him.

Part VIII

Sparks in the Dark

From a doomed tower on Long Island to the smart grid and the electric future — the last years of the three rivals, the verdict of history, and the current that still runs through everything.

Chapter 59: Tesla's Tower

On a stormy night in the summer of 1899, on a high plateau outside Colorado Springs, a tall gaunt man stood in the doorway of a strange wooden laboratory and ordered his assistant to throw a switch, and then the world went mad. From the top of a mast rising through the roof of the building, bolts of artificial lightning erupted — vast crackling arcs of electricity, some of them reported to be more than a hundred feet long, tearing through the mountain air with a thunder that could be heard fifteen miles away. The ground itself seemed to hum. Sparks leapt between metal objects; people walking nearby felt tingling through the soles of their shoes; horses at a livery stable bolted in terror. Butterflies, it was said, were held fast in the air, wreathed in blue coronas of static, their wings crackling. The man in the doorway was Nikola Tesla, and he was, in that moment, generating and hurling lightning larger than any human being had ever made. He believed he was on the verge of the greatest discovery in the history of the world: a way to send electrical power through the earth itself, without wires, to any point on the planet. He was wrong, but the grandeur of the error would consume the rest of his life.

To understand how Tesla came to this remote mountaintop, we have to pick up his story where his triumph left it. In the early 1890s Tesla had been at the summit of his fame and powers. His polyphase system had won the War of the Currents; his patents had electrified the White City and harnessed Niagara; his lectures had made him a celebrity on two continents. He had invented, in the **Tesla coil** of 1891, a device that remains fundamental to this day: a resonant transformer that could generate extraordinarily high voltages at high frequencies, producing the spectacular sparks and glowing tubes that became his signature. Let us define what he was chasing plainly, because it is the key to his tragedy. Tesla had become convinced that if he could make electricity oscillate at just the right frequency, he could pump energy into the earth and the atmosphere and set the whole planet ringing like a struck bell — a phenomenon he called resonance. And if the planet could be made to ring electrically, then anyone, anywhere, could tap that ringing and draw power or signals from it, with a simple receiver, out of the very ground and air. No wires. No power lines. Free energy and free communication for all humanity, delivered through the earth itself. It was one of the most audacious dreams any inventor has ever entertained, and Tesla pursued it with the total, unbending conviction of a prophet.

The analogies he used were revealing. He compared the earth to a vast reservoir of water: if you set up a pump and push and pull the water rhythmically, you can create waves and pressure changes that a person far across the reservoir can detect and use, without any pipe connecting the two of you. Push at the right rhythm — the resonant rhythm — and even small pushes build into great waves, the way a child pumping her legs at just the right moment sends a swing ever higher. Tesla believed the planet had such a rhythm, and that he could find it and drive it. At Colorado Springs, on money lent by a wealthy admirer, he built the equipment to try, and there he did genuinely remarkable things: he generated millions of volts, produced the largest artificial lightning yet made, lit lamps at a distance without wires, and made measurements of atmospheric electricity that were decades ahead of their time. He also, in his notebooks, recorded receiving mysterious repetitive signals that he speculated might come from another planet — an episode his critics seized on as proof of crankery, though Tesla more likely detected natural electrical phenomena he could not yet explain. Here is the first surprising fact: much of the physics Tesla explored on that mountaintop was real and important, and some of his measurements anticipated genuine discoveries about the electrical resonances of the earth and atmosphere. His error was not that his phenomena were fake. It was that he wildly overestimated what they could be made to do.

Fired by the Colorado results, Tesla returned east in 1900 determined to build the real thing — a full-scale wireless power and communication station that would prove his system to the world. He needed money, and he went to the one man in America who could supply it on the scale he required: the financier J. Pierpont Morgan, the same Morgan who had bankrolled Edison and orchestrated the birth of General Electric. Morgan, intrigued chiefly by the prospect of wireless communication across the Atlantic, advanced Tesla one hundred fifty thousand dollars. With it, on a plot of land at Shoreham, on the north shore of Long Island, Tesla began to build **Wardenclyffe**. At its center rose a colossal tower, some one hundred eighty-seven feet high, capped by an enormous mushroom-shaped dome of metal and topped by a shaft that plunged deep into the earth to make the electrical contact with the planet that his theory demanded. It was to be the first node of a worldwide wireless system. Tesla imagined a future in which ships at sea received news, in which stock prices and private messages and even images flashed instantly around the globe, in which a traveler could carry a small device that would receive power from the air — a vision that reads, a century later, with an uncanny resemblance to the wireless, connected world we actually inhabit. He saw the shape of the future with astonishing clarity. He was simply, and catastrophically, wrong about the method.

Then the money ran out, and the dream began to die by inches. Tesla's ambitions had swollen far beyond Morgan's original bargain; costs mounted; the tower was expensive beyond all estimate. And in the middle of it all came the blow that broke the project's back. On the far side of the Atlantic, a young Italian named Guglielmo Marconi, using far simpler equipment, succeeded in transmitting radio signals across the ocean, reportedly sending the letter S from England to Newfoundland in December 1901. Marconi's practical, unglamorous radio did the one thing Morgan actually cared about — transatlantic communication — and it did it cheaply, now, without a monstrous tower and a fantasy of planetary resonance. When Tesla went back to Morgan for more money, Morgan refused. He had bargained for a wireless telegraph, not for the free electrification of the human race, and he had no interest in funding a scheme to give away power that no meter could measure and no company could bill. Here is a second surprising fact, and a bitter irony: Marconi's radio equipment used a number of principles and patents that traced back to Tesla's own earlier work. Tesla fought a long legal battle over the priority of the radio patents, insisting that he, not Marconi, was the true father of radio. The fight outlived him — and in 1943, a matter of months after Tesla's death, the United States Supreme Court issued a ruling that restored the priority of several of Tesla's radio patents over Marconi's, a vindication that came too late for him to know and is still argued over by historians. The man was robbed by history of an honor and then handed it back when he was past all knowing.

Without Morgan's money, Wardenclyffe starved. Tesla struggled on for years, seeking new backers, mortgaging the property, pouring in what little he had, unable to let the dream go. But the tower never operated as he intended, and it never could have — for the plain and painful truth, which honesty requires us to state, is that Tesla's central premise was mistaken. You cannot efficiently broadcast usable power through the earth and atmosphere to the whole world the way he imagined; the energy spreads out and dissipates, and there is no way to make a planet ring with power on demand for billions of receivers. Modern wireless power transfer exists, but only over short distances, and long-range broadcast power of the kind Tesla dreamed remains physically impractical to this day. This is the correction of the most stubborn and beloved myth about Tesla: the legend that he had invented a working system of free wireless energy for all mankind, and that greedy financiers and rival capitalists suppressed it to protect their profits. The romantic story is that Tesla's dream was stolen. The sadder, truer story is that his dream was, in the form he pursued it, impossible — that his genius had, for once, outrun physics itself, and that the tragedy of Wardenclyffe was not conspiracy but overreach. He was not defeated by villains. He was defeated by nature, which does not care how brilliant you are.

There is a further dimension to Tesla's tragedy that deserves an honest reckoning, because it explains why so brilliant a man could pour years and fortunes into an impossible dream. Tesla's greatest gift — the very faculty that had let him conceive the rotating magnetic field whole, in a flash of vision, and build working motors in his mind before touching any metal — was also, in the end, his undoing. He trusted his inner vision so completely that he could convince himself of the reality of things he had seen only in his imagination. When he pictured the earth ringing with power, he saw it so vividly, in such precise and glorious detail, that the vision felt like proof. This is the shadow side of a certain kind of genius: the same imaginative certainty that lets a mind leap ahead of its evidence to a true discovery can just as easily carry it past the evidence into a false one, and from the inside the two feel identical. Tesla had been spectacularly right about the induction motor when the experts doubted him, and that triumph may have taught him a dangerous lesson — that his visions were to be trusted over the objections of lesser minds. With Wardenclyffe, the vision was wrong, and he could not see it, because seeing it would have meant distrusting the very faculty that had made him great.

It is also worth correcting the impression, common in the romantic retellings, that Tesla was simply a naive dreamer with no head for practical affairs. In fact his failure had a great deal to do with the ordinary, unglamorous business of money and management, at which he was genuinely poor. He had signed away, years earlier, the lucrative royalty rights on his polyphase patents — releasing Westinghouse from a royalty agreement during an earlier financial crisis, in a gesture of loyalty that cost Tesla an almost unimaginable fortune over the following decades. Had he held those royalties, he might have financed Wardenclyffe himself and never needed Morgan at all. He repeatedly let honor and grand vision override financial self-interest, gave away or failed to defend rights worth millions, and managed his affairs with a lordly disregard for the ledger that would have horrified a man like Insull. Tesla's poverty at the end was not simply the world's injustice done to him; it was, in painful part, the sum of his own choices — the choices of a man who cared for ideas and honor far more than for money, and who paid for that preference with the ruin of his final decades.

The end was grim. In 1917, with Tesla long since bankrupt and the property lost to creditors, the great tower at Wardenclyffe was dynamited and pulled down, its metal sold for scrap during the First World War, and the monument to Tesla's most sweeping ambition collapsed into rubble on the Long Island shore. It is impossible not to feel the weight of that image: the inventor who had harnessed Niagara, who had lit a city of a hundred thousand lamps, who had conceived the system that powered the modern world, reduced to watching from afar as his last great work was blown apart for scrap. He was in his sixties, his fortune gone, his greatest gamble a smoking ruin. He would live on for another quarter century, but the tower's fall marked the true end of Tesla the world-builder. What remained was the long twilight.

You meet Tesla's tower today more than you might think,

and in more than one way. You meet its failure in the honest limits of physics — every time your phone charges on a wireless pad only when it is touching the pad, you are meeting the boundary Tesla refused to accept. But you meet its vision, uncannily fulfilled by other means, every single time your phone pulls signal and data out of the empty air, every time you stream music with no wire in sight, every time information flashes wirelessly around the globe in an instant. Tesla dreamed of a world swimming in invisible signals and power, a planet knit together through the air, and that world exists — not through his ringing earth, but through the radio waves whose paternity he spent his last years fighting to claim. And the site itself endures: after decades of neglect, the grounds at Shoreham were preserved and are being made into a museum and science center in his honor, a quiet resurrection of the place where his greatest dream fell down. The tower is gone, but the man who built it has, in death, become larger than he ever was in the ruin of his final years.

For the ruin of Wardenclyffe was not the end of Tesla's story, nor of the story of the three rivals whose contest lit the world. There remained the long reckoning of their final years, their deaths, and the strange, still-unfolding verdict of history — the last accounting of who won, and what it all finally came to.

Chapter 60: The Long Twilight and the Electric Future

On the evening of January 7, 1943, a maid entered a room on the thirty-third floor of the Hotel New Yorker, ignoring the "Do Not Disturb" sign that had hung on the door for two days, and found the old man dead in his bed. He was eighty-six, and he had died alone, as he had lived for many years, penniless and largely forgotten by the public that had once thronged to see him throw lightning from his fingertips. On the windowsill and in the parks nearby, the pigeons he had loved and fed and sheltered — the creatures he had said he loved as a man loves a woman — waited for a hand that would not come again. Nikola Tesla, who had conceived the system that electrified the world, died owing money he could not pay, his rooms full of notebooks and unbuilt dreams, his greatest tower long since dynamited into scrap. The government, wary in wartime that his papers might contain military secrets, impounded his belongings; they were found to contain no death ray, only the residue of a magnificent and ruined mind. It was an ending of almost unbearable poignancy for a man whose work you were using at the very moment of his death, in the lamp that lit your room and the motor in your refrigerator, though you did not know his name.

The deaths of the three men who had fought the War of the Currents came, fittingly, in the reverse order of their fame, and each ending carried the shape of the life that preceded it. George Westinghouse went first, and his end was the quiet dignity of a builder undone by finance rather than engineering. He had been, of the three, the one who actually made things work at scale — the industrialist who bought Tesla's patents, armed the AC revolution, underbid to light the White City, and built the giant generators that harnessed Niagara. He was also, by the standards of the Gilded Age, a genuinely humane employer, granting his workers a half-day off on Saturdays, pensions, and decent conditions decades before such things were common. But in the financial panic of 1907, when the credit markets seized and the money dried up, Westinghouse — like Insull after him — found that his sprawling enterprises rested on too much borrowed money, and the bankers who rescued his companies did so at the price of his control. He lost command of the very firm that bore his name, eased out of the board of Westinghouse Electric by the financiers who now held the purse. He died on March 12, 1914, at sixty-seven. Three years before his death, his old adversaries in the profession had awarded him the Edison Medal — the highest honor of American electrical engineering, named, with an irony no one could miss, for the man whose current he had defeated. Westinghouse was the decisive figure of the whole story, and he is the one the popular memory has most unjustly forgotten.

Thomas Edison died on October 18, 1931, at eighty-four, in the New Jersey home he had lived in for decades, and his passing was mourned as the death of a national god. He had long since made peace, more or less, with losing the current war and his company; he had gone on to the phonograph and the movies and a hundred other ventures, and he remained, to the end, the most celebrated inventor in American history, the Wizard grown old, visited by presidents and industrialists, his birthday a kind of national holiday of ingenuity. When he died, the tributes were immense, and the nation reached for a gesture equal to the man. It was proposed that the entire country switch off its electric lights for a minute or two in his honor — a fitting salute to the man who had done more than any other to put those lights there. But the proposal ran headlong into the very world Edison had helped build: by 1931 the electric grid had become so essential, so woven into the machinery of life and safety, that switching it off, even for a minute, was judged too dangerous and disruptive to attempt. Here is a fact that says everything about what these men wrought: Edison could not be honored by darkness because the light he had helped kindle could no longer safely be extinguished. And so, on the evening of his funeral, a gentler tribute was made. At the request of the President, Americans were asked to dim their lights for a moment in his memory, and across the country, in millions of homes and on great public buildings, the lights dropped low and then rose again — a nation blinking its lamps in farewell to the man of light. It is one of the most beautiful gestures a country has ever made, and it was possible only because of him.

For a long while after their deaths, history's verdict on the three seemed settled and simple. Edison was the immortal, the household name, the byword for invention. Westinghouse faded into the background, a company logo more than a man. And Tesla was very nearly forgotten — a strange, tragic footnote, the eccentric who died with his pigeons, remembered, if at all, as a cautionary tale about genius unmoored from good sense. But history is not finished with anyone as long as the world keeps using their work, and over the second half of the twentieth century and into the twenty-first, the verdict was dramatically revised, and no reputation was more spectacularly resurrected than Tesla's. It began with a formal scientific honor: in 1960, the international body that governs scientific units adopted the **tesla** as the standard unit of magnetic flux density — the measure, roughly, of the strength of a magnetic field. Every time a physicist or an engineer today speaks of a magnetic field of so many teslas, or notes that a hospital's MRI scanner runs at one and a half or three teslas, they are speaking the name of the man who died alone in the Hotel New Yorker. It was the scientific establishment's quiet acknowledgment that the rotating magnetic field, Tesla's central insight, sat at the foundation of the electrical age.

Then, in the decades that followed, Tesla's rehabilitation swelled from a scientific footnote into a full-blown cultural phenomenon, and it is worth being honest about both its justice and its excesses. The just part is real and overdue: Tesla genuinely did conceive the polyphase alternating-current system that powers the modern world, genuinely was a pioneer of radio, robotics, and high-frequency electronics, and was genuinely sidelined in the popular story by the more famous Edison. Restoring him to his rightful place among the giants was an act of historical fairness. But the resurrection also curdled, in places, into myth — the internet legend of Tesla as a flawless martyr-saint of science, a wizard who invented everything, discovered free energy, and was robbed of it all by the villainous Edison and greedy bankers. We have already seen how much of that is embellishment: the falling-out with Edison was more ordinary than the legend claims, the free-energy tower failed because of physics and not conspiracy, and Edison, for all his ruthlessness in the current war, was a titan in his own right rather than a mere villain. The truest verdict honors all three men without cartoons: Tesla the visionary who saw furthest, Edison the systematizer who built the first working whole, and Westinghouse the industrialist who married vision to power and made it real. It took all three, and their bitter rivalry, to light the world. The war made them enemies; history has made them, at last, collaborators in a single monumental achievement none could have completed alone.

And what an achievement it was. In the year 2000, the National Academy of Engineering in the United States drew up a list of the twenty greatest engineering achievements of the twentieth century — ranking the automobile, the airplane, the computer, the telephone, spaceflight, all the marvels of the age. At the very top of the list, ranked first, ahead of them all, stood electrification: the building of the electric power systems that light and move the modern world. The engineers judged it the single most important engineering feat of the century, and their reasoning was simple. Nearly every other achievement on the list depends on it. The computer, the telephone, the appliances, the water and the machines and the light itself — strip away electrification and the whole edifice of modern life goes dark. The thing that Thales glimpsed in a rubbed stone, that Franklin drew from a storm, that Faraday coaxed from a spinning magnet, that Edison and Tesla and Westinghouse fought to distribute — that thing had become the indispensable substrate of civilization, the current running, quite literally, through everything.

The technology, meanwhile, kept evolving in ways that would have surprised the combatants. Alternating current won the war and has reigned ever since; the grid that spans the continents is, at its heart, still the polyphase AC system Tesla conceived and Westinghouse built. But in one of the finest ironies in the whole history of electricity, direct current — Edison's defeated current — has staged a quiet and remarkable comeback, and it is worth understanding why, because it corrects the lazy notion that AC was simply and eternally "better." Recall that AC won because, in the 1890s, only AC could easily change voltage, using the transformer, and so only AC could be sent long distances efficiently. But the twentieth century invented something the combatants lacked: power electronics, the semiconductor devices that can convert high-voltage DC to AC and back cheaply and efficiently. And it turns out that for sending truly enormous amounts of power over truly vast distances — across a continent, or under a sea — high-voltage direct current, or **HVDC**, is actually more efficient than AC, wasting less energy over the long haul and allowing grids to be tied together in ways AC cannot easily manage. So today, the longest and largest transmission links on earth — carrying power from remote dams and deserts to distant cities, connecting the grids of whole nations — increasingly run on high-voltage DC. Edison's current, beaten in the 1890s because it could not travel, now travels farther than AC ever could, riding technology he never imagined. And in the small world of your devices, DC never left at all: every battery, every computer chip, every phone and LED and solar panel runs on direct current, quietly converting the AC from your wall into the DC they actually use. The honest verdict on the great war is not that one current triumphed forever, but that each found its proper domain — AC for the grid, DC for the long haul and the electronic age — and that the world runs, as it always has, on both.

The frontier keeps moving, and here honesty demands that we hold present-day matters lightly, for they are recent and unsettled and may look different by the time you read this. The grid that the twentieth century built as a one-way river — power flowing from great central stations out to passive consumers — is being remade, in our own time, into something more like a two-way network. Rooftop solar panels turn ordinary homes into small power stations that feed electricity back into the wires. Wind farms and vast solar arrays are displacing the coal that Parsons's turbines were built to burn, though the transition is uneven, contested, and far from complete. The old problem that haunted Insull — that electricity could not be stored and had to be used the instant it was made — is being softened, at last, by grid-scale batteries that bank power when it is abundant and release it when it is scarce, a development still young and rapidly changing. Engineers speak of a **smart grid**, woven through with sensors and computers and communication, that can balance this fluctuating, decentralized flow far more nimbly than the dumb grid of the past — an idea whose full realization still lies ahead. And electricity is reaching back into the one great domain that fossil fuels long ruled: transportation. The electric car, an idea as old as the automobile itself and long left for dead, has returned in force, and with it, perhaps, the electrification of the very act of moving through the world. How far and how fast these changes will go, no honest writer can tell you; they are the live, uncertain frontier of the story, and you, reading this, may already know things about their outcome that could not be known as these words were written. What can be said with confidence is only this: the electric future is not finished, and the current is still rising.

Step back, now, all the way back, and consider the distance traveled. This book began some twenty-six centuries ago, on a sun-baked Greek coast, with a philosopher named Thales turning a warm yellow stone in his hands and watching a wisp of straw leap toward it across empty air, pulled by an invisible force he could name but never understand. From that clinging stone the story ran forward through Gilbert's careful experiments and Franklin's kite and the shock of the Leyden jar; through Galvani's twitching frogs and Volta's pile and Ørsted's trembling compass needle; through Faraday, the bookbinder's apprentice, spinning a wire near a magnet and, in that small motion, discovering the principle that would spin every generator on earth; through Maxwell's beautiful equations that revealed light itself to be electricity in flight. And then it ran through the three men whose rivalry gave this book its spine — the deaf, driving genius from Ohio who built the first working system; the ethereal visionary from a Balkan village who saw how to send power across a continent; the sturdy industrialist from New York who made the vision real — locked in a war of currents that was ugly and brilliant and world-changing all at once. From the amber to the grid; from a feather twitching in a Greek hand to a hundred thousand lamps blazing over Chicago, to the thunder of Niagara flung across the miles to Buffalo, to the light finally reaching the farthest darkened farm. Genius, this book has argued, was never enough by itself. It took the right questions, the right tools, the accumulated groundwork of a hundred generations, the money and the machines and the vast anonymous labor of countless hands — and it took, too, the strange alchemy of rivalry, three proud men pushing one another to build faster and better than any of them would have built alone.

We opened, in the prologue, with an image: the night the lights came on. A city holding its breath in the dark, and then, at the closing of a switch, the sudden blaze — windows leaping into gold, streets unspooling into rivers of light, a hush and then a roar as a crowd understood that the night had just been broken, that the oldest boundary of human life, the daily surrender to darkness, had been repealed. For all of history before that moment, when the sun went down the world went out, and human beings huddled around small flames and waited for the dawn. And then, in the span of a single long lifetime, that ended forever. The lights came on, and they have never gone out since. Somewhere as you read this, dusk is falling over a city, and without a thought, without a sound, a hundred million lamps are answering the dark — in kitchens and towers and hospitals and farmhouses, on bridges and streets and the small screen in your own two hands. It is the most ordinary thing in the world now, and the most easily forgotten, and it is the closest thing to a miracle that human beings have ever made with their own hands. Thales rubbed a stone and made a feather move. His descendants rubbed the same secret against the whole of the physical world, and made the night surrender. The current he touched runs, still, through everything — through the wires and the machines, through the cities and the century, through the light by which you are reading these final words. It is still rising. And the dark, for the first time in the long story of our kind, has learned to wait for our permission.

Epilogue: What the Light Cost and What It Gave

Somewhere near you, right now, a wire is humming. It runs behind the plaster of the wall you are leaning against, or under the floor, or overhead through the dark, and it is alive with a current that reverses itself sixty times every second — or fifty, if you are reading this in most of the world outside the Americas. You cannot hear it, though if the room is very quiet you may catch the faint sixty-cycle drone of a transformer or the sigh of a refrigerator cycling on. You do not think about it. That is the whole point. The greatest engineering achievement of the twentieth century, as the National Academy of Engineering officially ranked electrification, succeeded so completely that it made itself invisible. We notice it only when it fails.

It is worth pausing, at the end of this long story, to make it visible again — to look, as I asked you to look in the prologue, at the invisible thing running through your walls, and to reckon honestly with what it took to put it there and what it has done to us since.

Begin with what it gave, because the gifts are almost beyond counting. Before the light came on, the human day was ruled by the sun. For the whole of history until roughly a century and a half ago, darkness was a wall that fell across life every evening and could be pushed back only a few feet, at real expense and real danger, by fire — a candle, an oil lamp, a gas jet that hissed and leaked and now and then burned the house down. Work stopped. Reading stopped, unless you were rich. The night belonged to sleep, to fear, and to the small circle of the hearth. Edison's bulb, and the system Tesla and Westinghouse built to carry its power over the horizon, did something no god in any of the old myths had ever managed: they abolished the tyranny of the dark. A child in a farmhouse could now read after supper. A surgeon could operate at midnight. A factory could run three shifts. The day, for the first time, belonged to us.

And light was only the beginning — the least of it, really. The same current that lit the bulb could also turn a wheel, and here Tesla's rotating magnetic field changed the texture of ordinary life more profoundly than any single invention since the plow. The electric motor took the drudgery that had bent human backs for ten thousand years and simply lifted it away. It pumped the water, so no one had to haul it. It washed the clothes and cooled the food and moved the elevator that made the tall city possible. It ran the machines that made everything else. We are so accustomed to this that we forget it is a kind of miracle: that in the wall of nearly every home in the developed world there is a socket from which, on demand, a river of tireless mechanical strength will pour, cleanly and silently, for pennies. The Roman emperor, the medieval king, the great merchant prince of Florence — none of them commanded a fraction of the raw power that waits behind the cheapest wall outlet in the poorest apartment on your street.

From that same socket flowed everything that followed: the radio and the telephone and the television that stitched a scattered species into a single nervous conversation; the computer and the network of computers that now holds most of human knowledge and will, if you ask it, answer you; the hospital's whole humming apparatus of monitors and pumps and imaging machines that has added decades to the average life. None of it — not one line of it — runs without the current. Strip electricity out of the modern world and the modern world is simply gone; what remains is 1850, with worse teeth. When we say that these three men and the scientists before them changed the world, we are not reaching for a compliment. We are describing, plainly, what happened.

But this is a book that has tried to tell the truth, and the truth is that the light was not free, and I do not mean only the price on the meter.

It cost, first, in the mines and the smokestacks. The current has to come from somewhere, and for most of the electric age it has come from fire — from coal dug out of the earth by men who died young of it, burned in furnaces that fill the sky with the invisible exhaust we are only now learning to fear. The same combustion that spun Parsons's turbines and lit the White City has, over a century of accumulation, begun to change the climate of the whole planet. This is the shadow side of the gift, and it would be dishonest to end a celebration of electrification without naming it. The grid that freed us from the dark was built, for a hundred years, on burning the past — the compressed sunlight of ancient forests — and the bill for that fire is now coming due, in a currency our grandchildren will pay. It is a strange and sobering thought that the same abolition of darkness that let a farm child read after supper also began, invisibly and by the same act, to warm the whole atmosphere of the planet, and that the two consequences flowed from one and the same switch. Nothing in this book has a simple moral, and the light least of all.

It cost, too, in the smaller and more human coin of what the current did to the men who fought over it. The War of the Currents was not a clean contest of ideas. It was fought with staged electrocutions of dogs and horses, with a word — "to Westinghouse" — coined to make a rival's name a synonym for death, and, at its lowest, with the first human being ever strapped into an electric chair, a botched and monstrous killing that Westinghouse himself said could have been done "better with an axe." The greatest inventor of the age lent his laboratory and his prestige to a campaign of fear he knew, or should have known, to be dishonest. That, too, is part of what the light cost. Genius is not the same thing as goodness, and this story, if it teaches anything, teaches that the people who build the future are as small and frightened and self-interested as the rest of us, and capable of real cruelty in defense of a fortune.

And it cost the three of them everything, in the end, though in three different ways. Consider how they left the stage. Edison — American, self-made, relentless, holder of more than a thousand patents — was pushed out of the very company that bore his name, watched his beloved direct current lose the war he had poisoned to win, and died in 1931 an honored but diminished old man; on the night of his funeral, Americans dimmed their lights for a moment, and the country noticed how strange the near-dark had become. Westinghouse, the courtly Pittsburgh magnate who had actually electrified the world, lost control of his company in the financial panic of 1907, died in 1914 with his name already fading from the story, and is today the least remembered of the three though arguably the most consequential — the decisive man history filed under "and." And Tesla, who had given us the current that won, the motor that turned the wheels, the very shape of the modern power system, died alone and all but penniless in a New York hotel room in January of 1943, in the company of the pigeons he loved, his great tower long since dynamited for scrap, his fortune given away or frittered on dreams that never came. The unit of magnetic flux density bears his name now, and a car company, and a thousand internet legends. He would have preferred the money.

There is a temptation, looking at those three ruined endings, to draw a tidy moral — that the world uses up its geniuses, or that the visionary always loses to the businessman, or that no good deed goes unpunished. But the honest lesson is quieter and stranger than any of those. It is that the light does not care. The current that hums in your wall does not know whose idea it was. It does not remember Tesla's rotating field or Faraday's induction ring or the frog's leg that twitched on Galvani's table or the lump of amber a Greek philosopher rubbed against his sleeve twenty-six centuries ago and wondered at. All of that — the wonder, the mathematics, the greed, the cruelty, the decades of patient work and the weeks of ugly war — all of it has been poured into the invisible thing behind the plaster, and the thing simply flows, indifferent and immortal, doing its work.

Which brings us back, as I promised in the beginning, to your own lit room. Look up at the light. It is very likely alternating current, Tesla's and Westinghouse's victory, generated in a plant you will never see and carried to you across miles of wire at a voltage that would kill you instantly, then gently stepped down by a transformer to something your lamp can drink. But look closer — at the phone in your hand, the laptop on your knee, the little glowing brick charging in the corner — and you are looking at direct current, Edison's banished current, which never really died at all, and which now runs quietly inside every electronic thing you own and is staging, through the great high-voltage DC lines marching across continents, a comeback its inventor did not live to see. Both men are in your room. They are still, in a sense, arguing there, in the walls, in the wires, in the soft and steady and utterly ordinary light.

That light is the most extraordinary inheritance in human history, and we received it, as we receive most inheritances, without quite understanding what it had cost the people who left it to us. This book has been an attempt to understand — to trace the current back up the wire, through the grid, across the century, into the minds of the men who wrestled it into being, and further back still, into the long dark before them, when a person who wanted light after sundown had to strike a spark and hope. We do not have to hope anymore. We only have to reach out and touch the wall.

Go ahead. Turn the switch off, and sit for a moment in the dark that ruled every one of your ancestors. Then turn it back on. That small, casual, godlike gesture — that is what the light gave us. Everything in this book is the story of what it cost to put it under your fingertip.

Timeline of Electricity

The story in this book unfolds across more than twenty-six centuries, from a Greek philosopher's idle wonder at a lump of amber to the humming continental grid and the electric cars now recharging in the dark. The timeline below gathers the milestones — the discoveries, the inventions, the battles, and the turning points — into a single chronological thread. Dates before the modern era are approximate, and the most recent entries, marked as such, describe a story still being written and should be checked against current sources.

c. 600 BCE — Thales of Miletus records that a piece of amber, rubbed with fur or cloth, attracts light objects such as straw and feathers — the first known observation of static electricity. The Greek word for amber, *ēlektron*, gives us the word "electricity."

c. 250 BCE — The so-called "Baghdad battery," a set of Parthian-era clay jars sometimes claimed to be an ancient galvanic cell, is later attributed to this period; its purpose remains disputed and most likely non-electrical.

1600 — William Gilbert, physician to Queen Elizabeth I, publishes *De Magnete*, distinguishing magnetism from the "amber effect," coining the Latin term *electricus*, and founding the systematic study of electricity.

c. 1663 — Otto von Guericke builds an early friction machine, a rotating sulfur globe that generates static charge, making electricity reproducible in the laboratory.

1729 — Stephen Gray demonstrates the difference between conductors and insulators, showing that electric "virtue" can be carried along threads over a distance.

1733 — Charles du Fay proposes that electricity comes in two kinds, which he calls "vitreous" and "resinous" — later understood as positive and negative charge.

1745–46 — The Leyden jar, the first capacitor, is independently invented by Ewald von Kleist and by Pieter van Musschenbroek at Leyden; it can store a charge and deliver a violent shock, turning electricity into both spectacle and tool.

1752 — Benjamin Franklin performs (or proposes) the kite-and-key experiment, demonstrating that lightning is electrical. He introduces the one-fluid theory, the terms positive and negative, and invents the lightning rod.

1780s — Luigi Galvani observes that a frog's leg twitches when touched by two different metals, and proposes "animal electricity."

1800 — Alessandro Volta, disputing Galvani, builds the voltaic pile — the first true battery — producing a steady, continuous current for the first time. The unit "volt" honors him.

1808 — Humphry Davy demonstrates the electric arc lamp, the first practical electric light, using a giant battery and two carbon rods.

1820 — Hans Christian Ørsted notices that a current-carrying wire deflects a compass needle, proving that electricity and magnetism are linked. André-Marie Ampère rapidly formalizes the mathematics of the force between currents; the unit "ampere" honors him.

1821 — Michael Faraday builds the first primitive electric motor, converting electrical energy into continuous mechanical motion.

1826–27 — Georg Ohm publishes Ohm's law, relating voltage, current, and resistance ($V = I \times R$).

1831 — Faraday discovers electromagnetic induction: a changing magnetic field induces an electric current. This single principle underlies every generator, transformer, and the entire power industry. Joseph Henry independently discovers self-induction in America; the unit "henry" honors him.

1837 — Samuel Morse and others develop the electric telegraph, the first great practical application of electricity, soon spanning continents and oceans.

1860s — James Clerk Maxwell unifies electricity, magnetism, and light in four equations and predicts electromagnetic waves — the theoretical capstone of the field.

1866–67 — The self-exciting dynamo is developed (Werner von Siemens, Charles Wheatstone, and S. A. Varley), making large-scale electrical generation practical.

1876 — Edison opens his laboratory at Menlo Park, New Jersey — the world's first industrial research and development facility. Alexander Graham Bell patents the telephone the same year.

1877 — Edison invents the phonograph, recording "Mary had a little lamb" and becoming world-famous as the "Wizard of Menlo Park."

1878–79 — Joseph Swan in Britain and Edison in America race toward a practical incandescent lamp.

October 1879 — Edison achieves a practical, long-lasting incandescent bulb using a high-resistance carbonized filament, burning for hours and, with later refinements, hundreds of hours.

September 4, 1882 — Edison's Pearl Street Station, the first commercial central power plant, opens in lower Manhattan, supplying direct current at 110 volts to some four hundred lamps and eighty-five customers — the birth of the electric utility.

1884 — Nikola Tesla arrives in the United States with a letter of introduction to Edison. Charles Parsons invents the steam turbine, which will transform electrical generation.

1885 — Tesla leaves Edison's employ after the famous falling-out.

1886 — George Westinghouse founds Westinghouse Electric and bets on alternating current. In Great Barrington, Massachusetts, William Stanley demonstrates a practical AC transformer system, stepping voltage up for transmission and down for use.

May 1888 — Tesla is granted patents for his AC induction motor and polyphase system and delivers his landmark lecture to the American Institute of Electrical Engineers. He sells his polyphase patents to Westinghouse, arming the AC camp for the coming war.

1888–89 — Edison's camp wages the anti-AC "War of the Currents," a fear campaign branding alternating current the "executioner's current." Harold P. Brown stages public electrocutions of animals to dramatize AC's danger.

August 6, 1890 — William Kemmler becomes the first person executed in the electric chair, at Auburn Prison, New York — a botched killing, deliberately AC-powered to taint Westinghouse's brand.

1891 — Tesla patents the Tesla coil and is naturalized as a US citizen. The Lauffen-to-Frankfurt demonstration in Germany transmits three-phase AC roughly 175 kilometers, proving long-distance AC transmission. Edison General Electric is formed.

1892 — General Electric is created by merging Edison General Electric with Thomson-Houston; Edison's name is retained but he is pushed out of control, and GE soon adopts AC.

1893 — The World's Columbian Exposition in Chicago is lit by Westinghouse's alternating current — the dazzling "White City," with some hundred thousand lamps — a public vindication of AC.

1895 — The Adams Power Plant at Niagara Falls begins generating polyphase AC using Tesla's system.

November 1896 — Niagara power is transmitted about twenty-six miles to Buffalo, the decisive demonstration that alternating current had won the War of the Currents.

1897 — J. J. Thomson discovers the electron, identifying the particle that actually carries electric current.

1899–1900 — Tesla conducts high-voltage experiments at Colorado Springs, pursuing wireless power transmission.

1901 — Guglielmo Marconi transmits a radio signal across the Atlantic, launching commercial wireless and a long dispute with Tesla over priority.

1901–1917 — Tesla builds and then abandons the Wardenclyffe tower on Long Island, his grand scheme for global wireless power; it is demolished in 1917.

1907 — A financial panic costs Westinghouse control of his company.

1911 — Westinghouse receives the AIEE's Edison Medal, a poignant late honor from his rival's namesake award.

1914 — George Westinghouse dies on March 12.

1920s–1930s — Samuel Insull builds a vast utility empire, pioneering the regulated monopoly, load diversity, and mass electrification; interconnected grids and holding companies spread across the country before Insull's empire collapses in the Depression.

October 18, 1931 — Thomas Edison dies; on the night of his funeral, Americans briefly dim their lights in tribute.

1935 — The Rural Electrification Administration, a New Deal program, begins bringing power to American farms.

January 7, 1943 — Nikola Tesla dies, alone and nearly penniless, in the Hotel New Yorker. Later that year the US Supreme Court's radio-patent ruling partly favors Tesla over Marconi.

1951 — The first usable electricity is generated from nuclear fission (at the EBR-I reactor in Idaho), opening the atomic era of power.

1954 — Bell Labs demonstrates the first practical silicon solar cell.

1960 — The SI unit of magnetic flux density, the "tesla," is formally adopted, honoring the man who lost the war but shaped the grid.

1954 onward — High-voltage direct current (HVDC) transmission enters commercial service (the Gotland link in Sweden), quietly reviving Edison's current for efficient long-distance bulk power.

Late 20th century–present (*recent and ongoing; verify against current sources*) — Smart grids, grid-scale battery storage, large-scale integration of wind and solar power, the electrification of transport through electric vehicles, and the return of DC in electronics, data centers, and long-distance transmission reshape the electrical world that Edison, Tesla, and Westinghouse began. Electrification is named the greatest engineering achievement of the twentieth century, and the current still runs.

Who's Who of the Electrical Age

Electricity was not discovered by one person, nor tamed by three. It was the work of a long relay of curious minds, each handing the spark to the next across the centuries — philosophers, showmen, bookbinders, priests' sons, and hard-nosed industrialists. What follows is a gallery of the people who move through this book, arranged in rough chronological order so that the reader can watch the baton pass from hand to hand. Each entry gives the essential dates, the essential contribution, and, where I can, a detail worth remembering.

Thales of Miletus (c. 624–546 BCE). A Greek philosopher of the Ionian coast, traditionally counted among the Seven Sages. He is the first person recorded to have noticed that amber, when rubbed, attracts light objects — the founding observation of the entire science. Whether he theorized about it or merely remarked on it we cannot say, but the Greek word for amber, *ēlektron*, became our word "electricity," making Thales the accidental godfather of the whole field.

William Gilbert (1544–1603). Physician to Queen Elizabeth I and the true founder of electrical science. His 1600 masterwork *De Magnete* separated magnetism from the "amber effect," introduced the Latin term *electricus*, and insisted on experiment over ancient authority. He also, less happily, believed the Earth itself was a giant magnet — which, as it turns out, is roughly true.

Otto von Guericke (1602–1686). A German scientist and mayor of Magdeburg, better known for his dramatic vacuum experiments with hemispheres and teams of horses. Around 1663 he built one of the first friction machines, a spinning sulfur globe that produced static charge on demand, turning electricity from a rare curiosity into a repeatable laboratory effect.

Stephen Gray (1666–1736). An impoverished English dyer and amateur experimenter who, late in life, made the crucial distinction between conductors and insulators, showing that electric "virtue" could be sent hundreds of feet along a suspended thread. He famously electrified a charity-school boy, suspended on silk cords, in a demonstration called the "flying boy."

Pieter van Musschenbroek (1692–1761). A Dutch professor at Leyden who, in 1745–46, helped invent the Leyden jar — the first capacitor — and received such a violent shock from it that he wrote he would not repeat the experiment "for the whole kingdom of France." The jar made electricity a public sensation.

Benjamin Franklin (1706–1790). Printer, statesman, and America's first great scientist. His kite experiment of 1752 proved that lightning is electrical; he gave us the one-fluid theory, the terms positive and negative, and the lightning rod, which for the first time let humanity defend its buildings against the sky. He refused to patent the rod, wanting it freely available to all.

Luigi Galvani (1737–1798). An Italian anatomist at Bologna whose frog legs twitched when touched by metal, leading him to propose "animal electricity." He was wrong about the cause but right that something electrical was at work; his name survives in "galvanize" and "galvanic," and his dispute with Volta launched the science of the battery.

Alessandro Volta (1745–1827). An Italian physicist who, disputing Galvani, built the voltaic pile in 1800 — the first true battery, and the first source of steady, continuous electric current. This single invention turned electricity from a fleeting spark into a usable flow. The unit of electrical "push," the volt, honors him.

Hans Christian Ørsted (1777–1851). A Danish physicist who, during a lecture in 1820, noticed a compass needle jump when he switched on a nearby current — the accidental discovery that electricity and magnetism are one linked force. Electromagnetism, the principle behind every motor and generator, begins with that twitching needle.

André-Marie Ampère (1775–1836). A French mathematician and physicist who, within weeks of hearing of Ørsted's discovery, worked out the mathematics of the force between currents and founded the science of electrodynamics. A self-taught prodigy who had endured great personal tragedy, he is honored in the ampere, the unit of current.

Georg Ohm (1789–1854). A German schoolteacher who established, in 1827, the simple and fundamental relationship between voltage, current, and resistance — Ohm's law. His work was ignored and even ridiculed for years before it became the bedrock of electrical engineering. The unit of resistance bears his name.

Michael Faraday (1791–1867). A blacksmith's son and self-taught bookbinder's apprentice who became the greatest experimental physicist of the age. He built the first electric motor (1821) and discovered electromagnetic induction (1831) — the principle behind every generator and transformer, and thus the foundation of the entire power industry. He also gave us the concept of the field. He declined a knighthood and the presidency of the Royal Society, preferring to remain "plain Michael Faraday."

Joseph Henry (1797–1878). An American physicist, first

secretary of the Smithsonian Institution, who discovered self-induction independently of Faraday and built powerful electromagnets. The unit of inductance, the henry, honors him. Like Franklin, he declined to patent his discoveries, believing science should be freely shared.

James Clerk Maxwell (1831–1879). A Scottish physicist of extraordinary reach who, in the 1860s, unified electricity, magnetism, and light into four elegant equations and predicted the existence of electromagnetic waves. His theory is the capstone of classical physics and the direct ancestor of radio, radar, and wireless. Einstein kept a photograph of him on his study wall.

Heinrich Hertz (1857–1894). A German physicist who, in 1887, experimentally produced and detected the electromagnetic waves Maxwell had predicted, proving the theory correct and opening the door to radio. He died at thirty-six; the unit of frequency, the hertz, is his memorial, and we invoke it every time we tune a radio.

Werner von Siemens (1816–1892). A German engineer and industrialist who improved the self-exciting dynamo and built one of the great electrical firms of Europe. He helped turn electricity from a laboratory phenomenon into an industry, and the company bearing his name endures to this day.

Joseph Swan (1828–1914). An English chemist and physicist who developed an incandescent lamp in parallel with — and in some respects ahead of — Edison. Rather than fight an endless patent battle, the two men merged their British interests into the "Ediswan" company. Swan is a standing reminder that Edison did not invent the light bulb alone.

Charles Brush (1849–1929). An American inventor who dominated arc lighting in the 1870s and 1880s, illuminating streets and public squares with brilliant, harsh electric arcs before the softer incandescent bulb took over indoor lighting. His dynamos and arc lamps were the first widespread electric lighting in American cities.

Thomas Edison (1847–1931). The Wizard of Menlo Park, and one of the three central figures of this book. Largely self-educated and partly deaf from childhood, he built the world's first industrial research laboratory, invented the phonograph, produced the first practical incandescent lighting system, and opened Pearl Street Station in 1882. His genius was for whole systems, not single gadgets; his tragedy was his stubborn, ultimately losing faith in direct current. He held more than a thousand US patents.

Nikola Tesla (1856–1943). Born in Smiljan, in present-day Croatia, the son of a Serbian Orthodox priest, and the visionary at the heart of this book. Gifted with a near-photographic imagination, he conceived the rotating magnetic field and the AC induction motor, whose patents he sold to Westinghouse in 1888 — the invention that decided the War of the Currents. He also gave us the Tesla coil and pursued a lifelong dream of wireless power. Brilliant and eccentric, he died alone and nearly penniless; the unit of magnetic flux density, the tesla, honors him.

George Westinghouse (1846–1914). A Civil War veteran and prolific inventor from New York who made his first fortune with the railway air brake, then founded Westinghouse Electric in 1886 and became the decisive champion of alternating current. He bought Tesla's patents, won the Chicago fair and Niagara, and was, for the era, an unusually humane employer. The least remembered of the three rivals, he was arguably the most consequential — the industrialist who actually electrified the world.

William Stanley Jr. (1858–1916). An American engineer working for Westinghouse who, in 1886, built a practical transformer and used it to light the town of Great Barrington, Massachusetts — an early public proof that AC could be stepped up for transmission and down for use. His transformer made the whole AC system feasible.

Lucien Gaulard and John Dixon Gibbs. A French inventor and his English business partner who developed and patented an early AC transformer ("secondary generator") in the early 1880s. Westinghouse licensed their design, which Stanley then improved into a workable form. They are a reminder that the transformer, like the light bulb, had many parents.

Harold P. Brown. An engineer and self-styled electrical expert who, with quiet support from Edison's laboratory, led the ghastly public campaign against alternating current — electrocuting dogs, horses, and other animals to dramatize its dangers and helping engineer the AC-powered electric chair. His role is the moral low point of the current war.

William Kemmler (1860–1890). A Buffalo peddler convicted of murdering his companion with a hatchet, and the first human being ever executed in the electric chair, on August 6, 1890. His botched, prolonged killing horrified witnesses and led Westinghouse to remark that they could have done better with an axe. He was an unwilling pawn in a corporate war.

Charles Parsons (1854–1931). A British engineer who invented the steam turbine in 1884, a machine that could spin a generator faster, more smoothly, and more efficiently than any piston engine. His turbine became the heart of nearly every large power station and remains so today; most of the world's electricity is still generated by turbines descended from his design.

Samuel Insull (1859–1938). An Englishman who began as Edison's private secretary and became the true inventor of the modern electric utility. From Chicago he pioneered the regulated monopoly, economies of scale, load diversity, and mass electrification, bringing affordable power to millions before his vast holding-company empire collapsed in the Great Depression and he stood trial, and was acquitted.

J. P. Morgan (1837–1913). The most powerful American financier of his age, who backed Edison's ventures, engineered the merger that created General Electric, and briefly funded Tesla's Wardenclyffe tower before withdrawing his support. His money shaped the electrical industry as surely as any inventor's patent, and his decisions helped make and break the fortunes in this book.

Edward Dean Adams (1846–1931). An American financier who led the Niagara Falls power project, chairing the commission that ultimately chose Tesla's and Westinghouse's polyphase AC system to harness the falls. The Adams Power Plant, named for him, was the crowning triumph of alternating current.

Guglielmo Marconi (1874–1937). An Italian inventor and entrepreneur who commercialized radio and transmitted the first transatlantic wireless signal in 1901, winning fame and a Nobel Prize. His success came partly through patents that overlapped Tesla's earlier work, and their priority dispute was only partly resolved, in Tesla's favor, by a US Supreme Court ruling in 1943.

J. J. Thomson (1856–1940). A British physicist who, in 1897, discovered the electron — the tiny negatively charged particle that actually carries electric current through a wire. His work finally revealed what "electricity" is made of, decades after engineers had learned to put it to work. He won the Nobel Prize in 1906.

Charles Coffin (1844–1926). An American businessman, formerly a shoe manufacturer, who led Thomson-Houston into the 1892 merger with Edison General Electric and became the first president of the combined General Electric. It was Coffin, not Edison, who steered the great company that carried Edison's name into the AC age.

Glossary

The vocabulary of electricity is a museum of the people who built the science, so that to speak plainly about volts and amperes and farads is already to name Volta, Ampère, and Faraday; the definitions below aim to make every term in this book clear in a sentence or two, in the same order the alphabet gives them.

Adams Power Plant — The hydroelectric station at Niagara Falls, named for financier Edward Dean Adams, that began generating polyphase alternating current in 1895 using Tesla's system; its transmission of power to Buffalo in 1896 marked the decisive victory of AC.

Alternating current (**AC**) — Electric current that periodically reverses direction, typically fifty or sixty times per second. Because its voltage can be easily raised and lowered by transformers, AC became the standard for transmitting and distributing electric power.

Alternator — A generator that produces alternating current by rotating a magnetic field past coils of wire (or vice versa). Nearly all commercial electricity is produced by alternators.

Ampere (**amp**) — The unit of electric current, defined as one coulomb of charge flowing past a point each second. It is named for André-Marie Ampère.

Anode — The electrode through which conventional current enters a device; in a battery being discharged it is the negative terminal, and in electrolysis it is the positive one. Its meaning depends on the direction of current flow.

Arc (electric arc) — A sustained, luminous electrical discharge that leaps across a gap between two conductors, producing intense light and heat. It is the basis of the arc lamp.

Arc lamp — An early, very bright electric light that produces illumination from an arc between two carbon rods. Pioneered by Humphry Davy and commercialized by Charles Brush, it lit streets and public spaces before the incandescent bulb.

Armature — The winding of a motor or generator in which voltage is induced or which carries the main current; in many machines it is the rotating part that interacts with the magnetic field.

Base load — The continuous minimum level of electricity demand on a grid, met by plants that run steadily around the clock. It contrasts with peak load, the maximum demand at busy times.

Battery — A device that produces electric current from a chemical reaction, delivering direct current. Volta's pile was the first true battery.

Blackout — A complete loss of electrical power in an area, whether from equipment failure, overload, or damage to the grid. A partial reduction in voltage is instead called a brownout.

Blackstart — The process of restarting a power grid after a total collapse, using generators that can start without drawing power from the grid itself. It is one of the hardest problems in grid operation.

Brownout — A partial, deliberate or accidental reduction in electrical voltage or supply, often causing lights to dim, used to prevent a full blackout when demand outstrips supply.

Brush — A sliding contact, often made of carbon, that carries current to or from a rotating part of a machine by pressing against a commutator or slip ring.

Capacitance — The ability of a device to store electric charge, measured in farads. A capacitor, the modern descendant of the Leyden jar, stores energy in an electric field.

Carbon filament — The thin thread of carbonized material that Edison used as the glowing element in his first practical incandescent bulbs. It was later superseded by the more durable tungsten filament.

Cathode — The electrode through which conventional current leaves a device; in a discharging battery it is the positive terminal, and in electrolysis the negative one.

Central station — A power plant that generates electricity for many customers over a distribution network, rather than for a single building. Edison's Pearl Street Station of 1882 was the first commercial example.

Circuit — A closed loop of conductors through which electric current can flow. If the loop is broken, the current stops.

Circuit breaker — A safety device that automatically interrupts a circuit when the current becomes dangerously high, and which, unlike a fuse, can be reset rather than replaced.

Cogeneration — The simultaneous production of electricity and useful heat from the same fuel, also called combined heat and power, which greatly improves overall efficiency.

Commutator — A rotating switch, made of segments and brushes, that reverses the current connection at the right moments to keep a direct-current motor turning in one direction or to draw DC from a generator. Its absence is what makes Tesla's AC induction motor so simple and rugged.

Conductor — A material, such as copper or aluminum, through which electric current flows easily because its electrons move freely.

Coulomb — The unit of electric charge, equal to the charge carried by about 6.24 billion billion electrons, or the charge transported by one ampere in one second.

Current — The flow of electric charge through a conductor, measured in amperes. It may flow steadily in one direction (direct current) or reverse periodically (alternating current).

Direct current (DC) — Electric current that flows steadily in one direction. It is the current of batteries and of Edison's original system, and it powers nearly all modern electronics.

Distribution line — A lower-voltage power line that carries electricity from local substations to homes and businesses, as opposed to the high-voltage transmission lines that carry bulk power over long distances.

Diversity (load diversity) — The fact that different customers draw their peak power at different times, so that the combined demand on a utility is smoother and smaller than the sum of individual peaks. Samuel Insull built his utility empire by exploiting this principle.

Dynamo — An older term for an electrical generator, especially one producing direct current, that converts mechanical energy into electricity. Edison nicknamed his largest Pearl Street dynamo "Jumbo."

Earthing — See grounding.

Electric charge — A fundamental property of matter that comes in two kinds, positive and negative; like charges repel and unlike charges attract. The flow of charge is electric current.

Electric chair — A device for execution by electrocution, first used on William Kemmler in 1890 and deliberately powered by alternating current in an effort to associate AC with death during the War of the Currents.

Electric field — The region of influence around an electric charge, within which another charge experiences a force. It is one half of the electromagnetic field.

Electric motor — A machine that converts electrical energy into mechanical motion, using the force between magnetic fields and currents. Faraday built the first; Tesla's AC induction motor made the technology practical at scale.

Electrification — The historical process by which electric power spread through society, replacing older sources of light and mechanical energy. It was named the greatest engineering achievement of the twentieth century.

Electrocution — Death or severe injury caused by electric shock. The word was coined during the debates over the electric chair.

Electrolysis — The use of an electric current to drive a chemical change, such as splitting water into hydrogen and oxygen or depositing metal in electroplating.

Electrolyte — The conducting solution or paste in a battery or electrolytic cell through which ions move to carry current.

Electromagnetic induction — The generation of an electric current by a changing magnetic field, discovered by Faraday in 1831. It is the principle behind every generator and transformer, and thus the foundation of the power industry.

Electromagnetic spectrum — The full range of electromagnetic radiation, from radio waves through visible light to X-rays and beyond, all differing only in frequency and wavelength. Maxwell's theory revealed light to be part of it.

Electromagnetism — The unified force that links electricity and magnetism, first revealed by Ørsted and Ampère and fully described by Maxwell. A changing electric field produces a magnetic field, and vice versa.

Electromotive force (EMF) — The "push," measured in volts, that drives current around a circuit, supplied by a battery, generator, or other source. Despite the name, it is a potential difference, not a mechanical force.

Electron — The tiny negatively charged particle, discovered by J. J. Thomson in 1897, whose movement through a conductor constitutes electric current.

Electrostatics — The study of electric charges at rest, as opposed to the flowing charges of current electricity. It was the whole of electrical science before Volta's battery.

Excitation — The supply of current to the field windings of a motor or generator in order to create the magnetic field the machine needs to operate.

Farad — The unit of capacitance, named for Michael Faraday; a one-farad capacitor stores one coulomb of charge for each volt applied.

Field lines — Faraday's way of visualizing a field as a pattern of imaginary lines showing the direction and strength of the force at each point. The idea made the invisible field intuitive.

Field winding — The coils of a motor or generator that produce the machine's magnetic field, as distinct from the armature that carries the main current.

Filament — The thin thread of carbon or metal in an incandescent bulb that glows white-hot when current passes through it, producing light.

Frequency — The number of complete cycles an alternating current or wave completes each second, measured in hertz. Power grids run at 50 or 60 hertz.

Fuse — A safety device containing a thin wire that melts and breaks the circuit when the current grows dangerously high, protecting wiring and equipment. Unlike a circuit breaker, a blown fuse must be replaced.

Galvanic cell — A device that produces electricity from a chemical reaction; a battery is one or more galvanic cells. The term honors Luigi Galvani.

Generator — A machine that converts mechanical energy

into electrical energy through electromagnetic induction. See also dynamo and alternator.

Gigawatt (GW) — A unit of power equal to one billion watts, or one thousand megawatts, used to describe the scale of large power systems and national grids.

Great Barrington — The town in Massachusetts where William Stanley demonstrated a practical AC transformer system in 1886, an early proof that alternating current could be stepped up and down for transmission and use.

Grid — The interconnected network of generating stations, transmission lines, substations, and distribution lines that delivers electricity from producers to consumers across a region or nation.

Grid-scale storage — Large installations, increasingly batteries, that store electrical energy to balance supply and demand on the grid, buffering the variable output of renewable sources.

Grounding — The deliberate connection of an electrical system to the earth, providing a safe path for fault current and stabilizing voltages. Also called earthing.

Henry — The unit of inductance, named for Joseph Henry, measuring a coil's tendency to oppose changes in the current through it.

Hertz (Hz) — The unit of frequency, equal to one cycle per second, named for Heinrich Hertz. Household electricity in North America alternates at 60 Hz.

HVDC (high-voltage direct current) — A modern technology that transmits bulk electric power over very long distances as direct current at high voltage, with lower losses than AC for such distances — a partial revival of Edison's current.

Hydroelectric power — Electricity generated by the force of falling or flowing water turning a turbine, as at the Niagara Falls plant.

Impedance — The total opposition an alternating-current circuit offers to current flow, combining ordinary resistance with reactance. It is the AC counterpart of resistance.

Incandescence — The emission of light by a substance heated until it glows, the principle of the filament bulb. It differs from the arc and from modern light-emitting diodes.

Inductance — The property of a coil or circuit by which a changing current induces an opposing voltage, measured in henries. It is central to transformers and AC behavior.

Induction motor — An alternating-current motor, invented by Tesla, in which the rotating magnetic field of the stator induces current in the rotor, turning it without any electrical contact to the moving part. Its simplicity and ruggedness made AC practical for machinery.

Insulator — A material such as glass, rubber, or porcelain that strongly resists the flow of electric current and is used to prevent current from going where it should not. On power lines the porcelain or glass fittings that isolate the conductors from the poles are also called insulators.

Interconnection — The linking of separate power grids or systems so that they can share generation and support one another, improving reliability.

Inverter — A device that converts direct current into alternating current, the opposite of a rectifier. Inverters allow batteries and solar panels, which produce DC, to feed the AC grid.

Joule — The basic unit of energy in the international system; one watt of power sustained for one second delivers one joule.

Kilovolt (kV) — One thousand volts, the scale at which electricity is transmitted over distance; high-voltage lines may run at hundreds of kilovolts.

Kilowatt (kW) — One thousand watts, a convenient unit for the power drawn by homes and small machines.

Kilowatt-hour (kWh) — The standard unit of electrical energy for billing, equal to the energy used by a one-kilowatt load running for one hour. It is what the meter on your house counts.

Kinetoscope — An early motion-picture viewing device developed in Edison's laboratory, allowing one person at a time to watch a short film through a peephole.

Leyden jar — The first capacitor, invented in 1745–46, consisting of a glass jar lined inside and out with metal foil that could store a static charge and deliver a powerful shock.

Lightning rod — Franklin's invention: a pointed metal rod mounted on a building and connected to the ground, which safely conducts a lightning strike into the earth instead of letting it damage the structure.

Load — The amount of electrical power demanded by the devices connected to a circuit or system at a given moment.

Load factor — A measure of how steadily a system's capacity is used over time, comparing average demand to peak demand. A high load factor means efficient use of generating plant.

Magnetic field — The region of influence around a magnet or a current-carrying wire, within which magnetic materials and moving charges experience a force. Its strength is measured in teslas.

Maxwell's equations — The four fundamental laws, formulated by James Clerk Maxwell in the 1860s, that describe how electric and magnetic fields are generated and how they interact, uniting electricity, magnetism, and light.

Megawatt (MW) — One million watts, or one thousand kilowatts, a unit used to describe the output of power plants.

Menlo Park — The village in New Jersey where Edison built the world's first industrial research laboratory in 1876, earning himself the nickname "the Wizard of Menlo Park."

Morse code — The system of dots and dashes used to send messages over the electric telegraph, developed by Samuel Morse and Alfred Vail.

Ohm — The unit of electrical resistance, named for Georg Ohm; one ohm is the resistance that allows one ampere to flow under one volt of pressure.

Ohm's law — The fundamental relationship stating that the current through a conductor equals the voltage across it divided by its resistance ($V = I \times R$).

Overhead line — A power line strung on towers or poles above the ground, cheaper to build than an underground line but more exposed to weather and damage.

Patent — A legal grant giving an inventor exclusive rights to an invention for a limited time. Patents were the currency and the weaponry of the electrical age, central to the War of the Currents.

Patent pool — An arrangement in which rival companies agree to share or cross-license their patents, ending costly litigation, as the electrical giants eventually did.

Peak load — The maximum electrical demand on a system during a given period, which the grid must be able to meet even if only briefly.

Phase — The timing position of an alternating-current waveform relative to a reference. Systems that use several currents offset in time are called polyphase.

Phonograph — Edison's 1877 invention for recording and reproducing sound, the machine that first made him world-famous.

Polyphase — An alternating-current system, devised by Tesla, that uses two or more currents offset in phase, typically three, to produce a smoothly rotating magnetic field ideal for driving motors and transmitting power.

Power factor — In an AC system, the ratio of real power that does useful work to the apparent power supplied, reflecting how well the current and voltage are in step.

Power plant — A facility where electricity is generated on a large scale, also called a power station or generating station.

Rectifier — A device that converts alternating current into direct current, the opposite of an inverter. Rectifiers let AC-powered chargers feed DC devices.

Reactance — The opposition to alternating current caused by capacitance or inductance, as distinct from ordinary resistance. Together they make up impedance.

Regulated monopoly — A single company granted the exclusive right to provide electricity in an area in exchange for government oversight of its rates and service. Samuel Insull championed this model as the basis of the modern utility.

Renewable energy — Energy from sources that are naturally replenished, such as sunlight, wind, and flowing water, as opposed to finite fossil fuels.

Resistance — The opposition a material offers to the flow of electric current, measured in ohms and converting some electrical energy into heat.

Resonance — The tendency of a system to oscillate with large amplitude when driven at its natural frequency. Tesla exploited resonance in his coil to build up enormous voltages.

Right-of-way — The corridor of land secured for the route of a transmission line, along which the utility has the legal right to build and maintain its towers and wires.

Rotating magnetic field — The magnetic field, produced by polyphase alternating currents, that appears to sweep around in a circle, dragging a motor's rotor after it. Tesla's insight into this effect was the key to the AC induction motor.

Rotor — The rotating part of a motor or generator, turning inside the stationary stator.

Rural electrification — The extension of electric power to farms and countryside, accomplished in the United States largely through the New Deal's Rural Electrification Administration, founded in 1935.

Semiconductor — A material, such as silicon, whose ability to conduct electricity lies between that of a conductor and an insulator and can be precisely controlled. Semiconductors are the basis of all modern electronics.

Series circuit — A circuit in which components are connected end to end, so the same current flows through each in turn; if one fails, the whole circuit is broken. It contrasts with a parallel circuit.

Parallel circuit — A circuit in which components are connected side by side across the same voltage, so each receives full voltage and one can fail without stopping the others. Edison wired his lamps in parallel so that a single burned-out bulb would not darken the rest.

Short circuit — An unintended low-resistance connection that allows a sudden, dangerous surge of current, often causing sparks, heat, or fire. Fuses and breakers exist to interrupt it.

Single-phase — An alternating-current supply delivered on a single alternating waveform, the form in which most household electricity arrives.

Slip ring — A continuous conducting ring on a rotating shaft, contacted by a brush, used to pass current to or from a spinning coil without the reversing action of a commutator.

Smart grid — A power grid enhanced with digital sensors, communication, and automated control, allowing it to balance supply and demand more flexibly and to accommodate variable renewable sources. It is a still-evolving technology.

Static electricity — Electric charge at rest, accumulated on the surface of a material, as in the amber effect first noted by Thales. It is the oldest known form of electricity.

Stator — The stationary part of a motor or generator, within which the rotor turns; in an induction motor it produces the rotating magnetic field.

Steam turbine — A machine, invented by Charles Parsons in 1884, in which high-pressure steam spins a bladed wheel to drive a generator. It is the heart of most large power stations to this day.

Step-up transformer — A transformer that raises voltage, used to boost power to high voltage for efficient long-distance transmission. A step-down transformer lowers voltage for safe use.

Substation — A facility in the grid where transformers and switching equipment change voltage levels and route power between transmission and distribution networks.

Tesla (unit) — The unit of magnetic flux density, named for Nikola Tesla and adopted in 1960, measuring the strength of a magnetic field.

Tesla coil — A resonant, high-frequency, high-voltage transformer invented by Tesla in 1891, capable of producing dramatic sparks and important in the early study of radio.

Telegraph — The first great electrical communication system, sending coded messages as pulses of current along wires. Edison began his career as a telegraph operator.

Transformer — A device with two or more coils linked by a magnetic core that raises or lowers alternating-current voltage. Because it works only with changing current, it is the reason AC, not DC, became the standard for power distribution.

Transmission line — A high-voltage line that carries bulk electric power over long distances from generating plants toward the areas where it is used.

Tungsten filament — The durable metal filament, with a very high melting point, that replaced carbon in incandescent bulbs and gave them far longer life.

Turbine-generator set — A turbine coupled directly to a generator, the standard arrangement for producing electricity in large power stations.

Underground line — A power line buried below the surface, more protected and less visible than an overhead line but considerably more expensive to install.

Utility — A company that generates, transmits, or distributes electricity to the public, typically operating as a regulated monopoly within its service area.

Vacuum — The near-absence of air inside an incandescent bulb, necessary to keep the hot filament from burning up by reacting with oxygen.

Volt — The unit of electric potential difference or electromotive force, named for Alessandro Volta; it measures the "pressure" that drives current through a circuit.

Voltaic pile — Volta's first battery, of 1800, a stack of alternating metal discs separated by moist pads, which produced the first steady electric current.

Wardenclyffe — The unfinished tower and laboratory Tesla built on Long Island beginning in 1901 for his dream of global wireless power transmission, abandoned for lack of funds and demolished in 1917.

Watt — The unit of electrical power, or rate of energy use, equal to one joule per second; a device's wattage tells how fast it consumes energy.

Watt-hour meter — The instrument installed at a building that measures the electrical energy consumed, in kilowatt-hours, for billing.

Weber — The unit of magnetic flux, named for Wilhelm Weber, measuring the total magnetic field passing through a given area.

White City — The nickname for the 1893 World's Columbian Exposition in Chicago, whose gleaming buildings were lit by Westinghouse's alternating current in a spectacular public triumph for AC.

Wireless power — The transmission of electrical energy through space without connecting wires, a lifelong obsession of Tesla's that he was never able to realize on a large scale.

Essay A: How Electricity Actually Works

Through this whole book I have asked you to take a few ideas on trust — that current flows, that a moving magnet makes electricity, that one kind of current could cross a continent and the other could not. Here, in one place and from the ground up, is the science behind those ideas, told plainly and without a single equation you need to solve. If you understand this essay, the entire drama of Edison, Tesla, and Westinghouse will click into place, because their war was, underneath all the propaganda and money, an argument about the things described below.

Begin with the most fundamental fact, the one everything else is built on: the world is made of charged particles. Deep inside every atom sits a dense nucleus carrying positive electric charge, and around it swarm electrons, each carrying an equal and opposite negative charge. Charge is simply a property that some particles have, the way mass is a property they have, and it comes in these two flavors, positive and negative, with one iron rule governing them: like charges push apart, and unlike charges pull together. A positive and a negative attract; two positives, or two negatives, repel. That single rule of attraction and repulsion is the engine of the entire subject. Everything electrical, from the spark off a doorknob to the hum of the national grid, is charged particles obeying that rule.

In most matter the positives and negatives are balanced and locked together, so the object as a whole is neutral and nothing electrical seems to be happening. But in certain materials — metals above all — some of the electrons are only loosely held and can wander freely from atom to atom. These are the materials we call conductors, and copper is the great example, which is why the wires in your walls are made of it. Other materials — glass, rubber, dry wood, porcelain — hold their electrons tightly and refuse to let them move; these we call insulators, and we wrap them around conductors precisely to keep the wandering charge on its intended path. The whole physical apparatus of electrical technology, every wire and every bit of insulation, is just an arrangement of conductors to guide the charge and insulators to fence it in.

Now, static electricity — the oldest form, the one Thales saw in his amber — is what happens when charge builds up but has nowhere to go. Rub amber with fur, or your shoes across a carpet, and you scrape loose electrons from one surface onto the other. One object ends up with a surplus of electrons and a net negative charge; the other, robbed of them, is left positive. The two now yearn to reunite, and if you bring them close, the pent-up charge leaps the gap in a tiny spark — the same phenomenon as lightning, differing only in scale. Static electricity is dramatic but useless for running a civilization, because it discharges all at once and is gone. For two thousand years, from Thales to Franklin, electricity was almost entirely this: a curiosity of sparks and shocks, charge at rest, briefly and violently escaping.

The revolution that made everything else possible was the discovery of how to make charge flow steadily and continuously rather than jump all at once. A flowing stream of charge is called an electric current, and it is measured in amperes, or amps — one amp being a certain vast number of electrons drifting past a point each second. Here the most useful mental picture in all of electricity is water in a pipe. Imagine current as water flowing through a hose: the amount of water passing a given point each second is the current. This picture is not merely a teaching trick; it is genuinely close to how the mathematics behaves, and I will lean on it repeatedly.

But water in a pipe does not flow by itself. It needs pressure — a pump, or a height it can fall from — to push it along. The electrical equivalent of that pressure is voltage, measured in volts and named for Volta, whose battery first supplied a steady one. Voltage is the push, the difference in electrical "height" between two points that drives the charge from one to the other. A nine-volt battery pushes harder than a one-and-a-half-volt battery in the same way that a tall water tank pushes harder than a short one. And here is the crucial subtlety that trips up almost everyone: voltage is always a difference between two points, never an absolute quantity at a single point. A bird sits unharmed on a high-voltage wire because both its feet are at the same voltage; there is no difference across its body, and so no push, and so no current through it. It is only when you connect two points at different voltages — when you complete the path from high to low — that charge flows. This is why electricity always needs a complete loop, a circuit: the current must have somewhere to flow from and somewhere to flow to, out of the source, through the device, and back again. Break the loop anywhere, flip a switch open, and the flow stops instantly, everywhere, the way water stops in a hose the moment you crimp it.

The third quantity completes the picture, and it is resistance, measured in ohms after Georg Ohm. Resistance is how much a material fights the flow of current through it, turning some of the electrical energy into heat as the drifting electrons jostle their way past the atoms. A thick copper wire has very low resistance and lets current flow freely; a thin coil of high-resistance wire, or a carbon filament, chokes the flow and, in fighting it, grows hot — hot enough to glow. That glowing is precisely how Edison's bulb makes light: current forced through a high-resistance filament heats it white-hot, and it shines. Resistance is friction for electricity, and like friction it produces heat, sometimes as a nuisance and sometimes, as in a toaster or a light bulb, as the whole point.

These three quantities — voltage, current, and resistance — are bound together by the simplest and most useful rule in the field, Ohm's law, which says that the current equals the voltage divided by the resistance. In the water picture it is pure common sense: more pressure gives more flow, and a narrower, more obstructed pipe gives less. Raise the voltage and the current rises; raise the resistance and the current falls. Nearly every practical electrical calculation begins here. And it explains, incidentally, why electricity can be dangerous in ways that surprise people: it is current passing through your body, not voltage alone, that harms you, and how much current flows depends on your body's resistance and on the voltage pushing against it. A high enough voltage overcomes even the skin's resistance and drives a lethal current through, which is why the high-voltage transmission lines in this book must be kept so scrupulously out of reach.

There is one more piece of the picture that separates the merely interesting from the world-changing, and it is the marriage of electricity and magnetism. For most of history these seemed like two unrelated mysteries — the amber that grips a feather and the lodestone that grips iron. Then, in 1820, Ørsted noticed that a wire carrying a current deflects a nearby compass needle, and the two mysteries turned out to be one. An electric current, it emerged, always creates a magnetic field around itself, a region of magnetic influence circling the wire. Coil that wire into a spiral and the fields of each loop add together, so that the coil becomes a magnet you can switch on and off at will — an electromagnet. This is not a minor curiosity. It means that electricity can produce magnetism, and, as Faraday would show, that magnetism can produce electricity, and between those two facts lies every motor and every generator ever built.

Faraday's discovery, in 1831, is the true hinge of the modern world, so it is worth stating carefully. He found that moving a magnet near a coil of wire — or moving the coil near a magnet, it makes no difference — causes a current to flow in the coil. Not a steady magnet; a moving one, a changing magnetic field. The key word is *change*: only while the field is changing does a current appear, and the faster the change, the larger the current. This is electromagnetic induction, and it runs the world in both directions. Run it one way and you have an electric motor: feed current into coils, let their magnetism push against other magnets, and the thing spins, converting electricity into motion — the principle in every fan, pump, elevator, and electric car. Run it the other way and you have a generator: spin a coil near a magnet by any mechanical means — falling water, a steam turbine, a windmill — and the changing field induces a current, converting motion into electricity. Essentially all the electricity in your walls was made this way, by spinning magnets past coils, whether the spinning was driven by the Niagara River, a coal fire boiling water into steam, or a nuclear reactor doing the same. Faraday's induction is the faucet from which the entire electrical age pours.

It also, at last, explains the great quarrel at the heart of this book — the difference between direct and alternating current. Direct current, DC, is the simple kind: the charge flows steadily in one direction, out of the battery's positive terminal, through the circuit, and back to the negative, like water down a river. It is the current of Volta's pile, of Edison's system, and of every battery you own. Alternating current, AC, does something stranger: it flows first one way, then reverses, then reverses again, sloshing back and forth many times a second — sixty times in North America, fifty in most of the world. The charge in your wall does not actually travel to the lamp and stay there; it jiggles back and forth in place, and it is the *energy*, not the particular electrons, that flows to the bulb, the way a wave crosses the ocean while the water itself only bobs up and down. This seems, at first, like a bizarre and pointless complication. Why would anyone want a current that can't make up its mind which way to go?

The answer is the single most important fact in the practical history of electricity, and it follows directly from Faraday. Remember that induction responds only to *changing* magnetic fields. A steady direct current in a coil makes a steady magnetic field, which, being unchanging, induces nothing in a neighboring coil. But an alternating current, constantly reversing, makes a constantly changing field — and that changing field induces a current in a nearby coil without any wires connecting the two. This is the transformer, and it is the whole reason AC won. By winding two coils on a shared iron core, one with many turns and one with few, you can trade voltage for current or current for voltage almost without loss: a step-up transformer takes moderate voltage and high current and hands back very high voltage and low current, and a step-down transformer does the reverse. Why does this matter so enormously? Because the energy lost to resistance as power travels down a wire depends on the current, not the voltage — and it depends on the current squared, so that halving the current cuts the loss to a quarter. To send power a hundred miles without wasting most of it as heat in the wire, you must send it at high voltage and low current, and then, at the far end, step it safely back down for use. AC lets you do exactly this with a simple, silent, motionless transformer. DC, in the 1880s, offered no comparable way to change voltage, and so it could not be sent far. That, stripped of all the drama, is why Edison's stations had to sit within a mile of their customers while Tesla's and Westinghouse's system could reach across a continent.

The last quantity to name is the one you actually pay for: power, measured in watts, the rate at which energy is used. Power is simply voltage multiplied by current — the push times the flow — so a device drawing a lot of current at a high voltage, like a space heater, consumes a great deal of power, while a small LED sips only a little. Run a thousand watts for one hour and you have used one kilowatt-hour of energy, the unit the meter on your house counts and the electric company bills. Every appliance you own is rated in watts, and the whole grid, from a single turbine to the continental network, is just an arrangement for generating power at one place and delivering it, through Faraday's induction and Tesla's transformers, to the watt-hungry devices in your rooms.

That is the entire science, and it is a remarkably short list. Charge, which comes in two kinds and obeys one rule. Current, the flow of charge, pushed by voltage and resisted by resistance, the three of them tied together by Ohm's law. Magnetism, which current creates and which, when it changes, creates current in return — Faraday's induction, the source of all our power and the reason AC could travel when DC could not. And power, the rate of it all, which is what we generate, sell, and consume. Hold those few ideas in your mind and the invisible river in your walls is no longer a mystery but a machine you understand — the same machine a Danish professor glimpsed in a jumping compass needle, a London bookbinder built into the first motor, and three American rivals fought a war to control.

Essay B: AC vs DC — The Rematch

The story this book has told reaches its climax at Niagara Falls in 1896, when alternating current lit Buffalo and the War of the Currents was declared over. Edison had lost. Direct current, we are told, was beaten, buried, and forgotten, a technological dead end defended by a stubborn old man who could not see the future. It is a satisfying ending, clean and final, with a winner and a loser. It is also, viewed from our own century, badly incomplete — because the loser did not stay dead. Edison's current climbed out of its grave and is now running quietly through nearly everything you own, and, on the great transmission lines that increasingly stitch the continents together, it is winning back the very territory it lost. This essay is about that rematch. Some of what follows describes technologies still developing as this book is written, and the reader should treat the most recent figures and forecasts as approximate and subject to change; the broad reversal, however, is real and well established.

To understand the comeback, recall precisely what Edison lost and why. AC did not defeat DC because it was better in some absolute, eternal sense. It won a specific contest, under specific 1890s conditions, over a specific problem: how to move electric power over long distances cheaply. The answer, as the preceding essay explained, was the transformer, which works only with changing current and lets AC be stepped up to high voltage for efficient transmission and back down for safe use. In the 1880s there was simply no practical way to change the voltage of direct current. You generated DC at the voltage you needed and used it near where you made it, and that was that. Given the technology of the day, AC's ability to travel was decisive, and it deserved to win. But notice how narrow the victory actually was. It was a victory in *transmission and distribution* — in the wires between the power plant and the wall. It was never a verdict that direct current was inferior for the work electricity actually does once it arrives.

And this is the first crack in the tidy story, because a very large share of what electricity does today, it does as direct current. Consider the device most likely in your hand or pocket right now. Your phone runs on DC; so does your laptop, your tablet, your television's circuitry, your LED lightbulbs, your car's electronics, and every computer in every data center on earth. The reason is fundamental: the transistor, the tiny switch at the heart of all modern electronics, is a direct-current device. It works by controlling a steady one-way flow of charge, and the logic of computing — on and off, one and zero — maps naturally onto DC's steady presence or absence, not onto AC's endless sloshing back and forth. The entire digital revolution, the defining technology of our age, is built on Edison's current. Every time you charge a phone, the little brick in the wall is doing one specific job: it is a rectifier and converter, taking Tesla's alternating current from the socket and turning it back into Edison's direct current for the device to drink. We have wrapped our whole world in a thin skin of AC distribution over a vast body of DC consumption.

The battery, too, is inescapably a DC device, and batteries have never mattered more. A chemical cell produces a steady voltage in one direction; it cannot alternate. Volta's pile was DC, and so is the lithium-ion cell in your phone and the far larger pack in an electric car. As transport electrifies, the roads fill with machines that are, electrically, rolling reservoirs of direct current, charged from the grid and discharged into their motors. The grid delivers AC to the charger; the charger converts it to DC to fill the battery; and increasingly, when you brake, the car turns motion back into DC to top the battery up. The whole flourishing world of stored energy — the phone that lasts a day, the car that goes three hundred miles, the wall-mounted home battery, the grid-scale storage installations now being built to steady the flow of renewable power — is a world of direct current. Edison would have recognized every one of them.

Then there is the sun. A solar panel does not generate alternating current; it produces direct current, straightforwardly, when light knocks electrons loose in silicon. Every rooftop array and every solar farm generates DC at the source, which must then be run through an inverter — a device that does the opposite of your phone charger, turning DC into AC — before it can feed the existing alternating-current grid. As solar power grows into one of the largest new sources of electricity on the planet, an ever-larger fraction of the world's electricity is now *born* as direct current, converted to AC only because the grid it must join was built in Tesla's image. One can glimpse, in this, a strange possible future in which the conversions start to seem like needless waste, and stretches of the system quietly go over to DC end to end.

But the most striking part of the rematch is happening not in your devices or on your roof but on the largest wires in the world, and here Edison's current is not merely surviving in the margins — it is beating AC at the very game AC was invented to win. The technology is called high-voltage direct current, or HVDC, and its rise is one of the genuine ironies of electrical history. For very long distances — hundreds or thousands of miles of bulk power transmission — direct current has turned out to be *better* than alternating current, for reasons that were understood in principle long ago but became practical only when modern electronics gave engineers a way to change DC voltage at last. The thing that doomed DC in the 1880s, its stubborn refusal to be transformed, was finally solved not by a transformer but by banks of high-power semiconductor switches that can convert between AC and DC at enormous scale. Once you can step DC up to very high voltage, its old disadvantage evaporates and several quiet advantages appear.

The advantages are real and worth understanding, because they explain why utilities now spend billions on DC lines. Alternating current, it turns out, wastes energy in ways direct current does not. Because AC constantly changes, it sets up magnetic and electric effects in and around the conductors that consume power and, over long distances, actually limit how much a line can carry; a long AC line spends part of its capacity fighting itself. AC also insists that everything it connects march in perfect lockstep at the same frequency and phase, which makes tying together large or distant grids delicate and sometimes impossible. Direct current suffers none of this. A DC line carries power straight down the wire with lower losses over long hauls, needs fewer conductors, and can link two AC grids that are not synchronized — even two running at different frequencies — acting as a clean bridge between them. For undersea cables the case is overwhelming, since a long AC cable underwater is nearly unusable, and so essentially every major submarine power link, connecting islands to mainlands and one country's grid to another's across the sea, is direct current. The great long-distance lines now carrying hydroelectric and solar and wind power across continents — from remote dams and sun-baked deserts to distant cities — are more and more often HVDC. Edison's current, banished for being unable to travel, has become the current of choice for traveling farthest.

It is worth dwelling on why the solution took a full century to arrive, because the delay is the whole moral of the rematch. Edison's direct current was not defeated by a flaw in the idea of direct current; it was defeated by the absence of one specific tool — a practical way to change DC voltage. Alternating current had the transformer, a simple, silent, motionless device that had existed, in principle, since Faraday, and that trades voltage for current effortlessly. Direct current had no equivalent, and could have none, for a transformer works only on changing current and DC by definition does not change. To raise or lower a direct voltage efficiently you need to chop it, switch it, and reshape it thousands of times a second with electronic valves fast and rugged enough to handle enormous power — and no such valve existed in the 1890s, or the 1920s, or even the 1950s in any form that could move the output of a power plant. It waited on the invention of high-power semiconductors, the same family of devices that gave us the computer, and only when those matured did the door that had been locked against DC since Pearl Street finally swing open. Edison lost, in other words, not because he was wrong but because he was working two generations before the tools that would have vindicated him. That is a very particular kind of defeat, and a strangely hopeful one.

None of this means Tesla lost, and it is important not to overcorrect into a fresh myth. Alternating current remains, and will long remain, the right answer for the ordinary work of distribution — the web of medium and low-voltage lines that thread through every town and into every home. The transformer is still cheap, silent, reliable, and unbeaten for stepping voltage up and down at the neighborhood scale, and the AC induction motor Tesla invented still turns a huge share of the world's machinery. The grid you plug into is AC and will stay AC in its bones. What has happened is subtler and more interesting than a reversal of the old verdict: the sharp 1890s question of "AC *or* DC" has dissolved into a modern system that uses both, each where it excels. AC does the distributing; DC does the computing, the storing, the generating-from-sunlight, and, increasingly, the very-long-distance hauling. The two currents that went to war have made an uneasy peace inside the same system, often within the same building, sometimes within the same wall.

Which forces us to reframe, gently but firmly, the moral we inherited. "Edison lost the War of the Currents" is true as a matter of 1890s history and false as a verdict on the technology. Edison backed direct current for the wrong application — local urban distribution, where AC's transformer made it unbeatable — and he defended it dishonestly, with a fear campaign that remains the ugliest chapter in this story. He deserved to lose that fight, and he did. But his instinct that direct current was a fundamentally sound and valuable form of electricity was not wrong; it was merely a century early. The tools that could unlock DC's advantages — the power electronics that let it be converted and controlled at will — did not exist until long after both men were dead. Give Edison modern semiconductors and he might have built the HVDC line himself. The current he championed now runs the digital world, stores the world's portable energy, harvests the world's sunlight, and carries bulk power farther than AC ever could. That is not the record of a loser. It is the record of a good idea that arrived before the technology to redeem it, waited a hundred years, and came back to claim a share of the world its inventor never lived to see.

The honest verdict, then, is the one this book has tried to reach throughout: the War of the Currents was a real contest with a real outcome, but it settled a question of engineering economics in a particular decade, not the eternal superiority of one current over the other. AC won the war and built the grid. DC lost the war and, patiently, won back the future — the phone, the battery, the solar panel, and the longest wires on earth. Both men were right about their current; each was simply right about a different part of the world. And every time you unplug your charging phone, you are holding, in one small warm rectangle, the truce that ended their war and the proof that neither of them entirely lost.

Essay C: The Myths of Tesla

No figure in the history of technology has been more thoroughly reinvented by posterity than Nikola Tesla. In his own lifetime he was famous, then faded, then died nearly forgotten in a hotel room with his pigeons. In ours he has been resurrected as a kind of secular saint of science — the pure genius robbed and ruined by grubbier men, the prophet of free energy suppressed by greedy corporations, the true father of the modern world cheated of his due by the villainous Edison. There is a car company named after him, a currency, a thousand memes, and an entire internet subculture devoted to his martyrdom. Some of this rehabilitation is deserved and overdue; Tesla was a real genius who was genuinely neglected for most of the twentieth century, and correcting that neglect is a good thing. But the legend has now swollen so far past the documented record that the real man is in danger of vanishing behind it, and the real man is more interesting, and sadder, than the myth. This essay tries to separate the two, fairly and without cynicism, taking the most persistent legends in turn.

Begin with the famous origin story of the Edison feud: the fifty thousand dollars. As the tale is usually told, the young Tesla, newly arrived and working for Edison, was promised fifty thousand dollars if he could redesign and improve

Edison's inefficient dynamos. Tesla labored heroically, succeeded brilliantly, and went to collect his reward — whereupon Edison laughed in his face and said, "Tesla, you don't understand our American humor," refusing to pay a cent. Outraged, Tesla quit on the spot, and the war of the currents was born from this act of betrayal. It is a wonderful story, and it may even contain a kernel of truth, but it should be held at arm's length. Our main source for it is Tesla himself, recounting the episode decades later, and Tesla was an unreliable narrator of his own life — a man given to grand, self-dramatizing memories. Fifty thousand dollars was an enormous sum in the 1880s, far more than a company would plausibly dangle before a brand-new immigrant employee, and the likeliest reading is that a casual, perhaps joking remark by an Edison manager was remembered, or reshaped, by a proud and wounded young man into a solemn broken promise. Something soured between Tesla and the Edison operation, and Tesla left; that much is documented. But the crisp anecdote of the cheated fifty thousand belongs more to legend than to the record, and it is worth resisting the temptation to build a moral universe on it.

That leads to the largest myth of all, the one that frames everything else: "Edison stole from Tesla." In the popular imagination Edison is a thief and a fraud who robbed Tesla of his inventions and took the credit, and Tesla the true creator whose ideas Edison plundered. The documented reality is almost the opposite of a theft: the two men worked in different domains and largely did not overlap in their key inventions at all. Edison's central achievements — the practical incandescent lighting system, the phonograph, the DC central station — owed nothing to Tesla. Tesla's central achievement — the AC induction motor and polyphase system — owed nothing to Edison, was patented in Tesla's own name, and was sold by Tesla, quite lucratively at the time, to Westinghouse. Edison did not steal the AC system; he fought it, tooth and nail, as a business rival, and lost. There was no grand theft of Tesla's crown jewels. What there was, was a bitter commercial and technological war between two camps, waged with propaganda and patents, in which Edison behaved badly — the fear campaign against AC was genuinely disgraceful — but "behaved badly as a competitor" is a very different charge from "stole his rival's inventions." The saint-versus-villain framing flattens a complicated rivalry between two flawed, brilliant, and very different men into a comic book, and it does a disservice to both. Edison was not a fraud; he was one of the most productive inventors who ever lived, whose genius was for systems and relentless practical development. Tesla was not a helpless victim of Edison; his greatest triumph, the AC system, decisively beat Edison in the marketplace.

If Edison did not ruin Tesla, what did? Here the myth and the record part ways most poignantly, because the honest answer is that Tesla, to a large degree, ruined himself — not through any moral failing, but through the very qualities that made him a visionary. He was a spectacularly poor businessman. In a moment that the legend loves to cite as proof of his nobility, he tore up his enormously valuable royalty contract with Westinghouse when the company hit financial trouble, forgoing a fortune that would have made him one of the richest men in America. The legend presents this as pure selflessness; more likely it was a mix of genuine loyalty, poor judgment, and Tesla's lifelong indifference to money as anything but fuel for the next dream. Again and again he had success within his grasp and let it slip in pursuit of something grander and less attainable. He was not robbed of his fortune so much as he repeatedly gave it away or gambled it on visions that did not pay. This is a tragedy, but it is a tragedy of character and circumstance, not a crime committed against him by a mustache-twirling rival.

Now to the wilder legends, the ones that flourish online. First, the death ray. In his old age, impoverished and largely forgotten, Tesla issued press statements about a "teleforce" weapon — a beam that could destroy aircraft and armies at a distance, which the newspapers gleefully dubbed the "death ray." The myth treats this as a real, working superweapon that the government seized and suppressed after his death. The reality is sadder: there is no evidence Tesla ever built such a device, and the announcements came from a brilliant but declining old man, decades past his productive prime, seeking attention and funding for grand schemes he could no longer realize. When he died, government agents did examine his papers — that part is true, and understandable in wartime — but they found notes and speculations, not a working weapon. The death ray was a dream and a headline, not a suppressed invention.

Second, and most cherished, the myth of free wireless energy suppressed by big business. The story holds that Tesla had perfected a system to transmit unlimited free electricity through the air to everyone on earth, and that J. P. Morgan and other capitalists, realizing they could not put a meter on it and charge for it, deliberately cut off his funding and buried the technology to protect their profits. Wardenclyffe, the great unfinished tower on Long Island, is offered as the martyred monument to this suppressed miracle. The truth is more complicated and, alas, less flattering to the dream. Tesla genuinely did pursue wireless power transmission, seriously and for years, and it was a real and visionary goal. But the specific scheme behind Wardenclyffe — transmitting significant power around the globe through the earth and atmosphere — ran into physics that does not cooperate. Energy broadcast in all directions through space spreads out and dissipates; you cannot beam meaningful power to the whole planet the way you broadcast a faint radio signal, because power, unlike information, cannot be recovered once it has thinned out across the sky. Morgan did indeed withdraw his funding, and it is true that the lack of a way to meter freely transmitted power troubled the investors. But the deeper problem was that the system, as conceived, could not have worked at the scale Tesla promised, and Tesla's own shifting ambitions and financial troubles doomed the project as much as any conspiracy. Wardenclyffe was demolished for scrap in 1917. It is a genuinely moving ruin — the grave of a magnificent overreach — but it is not the tombstone of a working free-energy machine that greedy men stole from humanity. There was no such machine to steal.

There is a related family of legends worth touching on: the claim that Tesla invented, or single-handedly foresaw, nearly every marvel of the modern age — radio, radar, robotics, the smartphone, the internet, wireless communication, even directed energy — and was cheated of the credit for all of them. Like most durable myths, this one grows from real seeds. Tesla did do genuinely pioneering work in radio, to the point that a US Supreme Court ruling in 1943 recognized the priority of some of his patents over Marconi's; he did demonstrate remote control by radio with a small boat in 1898, a real ancestor of robotics; and he did speak, in interviews late in life, of a future of pocket-sized wireless devices through which people would send images and messages around the world, a genuinely striking anticipation of something like the smartphone. These are real achievements and real prophecies, and they are impressive enough on their own. But the meme inflates them into a claim that Tesla essentially invented the twentieth century by himself and that lesser men stole it piece by piece. The truth is that radio, radar, and computing each had many parents, that invention is almost always a collective and incremental thing rather than a lone genius's bolt from the blue, and that turning Tesla into the secret author of everything modern is finally a way of not taking his actual, specific, enormous contribution — the alternating-current system — seriously enough on its own terms.

One more legend deserves a plain answer, because it recurs constantly: the story that after Tesla died, government agents swept in and seized his papers, proof that he possessed secrets too dangerous or too valuable to release. The seizure is real; the sinister reading is not. In January 1943, in the middle of the Second World War, an elderly foreign-born inventor who had been making public claims about superweapons died in a hotel, and the government's Office of

Alien Property took custody of his effects and had them examined — a cautious and fairly routine wartime precaution, not a conspiracy. The examiner's report concluded that the papers contained speculative and theoretical material rather than any workable secret weapon, and the bulk of Tesla's estate was eventually released and now resides, fittingly, in a museum in Belgrade devoted to him. What the myth casts as a cover-up was, on the evidence, an anticlimax: the search for buried treasure that found, instead, the notebooks of a brilliant old man whose greatest work was decades behind him.

Why do these myths persist and grow? Partly because they answer a deep emotional need — the wish that pure genius should be rewarded and pure motives should triumph, and the suspicion that when they don't, dark powers must be to blame. Tesla makes an almost perfect vessel for that wish: brilliant, ascetic, unworldly, indifferent to money, dead poor while lesser men grew rich. It is comforting to believe he failed only because he was too good for this world, rather than because the world is complicated and even geniuses make ruinous mistakes and chase impossible things. And partly the myths persist as an overcorrection to a real injustice, for it is true that Tesla was unfairly neglected for most of the twentieth century while Edison was lionized, and the modern Tesla cult is in part a rebound from that neglect, swinging past the truth in its enthusiasm to restore him.

The man who emerges when the legends are peeled away is, I think, more compelling than the saint. He was a true original who gave the world the alternating-current system that powers it, a feat that alone secures his greatness. He was also a difficult, obsessive, socially isolated man of strange compulsions, a poor manager of money and people, prone to grandiose claims he could not deliver, who squandered real fortunes chasing visions that outran the physics of his time and died lonely and broke as a result. He was neither the helpless victim of a thieving Edison nor the suppressed messiah of free energy, but something harder to hold in the mind: a genuine genius whose gifts and flaws were tangled inextricably together, so that the same restless, uncompromising, unworldly mind that conceived the rotating magnetic field also led him, again and again, to reach past the possible and let the achievable fall from his hands. That is not a comic book. It is a human being, and a tragic one, and he deserves to be remembered as he was rather than as we wish he had been.

Essay D: The Business of Power

It is tempting to believe that the world was electrified by inventors — that once Edison had his bulb and Tesla his motor and Westinghouse his transformer, the light simply spread of its own accord, the way water finds its level. It did nothing of the kind. The inventions were necessary but nowhere near sufficient. What actually carried electric power into the homes and factories and farms of an entire civilization was not a machine but a business model: the electric utility, an institution as ingenious in its way as any dynamo, and one whose peculiar economics have shaped the wires in your walls as decisively as Faraday's induction. This essay is about that quieter revolution — how the *business* of electricity, more than any invention, determined how the light would reach you and who would control it.

Start with a puzzle that the founders of the industry had to solve before electricity could be anything more than a rich man's novelty. When Edison opened Pearl Street in 1882, electricity was appallingly expensive and almost useless to ordinary people, for a reason built into its nature: electricity, in that era, could not be stored. It had to be generated at the very instant it was consumed and used the very instant it was made. This single fact governs the entire economics of the industry to this day. A power company must build enough generating capacity to meet the highest demand its customers will ever place on it — the hottest afternoon, the darkest winter evening, the moment every light and motor is running at once — and then, most of the time, watch that expensive capacity sit idle. A plant sized for the peak is wildly underused during the trough. The great early problem of the electricity business was not making power; it was the crushing cost of building capacity that spent most of its life doing nothing.

The man who cracked this puzzle, and in doing so invented the modern utility, was not an inventor at all but a businessman: Samuel Insull, a sharp young Englishman who had come to America as Edison's private secretary and absorbed the trade at his master's elbow before striking out on his own in Chicago. Insull's genius was to see that the salvation of the industry lay in the very diversity of its customers. A single home wants power in bursts and leaves its equipment idle most of the day, which makes serving it ruinously inefficient. But a home peaks in the evening, a factory during the working day, a streetcar line in the rush hours, a shop at midday — and if you gather them all onto one system, their peaks and troughs fall at different times and partly cancel out, so that the combined demand is far steadier, and far smaller relative to the total connected load, than any one customer's alone. This is the principle of load diversity, and it was Insull's master stroke. The more varied customers he piled onto a single large generating plant, the more fully and cheaply that plant could be run, the lower his cost per unit fell, and the lower he could set his price, which drew in still more customers, which improved the diversity further, in a virtuous spiral. Insull chased volume relentlessly, cutting prices to pull in every possible user, because he understood that in this business bigger was genuinely cheaper, and cheaper made the whole thing possible.

That insight — that bigger is cheaper — is the key to everything that follows, because it means electricity is what economists call a natural monopoly, and understanding why explains the peculiar shape of the whole industry. Most businesses face competition comfortably: several bakeries can serve one town, and the rivalry keeps them honest. But some enterprises get cheaper and cheaper the larger they grow and the more they concentrate, so relentlessly that a single large provider can always undercut several small ones, until competition naturally collapses into monopoly. Electricity is the classic case, for two reasons. First, the load-diversity economics we have just seen reward scale enormously: one big plant serving a whole city beats a dozen small ones. Second, and even more decisively, the delivery system does not tolerate duplication. It would be absurd — and it was tried, briefly, in the early chaotic years — to have three or four rival companies each stringing its own separate set of wires down the same street, each digging up the pavement, each planting its own poles, to compete for the same houses. The wires are a single shared road, and it makes sense to build that road only once. An industry whose costs fall with size and whose infrastructure cannot sensibly be duplicated is one that will always tend toward a single provider. Electricity wanted, by its very nature, to be a monopoly.

This posed a genuine dilemma, and how different societies resolved it has shaped the delivery of power ever since. A monopoly is efficient — it avoids wasteful duplication and reaps the savings of scale — but an unregulated monopoly is dangerous, free to charge whatever it likes to captive customers who have nowhere else to turn. Here again Insull saw further than his rivals, and made a bargain that many businessmen of his era thought traitorous. He actively *invited* government regulation. Rather than fight to keep his monopoly free and unaccountable, he proposed a deal: let the utility be granted an exclusive territory, an official monopoly with no competitors, in exchange for submitting its prices and service to oversight by a public commission that would guarantee customers a fair rate and the company a fair, guaranteed return on its investment. This is the regulated monopoly, the bargain that governed the American electricity industry for most of the twentieth century and still governs much of it. It is a strange and clever compromise: the company gets certainty, protection from competition, and a dependable profit; the public gets universal service at supervised prices; and society gets the efficiency of a single large system without the extortion of an unchecked monopolist. Insull grasped that a secure, regulated, profitable monopoly was worth far more than a precarious, cutthroat, unregulated one, and he was right.

There was, however, a darker side to the empire-building, and it is part of the honest story. As utilities grew and consolidated, financiers assembled them into towering pyramids called holding companies — corporations that owned the controlling shares of other corporations that owned still others that owned the actual operating utilities at the bottom. By stacking these layers, a man at the top could control a vast electrical empire with a relatively small investment, using each layer to borrow against the next, and skim fees and profits at every level. Insull himself built such a pyramid, and in the boom of the 1920s it made him a titan; but the structure was dangerously top-heavy, riddled with debt and inflated valuations, and when the stock market crashed in 1929 the whole edifice came down in ruins. Insull's empire collapsed spectacularly, he was hounded into exile and put on trial for fraud — and, tellingly, acquitted, for his sins were more of overreach than of theft. The holding-company abuses of that era so alarmed the public and the government that they provoked a wave of reform legislation in the 1930s designed to break up the pyramids and rein in the financiers, a lasting mark the business of power left on the law.

The collapse of the holding companies also sharpened an old argument that runs through the whole history of electricity and has never been fully settled: should power be a private business at all, or a public service owned by the people? If electricity is a natural monopoly and a necessity of modern life, some argued, why let private owners profit from it? Why not have the government or the community own the wires and the plants directly, running them at cost for the public good? This is the case for public power, and it won important victories — municipal utilities owned by cities, rural cooperatives owned by their members, and, most dramatically, great federal projects that built dams and power systems to develop whole regions and bring electricity where private companies had found no profit in going. The opposing case, for private, investor-owned utilities under public regulation, held that private ownership brought discipline, capital, and efficiency that government operation would squander. Different countries and different regions struck the balance differently, and to this day the world's electricity is delivered by a patchwork of investor-owned utilities, municipal systems, cooperatives, and state enterprises, each embodying a different answer to the same old question of who should own the light.

It is worth pausing on the human meaning of what the utility model achieved, because it is easy to lose the marvel in the machinery of rates and regulation. When Insull began, electricity was a luxury of the rich and the commercial districts — a badge of wealth, priced beyond the reach of ordinary households. Within a generation, by driving down the cost through scale and diversity and volume, the utilities turned it into a near-universal necessity, something a working family simply expected to find behind the wall. This was not a gift; it was a business strategy, and it enriched the men who pursued it. But the effect was a democratization of power in the most literal sense — the spreading of tireless mechanical energy and clean light to tens of millions of homes that had never had either. The refrigerator, the washing machine, the electric iron, the radio: these became fixtures of ordinary life not merely because someone invented them but because the utility business had made the electricity to run them cheap enough for everyone. The social transformation of the twentieth-century home, and the release of countless hours of human labor, was underwritten by an accounting insight about load diversity. Rarely has a dry idea about how to run a plant more fully had such consequences for how people actually live.

The distinction between generating the power and delivering it, which the deregulation experiments would later seize upon, was in fact present in the industry's economics from the start, and it is worth naming clearly because it explains why the business took the shape it did. An electric utility does two quite different things. It *makes* electricity, in plants that in principle several companies could build and operate in competition with one another. And it *delivers* electricity, over a physical web of wires that runs to every building and that no sane society would build several times over. The first activity can support competition; the second cannot, being a natural monopoly of the purest kind. For most of the twentieth century these two functions were bundled together inside a single vertically integrated utility that did everything from turning the turbine to reading the meter, all under one regulated roof. The whole later argument over reform, as we will see, turns on the attempt to pry those two functions apart — to let the making of power compete while keeping the delivering of it a regulated monopoly. The seam was there in the economics from the beginning; the reformers merely tried to cut along it.

One last chapter of the business story belongs to our own era, and it should be flagged as recent and still contested. Beginning in the late twentieth century, many places grew dissatisfied with the old regulated-monopoly bargain, suspecting that guaranteed monopolies had grown complacent and that competition might drive prices down. They experimented with deregulation — breaking the vertically integrated utility into separate pieces, letting independent companies compete to generate power and sell it, while keeping the wires themselves, the natural monopoly that cannot be duplicated, under regulation. The results have been genuinely mixed and remain debated. In some places competition brought efficiency and choice; in others it produced volatility, manipulation, and crisis, most infamously in episodes where market gaming or extreme conditions sent prices soaring or the lights out. The lesson many drew is the one Insull understood a century before: electricity is not an ordinary commodity, because it cannot be stored, cannot be easily done without, and travels on shared wires that no market can duplicate — and so it resists being turned into an ordinary competitive market. The precise boundary between what should compete and what should be regulated is still being argued over, and the reader should regard the current state of that argument as unsettled.

Step back and the shape of the whole thing comes clear. The inventors gave us the means to make and move electricity, but it was the businessmen — Insull above all — who gave us the means to *afford* it, by discovering that the way to make power cheap enough for everyone was to gather everyone onto one great shared system, run it as full as possible, and govern the resulting monopoly by public bargain rather than private whim. The grid in your wall is the child of that discovery as much as of Faraday's or Tesla's. Every time you pay your electric bill you are participating in a settlement worked out a hundred years ago in Chicago — a truce between the efficiency of monopoly and the protection of the public — and the reason the light is there, cheap and constant and taken for granted, is that someone once figured out not merely how to generate it, but how to sell it to a whole civilization at once.

Essay E: Electricity by the Numbers

A history is a story of people, but electricity is also a story of quantities — of volts and hertz and miles and megawatts — and some of those numbers are so striking, or so revealing, that they tell the tale as well as any anecdote. What follows is an appendix of figures: the measurements, records, and comparisons that give the electrical age its true scale. Where a number is fixed by history or by definition, it is exact. Where it describes the present-day world — the size of a grid, the output of a source, the reach of a technology — it should be read as approximate and subject to change, since these quantities grow and shift from year to year and the reader will encounter this book at some unknown remove from its writing. Treat the modern figures as orders of magnitude, not gospel.

Begin with the units themselves, because each one is a small monument to a person in this book:

- **The volt**, the measure of electrical "push," honors Alessandro Volta, whose 1800 battery first supplied a steady one. A single flashlight cell gives about 1.5 volts.
- **The ampere**, the measure of current or flow, honors André-Marie Ampère. A bright household bulb draws well under an ampere; a hair dryer, perhaps

ten.

- **The ohm**, the measure of resistance, honors Georg Ohm, and is bound to the other two by his law: volts equal amperes times ohms.
- **The watt**, the measure of power, honors James Watt of steam-engine fame and equals one joule of energy per second — volts multiplied by amperes.
- **The farad** (capacitance) honors Faraday; **the henry** (inductance), Joseph Henry; **the hertz** (frequency), Heinrich Hertz; **the coulomb** (charge), Charles-Augustin de Coulomb; **the tesla** (magnetic flux density), Nikola Tesla, adopted in 1960; and **the weber** (magnetic flux), Wilhelm Weber. To speak the language of electricity is to recite a roll call of its founders.

Consider next the numbers that decided the War of the Currents, for the whole conflict turns on a few decisive quantities. Edison's Pearl Street Station distributed direct current at **110 volts**, a level chosen to be safe and to suit his lamps — and a level that, for the same reason, could not travel. The practical limit of his DC system was roughly **one mile**: beyond that, the voltage sagged and the lamps dimmed to a useless glow. That single number, one mile, is the reason Edison needed a power station in every neighborhood and the reason he lost. Alternating current shattered that limit through the transformer, and the decisive demonstration came at Lauffen in 1891, when three-phase AC was sent about **175 kilometers** — over a hundred miles — across Germany to Frankfurt, more than a hundred times DC's reach. Five years later, Niagara's power ran the roughly **26 miles** to Buffalo. The physics behind the rout is captured in one relationship: the power lost heating a transmission wire rises with the *square* of the current, so that raising the voltage tenfold and dropping the current tenfold cuts the loss to a hundredth. That squared relationship, more than any speech or patent, is why AC won.

The frequencies and voltages that emerged from these battles are still humming in your walls, standardized into a global patchwork:

- **60 hertz** — the frequency of alternating current in North America and parts of the world, meaning the current reverses direction 120 times and completes 60 full cycles every second.
- **50 hertz** — the frequency across most of Europe, Asia, Africa, and Australia.
- **110–120 volts** — the standard household voltage in North America, a descendant of Edison's original 110.
- **220–240 volts** — the standard household voltage in most of the rest of the world, which delivers the same power through less current.

These divisions are not laws of nature but frozen accidents of history — early choices by early companies that hardened into national standards and now cannot easily be changed, so that a traveler still packs adapters as a small souvenir of the industry's chaotic birth.

Now scale up from the wall to the grid, and the numbers grow vast. Household wiring runs at that 120 or 240 volts; the distribution lines down your street carry several thousand volts; but the great transmission lines that march across the landscape on steel towers run at staggering pressures:

- **Hundreds of thousands of volts** — the range of high-voltage transmission, commonly cited at levels such as 345,000 or 500,000 volts, with the largest lines reaching beyond 750,000 and, on the most extreme modern HVDC links, past a million volts. These voltages are lethal at a distance of feet, which is why the towers hold the conductors so far out in the open air.
- **Generating units** in a large power station are each rated in the hundreds of **megawatts** — millions of watts — and a big plant may total **1,000 megawatts (one gigawatt)** or more, enough to supply a substantial city.
- **National grids** are measured in hundreds of **gi-**

gawatts of capacity, and the largest, serving the biggest economies, in the neighborhood of a thousand gigawatts — a **terawatt** — at peak. (These totals grow year by year; treat them as approximate.)

To feel the reach of a century of growth, set the beginning against the present. Edison's Pearl Street plant, the birth of the whole industry in 1882, served about **85 customers** and some **400 lamps** from dynamos of modest size. A single large modern generating unit produces roughly a thousand times the power of that entire pioneering station, and the world now generates, all told, on the order of **tens of thousands of terawatt-hours of electricity a year** — a quantity with so many zeros that it defeats intuition. In a century and a half the industry grew from lighting a few square blocks of lower Manhattan to powering essentially the whole of human civilization, one of the steepest growth curves in the history of any technology.

A few individual landmarks deserve their own figures, because scale becomes real when it is attached to a place:

- **Niagara's Adams Power Plant**, the crowning triumph of AC, began with generators of about **5,000 horsepower** each — enormous for their day, and proof that polyphase AC could be built at industrial scale.
- **The 1893 White City** in Chicago blazed with roughly **100,000 incandescent lamps** powered by Westinghouse's alternating current, a display of electric light on a scale no one had ever witnessed, seen by millions of fairgoers.
- **Tesla's Wardenclyffe tower** rose about **187 feet**, its great mushroom cap intended to broadcast power across the globe; it was never completed and was dynamited for scrap in 1917.
- **Edison's patent count** stood at more than **1,000 US patents**, long a record; **Westinghouse held 361**, and **Tesla some 300 worldwide** — three of the most prolific inventors in history, measured by the paper trail of their ideas.

It is worth translating the abstract units into things you can feel, because the numbers on your electric bill hide a genuinely astonishing bargain. A **kilowatt-hour** — the unit the meter counts, a thousand watts running for an hour — is the energy that would keep ten old-fashioned hundred-watt bulbs lit for an hour, or run a typical refrigerator for the better part of a day, or boil many kettles of water. That same kilowatt-hour, delivered to your home for a price usually measured in small change, contains roughly the mechanical work that would take a fit human laborer the better part of a full day to produce by hand. Every unit of electricity you buy is, in effect, a day of tireless labor sold for pennies. The cheapness is the miracle. When Insull and his successors drove the price of electricity down and down across the twentieth century, they were quietly handing every ordinary household a workforce of invisible servants that no emperor of antiquity could have afforded.

The frequencies hide their own hidden precision. The 50 or 60 hertz of the grid is not merely a nominal figure; across an entire interconnected network, every generator, from the largest to the smallest, must spin in near-perfect synchrony, all their alternating currents rising and falling together in lockstep, or the system tears itself apart. A continent-spanning AC grid is thus a single vast machine whose thousands of spinning parts, hundreds of miles apart, all turn in time with one another to a tolerance of a fraction of a cycle — arguably the largest and most precisely coordinated machine humanity has ever built, and one that runs, unwatched, every second of every day. When that synchrony fails, the result is a **blackout**, and the great historical blackouts — cascading failures that darkened whole regions and tens of millions of people in minutes — are reminders of how tightly the entire structure is wound.

A closing set of comparisons places the electrical body of our civilization against the natural world it imitates. The signals in your own nerves are electrical, traveling at a tiny fraction of a volt; a single lightning bolt, by contrast, briefly reaches into the **hundreds of millions of volts** and carries a current of tens of thousands of amperes for a few thousandths of a second — a natural discharge so violent that all of human electrical engineering is, in a sense, the patient effort to tame and miniaturize what the sky does carelessly in an instant. Franklin's lightning rod was the first small victory in that long campaign; the continental grid, running at a millionth of the lightning's voltage but every second of every day, is its culmination.

These, then, are the numbers of the electrical age: a few exact units named for a few great men; the one-mile limit that doomed DC and the hundred-mile leap that crowned AC; the 50 and 60 hertz and the 120 and 240 volts frozen into the world's walls; the climb from 85 customers to a civilization; and the astonishing arithmetic by which a day's labor is delivered to your home for pennies. The story of electricity is finally a human drama, but it is written, underneath, in these figures — and the figures, rightly read, are as dramatic as the men.

Essay F: Where You Meet Electricity Today

The men in this book have been dead for the better part of a century, and yet you cannot get through an ordinary morning without shaking hands with every one of them. Their inventions and the principles behind them are not museum pieces; they are the invisible furniture of your daily life, so woven into the fabric of an ordinary day that you no longer see them at all. This essay is a tour of that day — a walk from waking to sleeping through a modern life — tracing each everyday device back to the people and the principles you have met in these pages. The point is not nostalgia. It is to make the invisible visible again, so that the next time you flip a switch or charge a phone you feel, for a moment, the long history running through your fingertips.

You wake, most likely, to an electric sound and an electric light. The alarm on your phone or clock is timed by the steady beat of alternating current or a quartz crystal, and the first thing you do — reach out and turn on a lamp — is the gesture that a whole century of struggle was fought to put under your thumb. That lamp is the direct descendant of the bulb that glowed over the counting-houses of lower Manhattan on the night in 1882 when Edison switched on a city. If it is an old-fashioned incandescent bulb, it works on the exact principle Edison perfected: current forced through a thin high-resistance filament until it glows white-hot. If it is one of the newer LED bulbs, as it increasingly is, it works on a principle Edison never knew — light coaxed directly out of a semiconductor, a child of the twentieth-century physics of the electron that J. J. Thomson discovered in 1897 — but it still runs on the grid Tesla and Westinghouse built, and it still answers to the same switch. Either way, the abolition of darkness that these men achieved is the first fact of your day.

You go to wash, and the water that comes hot from the tap has very likely been heated, or at least pumped and pressured, by electricity. The motor in the pump, the motor in the fan that clears the steam, the motor that will shortly spin your toothbrush or your razor — every one of them is a small monument to Tesla, whose rotating magnetic field made the alternating-current motor practical, and behind him to Faraday, who built the first crude electric motor in 1821 and discovered the induction that all of them depend on. Count the electric motors in a modern home and you will lose track past a few dozen; they hide in the refrigerator, the washing machine, the dishwasher, the ventilation fans, the garage door, the hard drives, the toys. Each one is Faraday's principle and Tesla's design, converting the current in the wall into silent, tireless motion. The drudgery that bent the backs of every generation before the twentieth century — the hauling, the grinding, the scrubbing, the pumping — has been quietly lifted away by these spinning ghosts, and you never think of them at all.

Breakfast is a gallery of resistance. The toaster, the kettle, the electric stovetop, the coffee maker — these do the opposite of the light bulb's neat trick, deliberately wasting electrical energy as heat by forcing current through elements that fight it. This is Ohm's law made edible: current through resistance yields heat, and the whole cheerful apparatus of a modern kitchen runs on Georg Ohm's simple relationship, exploited on purpose. The refrigerator that kept the milk cold overnight is subtler still, a motor-driven pump moving heat from inside to outside against its natural inclination — a machine that would have seemed like sorcery to every human being who lived before the electrical age, when keeping food cold meant ice cut from winter lakes and hoarded through the summer in sawdust.

You reach for your phone, and here the tour arrives at the strangest twist in the whole story. That glowing rectangle runs on direct current — Edison's current, the loser of the great war — drawn from a battery that is the lineal descendant of Volta's pile of 1800. The little charger you plug it into is quietly refighting the War of the Currents in miniature every time you use it: it takes Tesla's alternating current from the wall socket and rectifies it back into Edison's direct current for the phone to drink. Inside the phone, billions of transistors switch that direct current on and off to do their computing, each one a descendant of the electron physics that began with Thomson and of Maxwell's equations, which govern every radio wave the phone sends and receives. When you check the weather or send a message, you are using the electromagnetic waves that Maxwell predicted on paper in the 1860s and Hertz confirmed in his laboratory in 1887 — waves that Tesla and Marconi then raced to turn into radio. Your phone is a séance in which Volta, Faraday, Maxwell, Hertz, Edison, and Tesla are all summoned at once, several times a minute, without your ever noticing.

Step outside and the history goes vertical. The elevator that carried you down from a high floor exists because the electric motor made it possible to lift people safely and tirelessly, and without the electric elevator the skyscraper is literally unbuildable — no one will climb forty flights of stairs. The modern city, that forest of tall towers, is a physical consequence of Faraday's and Tesla's motor as surely as it is of steel and concrete. Overhead, if you pass beneath them, run the wires: the fat conductors on their tall steel towers carrying power at hundreds of thousands of volts, the whole visible skeleton of the grid that Insull's business model filled with customers. Most people walk under those lines their whole lives without once thinking that they are looking at the arteries of civilization, the descendants of the copper Edison buried under Pearl Street, grown continental.

The car you drive, or ride in, is electrical whether or not it is an "electric car." Every gasoline engine is started by an electric motor and fired by electric sparks timed by electronics; every car is a rolling nest of motors and computers. But increasingly the car is electric through and through — a battery of Volta's descendants storing direct current, feeding motors of Tesla's design, recharged from the grid of Tesla and Westinghouse through a charger that, once again, converts AC to DC. It is a fitting emblem of the modern truce between the currents, and it is no accident that the most famous electric-car company took Tesla's name, though one suspects Tesla himself would have preferred a royalty to a tribute.

Through your working day, the machine you rely on most is the one that would have astonished all three of our protagonists equally: the computer, and the vast invisible web of computers behind it. Every keystroke, every search, every stored photograph, is the switching of transistors by direct current, governed by the physics of the electron and the semiconductor. The "cloud" that holds so much of modern life is not a cloud at all but immense buildings full of humming machines, drinking electricity by the megawatt and pouring out heat that must itself be carried away by electric cooling. The data center is the cathedral of the electrical age, and it is growing so fast that it is beginning to reshape the grid itself, one of the newer pressures on a system its founders could never have imagined. That last point is a present-day and shifting matter, and the reader should take the details as current only as of this book's writing.

Nowhere do you meet the whole cast more completely than in a hospital, and it is worth pausing there, because it shows how much of modern survival now rests on the current. A century and a half ago a hospital was lit by lamps and candles, and everything in it that mattered was done by hand

and eye. Today the same building is a single vast electrical organism. The lights that let a surgeon see are Edison's; the monitors tracing a heartbeat and the imaging machines that peer inside the body are pure applications of the electron physics and electromagnetic theory that Thomson and Maxwell founded; the motors in the pumps and ventilators are Faraday's and Tesla's; the computers coordinating it all are Edison's direct current switching billions of times a second. And because a failure of the current here means death rather than mere inconvenience, the hospital keeps its own generators and batteries against a blackout — a private, miniature grid, standing guard over the one place where the abolition of the dark is not a convenience but a matter of life. Every operation performed after sundown, every premature infant kept alive in an incubator, every patient wheeled through a scanner, is a debt owed to the men in this book.

Evening comes, and the tour turns domestic again — the television and its descendants, all of them electric light and Maxwell's waves and the electron's dance; the lamps that let you read this sentence after sundown, the single most transformative gift of the whole enterprise, the abolition of the ancient rule that darkness ended the human day. When you finally go to bed and plug in your phone to charge overnight, you close the loop you opened at dawn: the grid pours its alternating current into your home, the charger turns it to direct current, the battery stores it as chemistry, and in the morning it will pour back out to wake you. The whole silent cycle runs on the accumulated genius of every figure in this book, none of whom you will thank, because the supreme achievement of the electrical age is precisely that it asks nothing of your attention.

That is the deepest thing to understand about where you meet electricity today: you meet it everywhere and notice it nowhere. The measure of how completely Edison, Tesla, and Westinghouse succeeded is that their revolution has become invisible — that a child today no more marvels at instant light and tireless motion than at air. But invisibility is not absence. The current is running through your walls right now, through the device in your hand, through the city outside your window, carrying in it the amber and the frog's leg, the battery and the field, the bulb and the motor and the transformer, and the long, strange, human story of how three rivals wired the world. Look up at the light once more. Everyone in this book is in the room with you.

Essay G: The Grid of the Future

Every previous chapter of this book has been history, grounded in the documented record of what happened. This one is different, and the reader deserves fair warning at the outset: it looks forward, into a future that has not yet been written, and much of what it describes is uncertain, contested, and changing even as these words are set down. The grid is in the middle of the largest transformation since Insull, and no honest writer can tell you exactly how it will end. What I can do is describe the forces now reshaping it, explain why they matter, and flag clearly where I am reporting established trends and where I am venturing into genuine uncertainty. Everything below should be read against the knowledge horizon of this book's writing; by the time you read it, some of these questions may be settled and some of these predictions overtaken. Treat this essay as a map of a journey still in progress, not a report from its destination.

Begin with the central fact that is forcing the change. The grid Tesla and Westinghouse built, and Insull filled with customers, rested on an assumption so basic it was rarely spoken: that electricity is generated in a relatively small number of large, controllable plants, burning fuel on demand, and flows one way — outward, from plant to home. The whole system was engineered around that one-way flow of dependable, dispatchable power, a plant that could be turned up when demand rose and down when it fell, matching supply to demand from moment to moment because electricity, as we have seen, cannot easily be stored. That assumption is now breaking down, and the breakdown is the root of nearly every challenge and opportunity the future grid faces.

The first great force is the rise of renewable energy, above all wind and solar power, which are growing rapidly across the world and in many places are now the cheapest new sources of electricity to build — a genuine and well-established trend, though the pace varies by region. Renewables are a profound good: they generate power without burning the fossil fuels whose accumulated exhaust is changing the planet's climate, addressing the great shadow cost of the electrical age. But they break the old grid's founding assumption in two ways. First, they are variable and uncontrollable: the sun sets, clouds pass, the wind drops, and unlike a coal plant you cannot simply turn the sun up when demand rises. The grid was built to follow demand with controllable supply; now it must cope with supply that follows the weather instead. Second, renewables are often distributed — scattered across thousands of rooftops and wind farms rather than concentrated in a few great stations — and they can push power *back* into a grid designed to carry it only one way. The one-way street is becoming a two-way network, and the whole system must be rethought to accommodate it.

The problem of variability points directly to the technology that may matter most, and it is the one thing the founders never solved: storage. For the whole history of electricity, the inability to store power cheaply has been the industry's defining constraint, the reason capacity must be sized to the peak and generation must instantly match demand. If electricity could be stored at large scale and low cost, that constraint would dissolve — solar power gathered at noon could light the evening, wind caught overnight could run the morning, and the grid could ride through the gaps in renewable supply. This is why grid-scale storage, above all large banks of batteries descended from Volta's pile, has become one of the most watched frontiers in the entire energy world. Battery costs have fallen dramatically and installations are growing fast — an established trend as of this writing — and other approaches, from pumped water reservoirs to novel chemistries and mechanical schemes, are being pursued in parallel. Whether storage will scale far enough, cheaply enough, to fully unlock a renewable grid is a genuinely open question, and one of the largest uncertainties in this essay. But its importance is not in doubt: storage is the missing piece the founders lacked, and the grid of the future will be shaped above all by how well it is supplied.

The second great force is a surge in demand from the electrification of things that used to burn fuel directly. The most visible is transport: electric vehicles, rolling batteries recharged from the grid, are being adopted at a pace that varies widely by place but is broadly rising, and a world of electric cars is a world that will draw far more power from the grid, and draw it in new patterns — millions of vehicles charging, ideally when power is cheap and plentiful, potentially even feeding power back when it is scarce. Heating, too, is electrifying, as electric heat pumps replace furnaces that burn gas or oil. The logic is clear: if the electricity itself can be made cleanly, then shifting cars and furnaces onto the grid removes their direct pollution. But it means the grid must grow substantially to carry loads it never had to carry, even as it juggles the new variability of its supply — a double challenge of more demand and less controllable generation at the same time. How large that growth must be, and how fast it can be built, are matters of active dispute.

To manage this vastly more complex system — two-way flows, variable supply, surging and shiftable demand — the grid is growing a nervous system it never had, and this is the meaning of the much-used phrase "smart grid." The old grid was largely dumb in the sense that mattered: it had little moment-to-moment awareness of itself, and it balanced supply and demand mostly by adjusting a few big plants. The smart grid layers digital sensing, communication, and automated control across the whole system, so that it can see itself in detail and respond in real time — shifting loads, routing power around trouble, coordinating thousands of scattered generators and storage units and flexible demands into a coherent whole. In principle this lets the grid absorb variable renewables and shiftable loads gracefully, turning what would be chaos into a managed dance. In practice the smart grid is a work in progress, unevenly deployed, and it brings its own new worry: a grid controlled by computers and networks is a grid that can, in principle, be attacked through them, and the security of this digital nervous system is a live and serious concern. This is very much a developing area, and the reader should regard specifics as provisional.

There is a subtler shift buried in all this that is worth drawing out, because it overturns the founding logic of the whole industry. The grid Insull built was engineered on the principle that supply should follow demand: customers used what they wanted when they wanted it, and the utility's job was to adjust its plants, moment by moment, to match. In a grid dominated by variable renewables and empty of cheap storage, that logic may partly invert, so that demand is nudged to follow supply — appliances, water heaters, and above all electric cars encouraged, by price signals or automatic controls, to draw power when the sun is high and the wind is strong, and to ease off when they are not. A fleet of electric vehicles, in this vision, becomes not merely a new load but a vast distributed reservoir, charging when power is abundant and, in some schemes, feeding a little back when it is scarce, so that the cars parked across a city help steady the grid that fills them. Whether this flexible, two-way, demand-following grid arrives in full is uncertain, and it raises real questions about control, fairness, and privacy that are only beginning to be argued over. But the direction of travel is clear enough: the rigid one-way system of the twentieth century is giving way to something looser, more distributed, and more negotiated, in which the line between producer and consumer blurs and millions of small devices participate in balancing the whole.

A quieter thread running through all of this is the return of direct current, which the essay on the rematch of the currents described. The future grid will very likely be more Edisonian than the one Tesla built, not because AC is being replaced but because DC is reclaiming the roles it always suited. The long-distance backbone that will carry power from remote wind farms and desert solar arrays and distant hydro dams to the cities that need it is increasingly high-voltage direct current, which moves bulk power farther and more efficiently than AC and can knit together grids that AC cannot easily join. And at the small scale, the world of solar panels, batteries, LED lights, and electronic devices is natively a direct-current world, so that some envision local networks — a home, a building, a campus — running partly on DC to avoid the wasteful back-and-forth conversions the all-AC grid now requires. Whether such DC microgrids become common is uncertain, but the broad pattern is clear: the future grid will use both currents deliberately, and Edison's, unexpectedly, will hold more of it than at any time since 1896.

Behind all these specifics stands the largest and most contested question of all, the one that gives the whole transformation its urgency: the effort to move the world's energy system off the fossil fuels whose burning is warming the planet — the energy transition. The electrical age was built, for a century, on fire, and the bill for that fire is the shadow cost this book's epilogue names. The transition is the collective attempt to keep the light and lose the smoke: to generate the world's growing demand for electricity without combustion, from sun and wind and water and, in the view of many though not all, from the atom. It is one of the largest undertakings humanity has ever attempted, its pace and prospects are fiercely debated, and its outcome is unknown. I will make no prediction. I will only observe that it is, at bottom, the same problem the founders faced, returned in a new and vaster form: how to make electricity that a whole civilization can afford — but now with a constraint they never felt, that it must also be electricity the planet can afford.

One last quality the future grid must have deserves a word, because recent history keeps insisting on it: resilience. As electricity becomes ever more the foundation on which everything else rests — as transport, heat, industry, and the entire digital world pile onto the grid — the cost of the grid failing rises with every passing year. A blackout that was once an inconvenience becomes, in a fully electrified society, a catastrophe touching every part of life at once. And the pressures on reliability are mounting from several directions: more extreme weather straining the wires, the variability of renewable supply, the complexity of the smart grid, and the new vulnerability of a networked system to deliberate attack. Building a grid that can bend without breaking — that can lose a piece and keep running, restart itself after collapse, and withstand both storms and saboteurs — is among the central tasks of the coming decades. How well it will be accomplished, no one yet knows.

So this is where the current runs now, at the far edge of the story that began with a Greek philosopher and his lump of amber: into a grid being rebuilt in flight, absorbing the sun and the wind, learning to store what it cannot yet store, bracing for a surge of electric cars and electric heat, growing a digital nervous system, quietly welcoming Edison's current back onto its longest wires, and straining toward a future without the fire that built it. I cannot tell you how it ends, and I have tried throughout this essay to be honest about that. But I can tell you that the transformation now under way is as consequential as anything in this book, and that the men who fought the War of the Currents, could they see it, would recognize both the problem and the stakes — for it is still, after all this time, the same question they gave their lives to: how to carry the light to everyone, over the horizon, and keep it burning.

A Note on Sources

A book that ranges across twenty-six centuries and three continents, and that dares to correct a few beloved legends along the way, owes its readers an honest account of where its knowledge comes from and how far that knowledge can be trusted. This note is that account. It is not a formal apparatus of footnotes — this is a narrative history for the general reader, not a monograph — but a plain statement of the kinds of sources I have leaned on, and, more importantly, of where the historical record is solid and where it grows contested, thin, or frankly legendary.

For the deep scientific prehistory — Thales and his amber, Gilbert and *De Magnete,* the Leyden jar, Franklin's kite, Galvani's frogs and Volta's pile, Ørsted and Ampère and Ohm, Faraday and Maxwell — I have relied on the settled consensus of the history of science, as recorded in standard reference works, the *Encyclopaedia Britannica,* and the master survey articles on the history of electromagnetic theory. This material is generally firm, though the reader should know that the further back one goes, the more the record softens into tradition. That Thales observed the amber effect is universally reported, but we have it only at second and third hand, from writers who lived centuries after him; the famous story of Franklin's kite rests largely on an account written years later; and dozens of such vignettes have been polished by generations of retelling. I have flagged the most doubtful of these in the text and have tried never to present a good story as a certain fact.

For the three central figures, the sources are richer but also more treacherous, because fame breeds legend. On Edison, the indispensable resource is the Thomas Edison Papers project at Rutgers University, which has digitized his notebooks, patents, and correspondence and made it possible to check the man against the myth; for the shape of his life and character I have also drawn on the major biographies, including Edmund Morris's *Edison*, and on the institutional records of the Smithsonian's National Museum of American History. On Tesla, the situation is more delicate, and I have tried to be especially careful. Tesla's own autobiography, *My Inventions*, is a primary source of enormous value and equally enormous unreliability: it is the origin of many of the best-loved Tesla stories, including the fifty-thousand-dollar anecdote, and it was written by a proud, self-dramatizing man decades after the events, with every incentive to burnish. I have treated it as I would treat any vivid memoir — as precious evidence of how Tesla saw himself, and as shaky evidence of what actually happened. For the corrective, sober reconstruction of Tesla's life and work I am indebted above all to W. Bernard Carlson's *Tesla: Inventor of the Electrical Age*, a model of myth-resistant scholarship, and to Marc Seifer's detailed biography. On Westinghouse, the least documented of the three in popular memory, I have leaned on the biographical literature, including Quentin Skrabec's study, and on the records of the company and the era.

For the War of the Currents itself — the propaganda, the animal electrocutions, Harold Brown, the electric chair, and Kemmler — the narrative backbone comes from Jill Jonnes's *Empires of Light*, the definitive popular account of the conflict, checked against contemporary newspaper coverage, the Edison Papers, and the summaries maintained by institutions such as the U.S. Department of Energy and the Smithsonian, as well as the standard reference article on the war of the currents. This is one of the better-documented episodes in the book, and yet it is also one of the most mythologized, and I have worked hard to separate the two. The electrocution of Topsy the elephant, for instance, is routinely folded into the current war and attributed to Edison as a deliberate anti-AC stunt; the sober record shows that Topsy was killed in 1903, well after the war was over, in a separate affair in which Edison's personal role is much overstated. I have said so plainly rather than repeat the tidy version.

For the build-out of the electrified world — Niagara and the Adams plant, the steam turbine, Insull and the invention of the utility, rural electrification, and the rise of the grid — I have drawn on the industrial and economic histories, including Maury Klein's *The Power Makers*, on the records of the Niagara projects, and on the historical and statistical materials published by the U.S. Department of Energy and the Energy Information Administration. For the cultural meaning of electric light I have found Ernest Freeberg's *The Age of Edison* especially illuminating. And for the modern grid and the energy transition that occupy the final chapter and several of the essays, I have leaned on Gretchen Bakke's *The Grid* and Richard Rhodes's *Energy: A Human History*, supplemented by current institutional sources.

That last body of material demands the frankest warning of all. Everything in this book that describes the present-day and near-future world — the state of renewables, the economics of storage, the spread of electric vehicles, the progress of the smart grid, the return of DC, the pace of the energy transition — is a moving target, and I have flagged it as such wherever it appears. These are not historical facts settled in the archive; they are the shifting conditions of a story still unfolding, and the figures and forecasts I give are approximate, drawn from the best current understanding at the time of writing, and certain to be out of date to some degree by the time you read them. Where a modern number matters, I have urged the reader to check it against current sources. The history in this book is meant to last; the futurology is meant only to orient.

A word, finally, on the reference tools of our own moment, since honesty requires it. In assembling and cross-checking the framework of dates, names, and facts, I have made use of the great collaborative reference works of the internet, including the master Wikipedia articles on the war of the currents and on each of the principal figures, alongside the more traditional *Britannica* and the specialist resources such as the Edison Tech Center. These collaborative sources are marvelous for orientation and for tracking down primary references, and they are also, by their nature, uneven and occasionally wrong; I have used them as starting points and signposts, not as final authorities, and have tried to anchor every significant claim in something more solid — a primary document, an institutional record, or the considered judgment of the serious biographers named above.

The result is a book that aims to be accurate without pretending to be exhaustive, and confident without pretending to certainties it does not possess. Where the record is firm, I have stated things firmly. Where it is contested, I have said so. Where legend has crept in and hardened into supposed fact — the fifty thousand dollars, the stolen inventions, the suppressed free-energy machine, the elephant — I have tried to set the documented record beside the meme and let the reader see the difference. History is not a collection of certainties; it is a disciplined argument with the evidence, and the honest historian's job is to show his work and mark his doubts. I have tried to do both, and where I have failed, the fault is mine and not my sources'.

Further Reading

This book has tried to tell the whole story in one volume, from amber to the smart grid, and that ambition necessarily means it lingers over some things and hurries past others. A reader whose appetite has been whetted — for the drama of the current war, for the inner life of one of the three protagonists, for the science, or for the modern grid — will want to go deeper, and the good news is that this subject has attracted some genuinely excellent writers. What follows is a guide to the best of them, with a sentence or two on what each one offers and why you might turn to it next. I have kept to books that a general reader can enjoy, and I have arranged them roughly in the order the story unfolds.

Jill Jonnes, *Empires of Light: Edison, Tesla, Westinghouse, and the Race to Electrify the World* **(2003).** If you read only one other book, read this one. It is the definitive popular narrative of the War of the Currents, and it shares the exact dramatic spine of the book you are holding — the three-cornered rivalry — but at greater length and with a novelist's eye for scene and character. Where my book gives you the whole sweep of electrical history, Jonnes gives you the current war itself in rich, immersive detail, and there is no better account of the propaganda, the personalities, and the stakes.

W. Bernard Carlson, *Tesla: Inventor of the Electrical Age* **(2013).** The Tesla biography for readers who want the real man rather than the internet legend. Carlson is a careful historian of technology, and his Tesla is neither the martyred saint nor the suppressed prophet of the memes but a genuine and flawed genius whose triumphs and self-inflicted disasters are traced with rigor and sympathy. This is the essential corrective to the myths, and the ideal next step for anyone drawn to Tesla by the essay in this book.

Marc J. Seifer, ***Wizard: The Life and Times of Nikola Tesla*** **(1996).** A fuller, more sprawling, more sympathetic Tesla biography than Carlson's, dense with detail about his experiments, his social world, and his long decline. It is more willing to take Tesla's grander claims seriously, so it makes an interesting companion and counterweight to Carlson — read the two together and you will have triangulated the man.

Edmund Morris, ***Edison*** **(2019).** A vast, ambitious biography of Edison by a Pulitzer Prize–winning master of the form, famous (and to some readers frustrating) for telling the life in reverse chronological order, decade by decade backward from death to birth. It is less a quick reference than an immersive literary experience, and it repays the reader who wants to live inside Edison's restless, relentless, contradictory mind at length.

Ernest Freeberg, ***The Age of Edison: Electric Light and the Invention of Modern America*** **(2013).** Where Morris gives you the man, Freeberg gives you the meaning — a cultural history of what electric light actually did to society, how people reacted to the abolition of darkness, and how the bulb remade cities, work, leisure, and the rhythms of daily life. If the epilogue and the "Where You Meet It Today" essay caught your interest, this is where to go for the human texture of the electrical revolution.

Quentin Skrabec, ***George Westinghouse: Gentle Genius*** **(2007).** The book to read for the forgotten third man. Westinghouse is the least served by popular memory, sidelined by the glamorous Tesla-versus-Edison framing, and this biography restores him as the decisive industrialist he was — the humane employer, the relentless inventor, the businessman who actually electrified the world. A necessary supplement to a story that too often leaves him out.

Maury Klein, ***The Power Makers: Steam, Electricity, and the Men Who Invented Modern America*** **(2008).** For the reader who wants the bigger industrial picture — how steam and electricity together built modern American power, and the entrepreneurs and engineers who drove the build-out. It sets the electrical story in the broader sweep of industrialization and is especially good on the machinery and the money.

Gretchen Bakke, ***The Grid: The Fraying Wires Between Americans and Our Energy Future*** **(2016).** The essential companion to the last third of this book and to the essays on the modern and future grid. Bakke writes vividly about the aging, fragile, and rapidly changing electrical system we actually depend on, its vulnerabilities, and the wrenching transition toward renewables. If "The Grid of the Future" left you wanting more, start here.

Richard Rhodes, ***Energy: A Human History*** **(2018).** A grand, sweeping account of humanity's long relationship with energy in all its forms, from wood and coal through oil to electricity and the atom, by one of America's finest writers of narrative history. It places the electrical story within the largest possible frame — the whole human quest for power — and is the book to read for perspective and scale.

Beyond these, the curious reader has a wealth of primary material now freely available. **Tesla's own autobiography,** ***My Inventions*** (1919), is short, strange, and revealing — best read, as the note on sources warns, as a window into how Tesla saw himself rather than as a reliable record. **The Thomas Edison Papers** at Rutgers University have digitized an astonishing trove of Edison's notebooks, patents, and letters, browsable online, for anyone who wants to watch the invention factory at work in the inventor's own hand.

And the historical and educational materials published by the **U.S. Department of Energy**, the **Smithsonian Institution**, and the **Edison Tech Center** offer reliable, accessible overviews of the science and the history alike. From these, and from the books above, a reader could happily spend years following the current wherever it leads — which is, in the end, very nearly everywhere.

Bibliography

The sources below informed the research and writing of this book. They are grouped into three categories — narrative and scholarly books; primary and archival sources; and reference and institutional works — and within each group arranged alphabetically by author or title. This is a working bibliography for the general reader, not an exhaustive scholarly apparatus; fuller guidance on how these sources were used, and on where the record is contested, appears in "A Note on Sources."

Books

Bakke, Gretchen. *The Grid: The Fraying Wires Between Americans and Our Energy Future.* Bloomsbury, 2016.

Carlson, W. Bernard. *Tesla: Inventor of the Electrical Age.* Princeton University Press, 2013.

Cheney, Margaret. *Tesla: Man Out of Time.* Prentice-Hall, 1981.

Freeberg, Ernest. *The Age of Edison: Electric Light and the Invention of Modern America.* Penguin Press, 2013.

Hughes, Thomas P. *Networks of Power: Electrification in Western Society, 1880–1930.* Johns Hopkins University Press, 1983.

Israel, Paul. *Edison: A Life of Invention.* John Wiley & Sons, 1998.

Jonnes, Jill. *Empires of Light: Edison, Tesla, Westinghouse, and the Race to Electrify the World.* Random House, 2003.

Klein, Maury. *The Power Makers: Steam, Electricity, and the Men Who Invented Modern America.* Bloomsbury Press, 2008.

McNichol, Tom. *AC/DC: The Savage Tale of the First Standards War.* Jossey-Bass, 2006.

Morris, Edmund. *Edison.* Random House, 2019.

Nelson, Craig. *The Age of Radiance: The Epic Rise and Dramatic Fall of the Atomic Era.* Scribner, 2014.

Rhodes, Richard. *Energy: A Human History.* Simon & Schuster, 2018.

Seifer, Marc J. *Wizard: The Life and Times of Nikola Tesla — Biography of a Genius.* Citadel Press, 1996.

Skrabec, Quentin R. *George Westinghouse: Gentle Genius.* Algora Publishing, 2007.

Primary & Archival Sources

Edison, Thomas A. Notebooks, patents, and correspondence. *The Thomas A. Edison Papers*, Rutgers University, New Brunswick, New Jersey (digital edition).

Tesla, Nikola. *My Inventions: The Autobiography of Nikola Tesla.* Originally published in *Electrical Experimenter*, 1919.

Tesla, Nikola. "A New System of Alternating Current Motors and Transformers." Lecture before the American Institute of Electrical Engineers, 1888. AIEE / IEEE archives.

United States Patent Office. Patent records of Thomas A. Edison, Nikola Tesla, George Westinghouse, and William Stanley Jr.

Niagara Falls Power Company / Niagara Parks. Records of the Adams Power Plant and the Niagara power development.

Tesla Science Center at Wardenclyffe. Historical materials on Tesla's Long Island laboratory and tower.

American Institute of Electrical Engineers (AIEE). Edison Medal citations, including the award to George Westinghouse, 1911. IEEE archives.

Reference & Institutional Sources

Encyclopaedia Britannica. Biographical and topical entries on electricity, Thomas Edison, Nikola Tesla, George Westinghouse, Michael Faraday, James Clerk Maxwell, and Alessandro Volta.

Edison Tech Center (edisontechcenter.org). Histories of alternating-current power and electrical engineering timelines.

History.com. "What Was the War of the Currents?" and related articles.

IEEE History Center / Engineering and Technology History Wiki (ethw.org). Biographies and technical histories of electrical pioneers and milestones.

National Academy of Engineering. "Greatest Engineering Achievements of the 20th Century," which ranks electrification first.

National Park Service. Thomas Edison National Historical Park, West Orange, New Jersey. Laboratory and biographical materials.

Smithsonian Institution, National Museum of American History. Electricity collections and exhibitions, Washington, D.C.

U.S. Department of Energy. "The War of the Currents: AC vs. DC Power," and historical energy timelines.

U.S. Energy Information Administration (EIA). "History of Energy" and electricity statistics.

Wikipedia. Master articles including "War of the currents," "Thomas Edison," "Nikola Tesla," "George Westinghouse," "History of electromagnetic theory," "History of electric power transmission," "Electric power industry," "Michael Faraday," "James Clerk Maxwell," "Alessandro Volta," "Benjamin Franklin," and "Niagara Falls (Adams Power Plant)." Used for orientation and cross-reference, anchored where possible to the primary and institutional sources above.

www.ingramcontent.com/pod-product-compliance
Lightning Source LLC
La Vergne TN
LVHW100132061126
851857LV00010B/61